HOSPITAL AIRBORNE INFECTION CONTROL

HOSPITAL AIRBORNE INFECTION CONTROL

WLADYSLAW KOWALSKI

CRC Press
Taylor & Francis Group
Boca Raton   London   New York

CRC Press is an imprint of the
Taylor & Francis Group, an **informa** business

CRC Press
Taylor & Francis Group
6000 Broken Sound Parkway NW, Suite 300
Boca Raton, FL 33487-2742

Printed in the United States of America on acid-free paper
Version Date: 20111020

International Standard Book Number: 978-1-4398-2196-1 (Hardback)

Visit the Taylor & Francis Web site at
http://www.taylorandfrancis.com

and the CRC Press Web site at
http://www.crcpress.com

I dedicate this book to all the doctors, nurses, health care workers, scientists, and engineers out there who are working in health care and who are seeking to save lives any way they can. I hope this book makes a small contribution to their understanding of this complex problem.

Contents

Preface

This book addresses a contemporary topic of considerable interest and some controversy in the field of health care—the subject of airborne pathogens. This book is not about debate, it is a compendium of information from the literature of science and medicine that is intended to assist medical professionals, health care workers, engineers, designers, and infection control professionals in dealing with hospital-acquired infections and outbreaks caused by pathogens that may spread by the airborne route as well as by other routes, these being primarily direct or hand contact and contact with fomites or contaminated equipment. A great many examples are presented in these pages that make it clear that airborne inhalation is not the only airborne route by which infections may be transmitted. The simple fact that a number of outdoor environmental spores and bacteria regularly show up as hospital-acquired or nosocomial infections, and that these pathogens can routinely be detected in the air, makes it abundantly clear that microbes may be transported by air currents inside hospitals regardless of whether they are respiratory diseases or not. Some new concepts are introduced in this book that may be of some benefit in understanding the complexities of airborne transmission and airborne transport of nosocomial pathogens. Foremost among these concepts is the idea that airborne transmission and fomite transmission are inextricably linked and therefore must be dealt with together and not separately. That is to say, the management of airborne nosocomial infections must involve strategies for surface disinfection and hand disinfection as well as air disinfection.

This book does not address certain topics that are commonly included in discussions of infection control such as vaccination programs, cohorting, management and procedural controls, physiology, and treatment. Although these important topics are mentioned in various parts of this book, it is not the intention of this book to repeat information on these subjects when so many excellent texts are available that address these topics in much greater detail than is possible here. It is the intent of this book to provide perspective on the matter of airborne nosocomial infections and enlighten readers as to the nature of the problem so that appropriate and effective solutions may be pursued.

Any readers who spot technical errors or typos in this book are encouraged to send them to me at drkowalski@aerobiologicalengineering.com. Errata will be posted at www.aerobiologicalengineering.com/HAIC/errata.html.

Acknowledgments

I most gratefully acknowledge all those who assisted me in the research and preparation of this text, including Dr. William Bahnfleth of the Pennsylvania State University Architectural Engineering Department; Dr. Jim Freihaut of the Pennsylvania State University Architectural Engineering Department; Dave Witham of Ultraviolet Devices Incorporated (UVDI) of Valencia, California; Dr. Ed Nardell of the Harvard School of Public Health; Richard Vincent of the Mount Sinai School of Medicine; Azael Capetillo of the University of Leeds; Mary Clancy, RN, of Environmental Dynamics, Inc., Sterling, Virginia; Dieter von Merhart of Masterpack Group, The Netherlands; Katja Auer of American Ultraviolet Company, Lebanon, Indiana; Fahmi Yigit of Virobuster GmbH, Windhagen, Germany; Chuck Dunn of Lumalier, Memphis, Tennessee; Fred Zander, Zander Scientific, Inc., Vero Beach, Florida; and Danielle Scher, RN. Thanks to the Alaskan Native Medical Center in Anchorage, and the Hershey Medical Center in Pennsylvania where I performed studies and took some of the photos in this book. Thanks to the Western Michigan University Waldo Library staff for their endless assistance. Thanks also to my dear Aunt Mary who passed away during the writing of this book, and to my friends Richard Gregg and Dr. James Kendig, who also passed away. I especially thank my father, Stanley J. Kowalski, for his support and for surviving the Second World War so that his son might write this book, and for living to be 91 years old so that he might see it published as well.

The Author

Wladyslaw Kowalski, PhD, PE, has authored numerous articles related to air cleaning technology, biodefense, and hospital air disinfection, and has authored books and contributed chapters on these subjects, including the *ASHRAE HVAC Design Manual for Hospitals and Clinics.* He consults with hospitals on indoor air quality issues, including mold and bacterial contamination problems, and investigates hospital infection problems using air and surface sampling. He designs air disinfection systems for hospitals and commercial buildings and speaks at international conferences on disease control and air cleaning topics. He is secretary of the Air Treatment Group of the International Ultraviolet Association (IUVA). He is currently engaged in research on ultraviolet germicidal irradiation (UVGI) and is developing new types of UVGI air and surface disinfection systems for the health care industry.

Dr. Kowalski earned his BS in mechanical engineering from the Illinois Institute of Technology and his MS and PhD in architectural engineering from The Pennsylvania State University. He currently lives on his farm in Michigan near Kalamazoo where he grows fruits and vegetables.

1

Airborne Nosocomial Infections

Introduction

Nosocomial, or hospital-acquired, infections have been a persistent problem in hospitals for over a century and they can have complex, multifaceted etiologies. The microorganisms that cause nosocomial infections have been well cataloged and the infections they cause are fairly well understood, but the traditional solutions, cleaning, scrubbing, disinfectants, ventilation, filtration, management and procedural controls, etc., have failed to completely eliminate the problem. Even the most modern hospitals with the best personnel and equipment find themselves stymied by the persistence of these insidious pathogens and their endless intrusions into hospital wards and operating rooms. A single nosocomial infection can wipe out the efforts of the most skilled surgeons and incur untold costs to the hospital, as well as result in tragic fatalities. Nosocomial infections can spread throughout hospital wards in epidemic-like fashion, even reaching rooms on floors far from the original source. The degree to which most of these infections are airborne is not well understood, but a growing body of evidence, much of which is presented in this book, indicates that the airborne component of nosocomial infections is not insignificant.

The majority of nosocomial infections have long been considered to be spread by direct contact or fomite contact as opposed to airborne transmission, and so the tendency has been to dismiss the airborne route as unimportant except for well-known exceptions like tuberculosis (TB). However, as more researchers have turned their focus on demonstrating or disproving the possibility of airborne transmission of any particular pathogen, they have found strong evidence that airborne transmission plays a role in many nosocomial infections. If we consider the fact that most fomite contamination of surfaces, especially horizontal surfaces like floors, must arrive via the air, then we must admit that airborne transport plays a role in surface contamination. The three traditional classifications of airborne transmission, direct contact, and indirect contact would seem to be inadequate to completely describe actual nosocomial infection etiology, and a new perspective may be in order. *Aspergillus* and other fungal spores, for example, hail from the outdoor environment and are naturally transported by the air both inside and outside hospitals. Bacterial spores like *Clostridium* are also well suited for airborne transport and even though they must be ingested to produce an infection, the evidence suggests that they may travel far from the infectious source patient via air currents. Staphylococci like methicillin-resistant *Staphylococcus aureus* (MRSA) are routinely inhaled in indoor environments, but of course they cause no problems except in nosocomial settings where they may settle directly on surgical sites or else may become fomites on surfaces. These are but a few of the examples that will be detailed throughout this book, and it will be seen that the majority of common nosocomial pathogens have the potential to be transported by the airborne route at one or more stages of the transmission process.

Any surface contamination that results from settled microbes is the result of airborne transport. The subjects of airborne infection and surface-borne infection are so intimately linked that it may be impossible to completely separate the two, and it is for this reason that both subjects are addressed together in this book. Numerous examples cited from the literature will be provided throughout this book that show the interrelationship of these two modes of transport, and it will become clear that the control of airborne infections depends not only on hospital airborne disinfection but also on hospital surface disinfection and that these two components must form part of an overall program designed to control airborne nosocomial pathogens.

The same holds true for hand disinfection—hand hygiene is also an integral part of any program intended to control airborne infections. Hands may become contaminated from fomites as well as from direct contact with infected patients, and this transmission mode cannot easily be separated from the other routes when considering airborne pathogens.

The subsequent chapters examine the epidemiology and etiology of all types of nosocomial infections that may be transmitted by the airborne route or otherwise be transported by air currents to ultimately contaminate hospital surfaces. The following sections introduce various general topics and background information that will be referred to throughout this book and address issues that will not be treated further.

Nosocomial Infection Epidemiology

Nosocomial infections are a major public health problem and represent a severe financial burden for hospitals (Davis and Shires 1991; Scott 2009). The cost of airborne infections is difficult to assess without a more detailed accounting of airborne epidemiology, and this limitation in the data affects the debate on how much should be spent on hospital engineering to reduce the risk (Hoffman, Bennett, and Scott 1999). Respiratory diseases account for

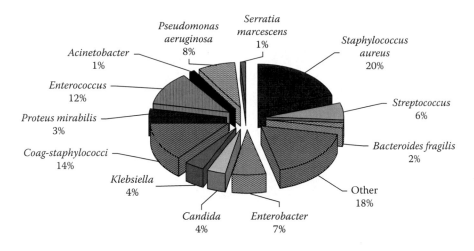

FIGURE 1.1
Primary causes of nosocomial infections. Based on data from NNIS (HICPAC 2003).

one-fourth of all medical costs in the United States today, and annual health care costs are estimated to total $30 billion (Fisk and Rosenfeld 1997). The probability of acquiring one or more nosocomial infections has remained relatively constant at about 5% of those hospitalized (Freeman 1999). It is estimated that surgical site infections (SSIs) occur in 2–5% of the patients undergoing surgery and that they account for almost one-quarter of all nosocomial infections (Wong 1999).

Figure 1.1 shows the variety of pathogens that are responsible for nosocomial infections. Almost all the pathogens shown are, in fact, potentially airborne, although only a fraction of them likely produce infections directly or indirectly via the airborne route.

Subdividing nosocomial infections into categories of airborne and nonairborne is problematic mainly due to a lack of data regarding airborne disease transmission. Various sources estimate that between 10% and 20% of hospital-acquired infections (HAIs) are airborne (Eickoff 1994; Durmaz et al. 2005; McDowell 2003; Bennett, Jarvis, and Brachman 2007). It is likely that the percentage of infections that can be attributed to the airborne route varies by species and even by the type of infection (SSI, respiratory, etc.), and perhaps even by the environmental conditions in any ward, room, or hospital, and therefore it would be difficult to precisely quantify the overall economic impact of airborne infections in hospitals today.

Table 1.1 summarizes the typical annual cases for known and potential nosocomial pathogens based on data from a variety of sources (see Kowalski 2006; Klevens et al. 2007) and excluding uncommon or rarely occurring pathogens. SARS virus is included based on a single year of outbreaks in Hong Kong, although it has rarely occurred since its first appearance. The annual cases cannot be subdivided into those caused by airborne transmission and

TABLE 1.1

Annual Cases of Potentially Airborne Nosocomial Pathogens

Microbe	Group	Airborne Class	Type	Source	Annual Cases
Influenza A virus	RNA Virus	1	Communicable	Humans	2,000,000
Measles virus	RNA Virus	1	Communicable	Humans	500,000
Streptococcus pneumoniae	Gram+ bacteria	2	Communicable	Humans	500,000
Streptococcus pyogenes	Gram+ bacteria	1	Communicable	Humans	213,962
Respiratory syncytial virus	RNA virus	1	Communicable	Humans	75,000
Varicella-zoster virus	DNA virus	1	Communicable	Humans	46,016
Parainfluenza virus	RNA virus	2	Communicable	Humans	28,900
Mycobacterium tuberculosis	Bacteria	1	Communicable	Humans	20,000
Bordetella pertussis	Gram− bacteria	1	Communicable	Humans	6,564
Rubella virus	RNA virus	1	Communicable	Humans	3,000
Staphylococcus aureus	Gram+ bacteria	1	Endogenous	Humans	2,750
Pseudomonas aeruginosa	Gram− bacteria	1	Noncommunicable	Environmental	2,626
Klebsiella pneumoniae	Gram− bacteria	1	Endogenous	Humans/ Environmental	1,488
Legionella pneumophila	Gram− bacteria	1	Noncommunicable	Environmental	1,163
Haemophilus influenzae	Gram− bacteria	2	Communicable	Humans	1,162
Histoplasma capsulatum	Fungal spore	1	Noncommunicable	Environmental	1,000
Aspergillus	Fungal spore	1	Noncommunicable	Environmental	666
Serratia marcescens	Gram− bacteria	2	Endogenous	Environmental	479
Acinetobacter	Gram− bacteria	2	Endogenous	Humans	147
Corynebacterium diphtheriae	Gram− bacteria	2	Communicable	Humans	10
SARS coronavirus	RNA virus	1	Communicable	Humans	(10)

those resulting from other routes, but even if it was assumed that only 10% of these infections had an airborne component, it can be seen that the problem is potentially significant.

Table 1.2 shows nosocomial pathogens that are definitively airborne (Airborne Class 1) based on the indicated references. Although these latter agents are occasionally fatal, it is the bacteria such as *Streptococcus, Staphylococcus, M. tuberculosis, Bordetella, Pseudomonas, Klebsiella, Legionella, Haemophilus,* and *Serratia* that cause the most fatalities. The fungi are predominantly a problem to the immunocompromised, who have insufficient immunity to resist microbes that would not pose a threat to healthy individuals. Many of the bacteria listed in Table 1.2 are endogenous, which means that they hail from human carriers. In most cases these endogenous microbes, like *Streptococcus* and *Staphylococcus*, are commensals, meaning that they live symbiotically in or on humans and cause no particular problems. It is usually only when these commensal microbes get into the wrong place that they cause serious infections. An endogenous microbe can cause an infection in the same patient from which they come and a surgical wound can become contaminated with a person's own endogenous microflora. Endogenous

TABLE 1.2

Airborne Class 1 Nosocomial Pathogens

Microbe	Type	Primary References
Aspergillus	Fungal spore	HICPAC 2003, 2007
Bordetella pertussis	Bacteria	HICPAC 1998
Clostridium difficile	Bacterial spore	Nielsen 2008; Best et al. 2010; Snelling et al. 2011; Wilcox et al. 2011
Coronavirus (SARS)	Virus	HICPAC 2007, ASHRAE 2009
Histoplasma capsulatum	Fungal spore	Ryan 1994; Heymann 2008
Influenza A virus	Virus	HICPAC 2003; HICPAC 1998; ASHRAE 2009
Legionella pneumophila	Bacteria	HICPAC 2007
Measles virus	Virus	HICPAC 2003, 2007; ASHRAE 2009
Mumps virus	Virus	ASHRAE 2009; NCIRD 2011
Mycobacterium tuberculosis	Bacteria	HICPAC 2003; CDC 2005; HICPAC 2007
Norwalk virus (Norovirus)	Virus	Sawyer et al. 1988; Chadwick and McCann 1994; Caul 1994; Gellert, Waterman, and Ewert 1990
Pseudomonas aeruginosa	Bacteria	Lowbury and Fox 1954; Ransjo 1979; Grieble et al. 1974; Kelsen and McGuckin 1980
Respiratory syncytial virus (RSV)	Virus	Hall, Douglas, and Geiman 1980; Hall and Douglas 1981; Baron 1996
Rubella virus	Virus	NRC 1998
Staphylococcus aureus	Bacteria	HICPAC 2003
Streptococcus pyogenes	Bacteria	CDC 1999; HICPAC 1998
Varicella-zoster virus (VZV)	Virus	HICPAC 2003; HICPAC 1998, 2007

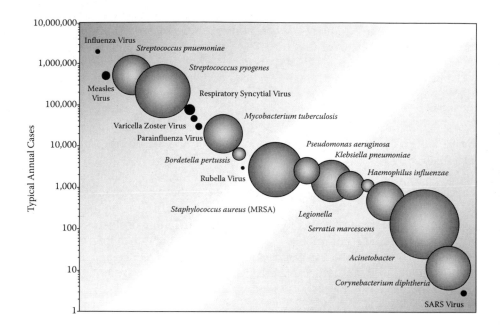

FIGURE 1.2
Total annual cases of potentially airborne nosocomial pathogens. The sphere volumes represent the relative size (aerodynamic diameter) of the indicated pathogen. Viruses are shown in black.

microbes may also come from personnel in the operating room. Several of the pathogens shown in Table 1.2 are environmental microbes, like *Aspergillus* and *Pseudomonas*, which hail from the environment and are common contaminants of hospitals. It is a fair question to ask why hospital ventilation systems can allow ambient environmental microbes to enter the building and cause infections when the requisite hospital air filters should easily remove them, and this matter will be evaluated in later chapters. Figure 1.2 graphically illustrates the annual cases with each pathogen shown in terms of its relative size.

Figure 1.3 illustrates the common types of nosocomial infections and how they break down by frequency, based on data from the National Nosocomial Infection Surveillance Report (CDC 1996). Although this breakdown does not indicate which infections may be airborne, it does help eliminate categories that are unlikely to be airborne. Bloodstream infections and urinary tract infections (UTIs) can be largely excluded from consideration as airborne infections, although it is possible that some of these infections may result from equipment contaminated by microbes settling on equipment. The degree to which intravascular devices and catheters become contaminated with fomites from air settling is not known, although many of the contaminating bacteria are potentially airborne (Farr 1999). The airborne route has a much larger role in respiratory infections (the Other category in Figure 1.3), and also pneumonia and surgical site infections. Pneumonia is the second most common nosocomial infection overall and the most common infection

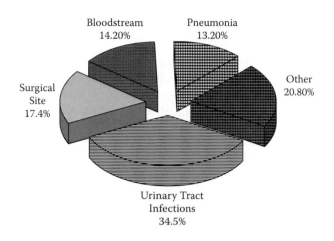

FIGURE 1.3
Types of nosocomial infections. Based on data from the NNIS report (CDC 1996).

in intensive care units (Bonten and Bergmans 1999). Two types of pneumonia can be distinguished, respiratory and ventilator-assisted, of which the
latter is considerably more likely to result from direct contact. Ventilator-
assisted pneumonia can therefore be largely excluded from consideration
as an airborne infection, although again there is a possibility that fomites
contaminating respirator equipment may have settled from the air. The fact
that some SSIs have been caused by airborne pathogens such as *S. aureus* and
Group A *Streptococcus* (GAS) has been well established, and data to support
this view is presented in later chapters.

Airborne Nosocomial Pathogens

Throughout this book a true *airborne nosocomial pathogen* will be defined as any
nosocomial pathogen in which airborne transmission or airborne transport contributes to the etiology of infection regardless of how minimal that contribution
may be, as long as evidence exists that the pathogen has been shown to be airborne in hospital environments and that it has been responsible for nosocomial
infections. *Potential airborne nosocomial pathogens* will refer to all other agents
that are suspected to have an airborne component in their etiology but have not
yet been absolutely proved to transmit by this route in hospitals. This latter category may include airborne pathogens that transmit via the airborne route in
the community but have not (yet) presented a problem in nosocomial settings.
In order to quantify this distinction the concept of nosocomial *Airborne Class*
is introduced. Very simply, Airborne Class 1 refers to true airborne nosocomial pathogens while Airborne Class 2 refers to potential airborne nosocomial

pathogens. Table 1.2 lists all Airborne Class 1 nosocomial pathogens and the primary references for airborne transport in hospitals. Additional supportive references are presented in Chapter 4 and elsewhere in this book.

The Class 2 pathogens are individually introduced in Chapter 4 and are also addressed in various other chapters. The entire array of Class 1 and 2 pathogens form an airborne nosocomial pathogen database that includes viruses, bacteria, bacteria spores, and fungal spores. This database will be used in later chapters as a basis for analysis of the filtration and disinfection capabilities of air cleaning equipment.

Several other fungal spores like *Mucor* and *Rhizopus* are indisputably airborne and are known to be a threat to the immunocompromised but have not yet caused a significant enough problem to warrant their inclusion in Class 1. This is typical of many of the Class 2 pathogens that have been known to cause airborne infections in the community or in specific geographic areas. If the airborne route offers advantages in transmission then someday these microbes may adapt and become a more manifest threat in hospital settings. In fact, many of the bacteria and viruses in Table 1.2 have evolved and adapted to indoor environments to the degree that they cannot survive in the outdoor air, and can only survive in indoor environments long enough to transmit to a new host. It is likely that eventually some Class 2 pathogens will adapt to the hospital environment and thereby become Class 1 pathogens. It is also likely that this adaptation will occur in the community first, because hospitals are relatively harsh environments for the survival of microbes.

Airborne Class 3 implicitly represents all other pathogens for which airborne transmission is not plausible and has not occurred under natural conditions in nosocomial settings. The unnatural condition of forced aerosolization may result in Class 3 pathogens transmitting infections by the air. This class includes microbes such as *Escherichia coli*, *Salmonella*, *Shigella*, *Candida*, and HIV. This class is not the subject of this book and these microbes will not be addressed further in this text except incidentally.

Airborne Infection Categories

An extensive review of the literature has indicated that airborne infections, suspected or otherwise, can be categorized into a series of distinct groups that provides perspective in understanding both the specific etiology and the array of pathogens that present themselves in different nosocomial environments. These categories are summarized in Table 1.3 along with examples of the type of airborne nosocomial pathogens that occur in these settings. The chapter that focuses on these particular infection categories is shown in the final column, although there is some overlap, especially for Category 2, and these topics are also partly addressed in other chapters throughout this book.

TABLE 1.3

Categories of Airborne Nosocomial Infections

No.	Category Title	Typical Pathogens	Chapter
1	Respiratory Infections	Influenza, TB, measles	11
2	Nonrespiratory Infections	*Clostridium difficile*, Norwalk	13
3	Surgical Site Infections (SSIs)	MRSA, GAS	12
4	Burn Wound Infections	MRSA, *Pseudomonas*	15
5	Immunocompromised Infections	Opportunistic bacteria and fungi	15
6	Pediatric Infections	Viruses and commensal bacteria	14
7	Nursing Home Infections	Viruses and pathogenic bacteria	16

These categories are invaluable in helping isolate both the specific pathogens that cause airborne infections and the aerobiological pathways by which they are transported and transmitted, and each of the indicated chapters specifically identifies the complete array of airborne nosocomial pathogens relevant to the category. Quantifying the specific pathogens for each category is of some help in planning disinfection measures and control strategies, and in eliminating nonairborne pathogens from consideration. There has been some tendency in the literature to assume that respiratory infections are the only airborne infections, and these seven categories should help resolve the confusion between airborne respiratory transmission and noninhaled infections that may be transported by the airborne route, thereby allowing more complete focus on each of the individual nosocomial problems. There are certainly additional categories that can be defined, but the categories above account for virtually all cases of airborne nosocomial infections that have been definitively identified in the literature. The categories could also conceivably be further subdivided, such as SSIs that could be broken down by type of surgery, but the categories shown are sufficient to begin an exploration of the means and modes of airborne infection transmission and possible control methodologies.

Respiratory infections (Category 1) are a well-defined class of pathogens that include bacteria, viruses, and fungi, and that are often transmitted via airborne inhalation. It seems clear, however, that most, if not all, respiratory pathogens can also be transmitted by direct and indirect contact, and therefore the control of airborne respiratory infections cannot be fully accomplished without also addressing surface disinfection and hand hygiene.

Nonrespiratory infections (Category 2) include mainly gastrointestinal infections and specifically exclude those in the SSI, burn wound, and immunocompromised categories. There is some overlap between this category and Categories 6 and 7, but this should present no undue confusion. The airborne transmission of gastrointestinal pathogens is a relatively recently recognized problem, and certain agents such as Norwalk virus and rotavirus are suspected of being airborne. Besides the fomite route, any airborne pathogens that are inhaled or otherwise enter the upper respiratory tract may be cleared by mucociliary action and then ingested, which may result in a gastrointestinal infection (Hoffman, Bennett, and Scott 1999).

Surgical site infections (Category 3) are a distinct and important class of airborne infections that involve mainly endogenous or commensal bacteria that may come from either the patient or the surgical team in the operating room. SSI infections involve mainly endogenous microbes, but some environmental microbes also pose infection risks in the operating room.

Burn wound infections (Category 4) are also a distinct class of infection in which a specific array of microbes can be identified as causative agents and these include both endogenous and environmental microbes.

Immunocompromised infections (Category 5) affect those with reduced immunity due to either AIDS or to immunodeficiency caused by other factors (i.e., cancer therapy). These infections may be due to both endogenous and environmental opportunistic pathogens.

Pediatric infections (Category 6) involve pathogens that can affect neonates, newborns, and children, including viruses like respiratory syncytial virus (RSV) and varicella-zoster virus (VZV). They occur primarily in neonatal intensive care units (NICUs), newborn nurseries, and childcare facilities. The array of Category 6 pathogens is fairly distinct and includes both respiratory and nonrespiratory nosocomial and community-acquired pathogens that may also occur as Category 1 and 2 infections. Childcare facilities and schools could also be included in a broader definition of this category, although they are not really nosocomial settings.

Nursing home infections (Category 7) include a wide array of pathogens that can affect the elderly. Many of these agents are also respiratory and nonrespiratory airborne infections. These pathogens apply to long-term care facilities (LTCFs) as well as home care.

Infection Control Precautions

There are four classes of infection control precautions defined by the Healthcare Infection Control Practices Advisory Committee (HICPAC 2007), and these are Standard Precautions, Droplet Precautions, Contact Precautions, and Airborne Precautions. Most of the pathogens treated in this book are subject to one or more of these precautions (as specified in Chapter 4).

Standard Precautions are recommended for the care of all patients and are designed to reduce the risk of transmission of pathogens from both recognized and unrecognized sources of infection in hospitals. Standard Precautions apply to (1) blood; (2) body fluids, secretions, and all excretions except sweat; (3) nonintact skin; and (4) mucous membranes. Depending on the situation, these precautions may include handwashing, gloves, facemasks, eye protection, face shields, gowns, patient-care equipment, linen, and environmental control. Infectious patients should be placed in private rooms.

Droplet Precautions apply to infections likely to produce droplets via coughing, sneezing, or other means. Droplet transmission involves droplets (larger then 5 microns) contacting the mucous membranes of the nose or mouth of susceptible persons. Droplets of this size do not remain suspended in the air and generally travel only about 3 feet through the air, and therefore transmission requires close contact between the infected patient and the recipient. Facemasks are required and patients should be placed in private rooms or cohorted.

Contact Precautions are intended to prevent direct contact transmission, which can involve skin-to-skin contact and physical transfer of pathogens from an infected patient to a susceptible host. Direct contact can occur between two patients by hand or other contact. Indirect contact transmission involves contact with a contaminated object or inanimate surface in the hospital environment. Gloves should be worn and changed between patients and procedures, and hands should be washed before regloving. Gowns should be worn if there is going to be substantial contact, and gowns should be removed before leaving the room. Patient placement should be in a private room. Patient care equipment should be dedicated to a single patient.

Airborne Precautions require special air handling and ventilation. Airborne transmission occurs by small droplets (less then 5 microns), droplet nuclei, or dust particles that remain suspended in air. Airborne transport of these particles may carry pathogens long distances from the patient source. Patient rooms should have negative air pressure and 6–12 air changes per hour (ACH) with no air recirculated to other areas unless properly filtered with high efficiency filters. Respiratory protection (N95 respirators) should be worn when entering the room.

For additional details of the various precautions see HICPAC (2007) or other authoritative references.

Infections in Health Care Workers

Health care workers (HCWs), including doctors, nurses, dentists, lab technicians, and any other workers who function in a support role in health care facilities, are subject to airborne microbiological hazards from patients, from the hospital environment, and even from other health care workers. Health care workers are at risk for contracting a variety of infections on the job and a number of these infections may be respiratory in nature such as TB, influenza, measles, mumps, cold viruses, RSV, GAS, MRSA, and other drug-resistant microorganisms (Charney and Fragola 1999). In most cases where medical workers have contracted respiratory infections from inhalation, the root cause was often inadequate local ventilation, malfunctioning systems and equipment, or administrative control problems (Castle and Ajemian 1987).

Exposure to respiratory pathogens can occur during the treatment of patients and by handling contaminated materials. Although inhalation is a possible infection route in all cases, close proximity to patients for extended periods is a major risk factor regardless of whether the pathogens are aerosolized or not. Measures can be taken to reduce the risks to HCWs in health care environments including engineering controls, work practice controls, administrative controls, and personal protective equipment. Engineering control strategies can be used to control the source of contaminants as in TB isolation rooms. Work practice controls include good hygiene and good housekeeping. Childcare providers are at risk of acquiring infectious diseases in the workplace (Cordell and Solomon 1999). Infections caused by VZV, parvovirus, and cytomegalovirus (CMV) pose a significant risk to pregnant women.

Multidrug Resistance

The emergence of multidrug resistance in nosocomial pathogens has reduced the arsenal of effective antibiotics available for treatment. A warning from the Centers for Disease Control and Prevention (CDC) states that more than 70% of the bacteria that cause infections are resistant to at least one antibiotic used to treat them. The most dangerous of these microbes are those that have the potential to spread by the airborne route and include MRSA and multidrug-resistant tuberculosis (XTB). Once these microbes develop immunity to antibiotic treatment they have a tendency to cause outbreaks in hospital environments, and some have, on occasion, spread into local communities.

Bacteria that cause respiratory infections have been developing increased drug resistance over the past decades. The drug resistance of streptococcal infections, which can cause scarlet fever, has increased from 0.8% to 28% in the past decade. MRSA, the so-called superbug, has shown up repeatedly outside hospital settings and has become a problem in athletics, where it may contaminate sports equipment and facilities (Kowalski 2007). MRSA is being isolated with increasing frequency, and according to one study, it is the most frequently isolated airborne microbe (Durmaz et al. 2005). MRSA has been reported in prisons, sports facilities, and residential homes. MRSA is no longer confined to hospitals, and new strains are infecting young, healthy people. MRSA may not cause noticeable problems in healthy people, because it is largely a commensal or endogenous microbe, but once a person ends up in the hospital, it can cause unwanted complications.

Multidrug-resistant tuberculosis has caused a resurgence in this disease worldwide, and close to a million people die from it each year. The CDC has warned that outbreaks from small pets like hamsters, mice, and rats have sickened up to 30 people in at least 10 states with dangerous multidrug-resistant bacteria. Infected farm animals have also been found. Studies indicate that when multidrug-resistant microbes develop in hospital settings, they often first spread to the immediate surrounding community then begin spreading outward and elsewhere as people move about, and silent outbreaks begin occurring in distant cities and other countries.

Toward Airborne Infection Control

The key to controlling airborne infections lies in the use of all resources, methodologies, and technologies for controlling microbial contamination in hospitals regardless of whether they are applied to air, surfaces, equipment, or hands. Any top-down structured approach to controlling airborne infections must address the same issues and use the same methods used for controlling other types of infections and this includes management of infections, isolation of patients, vaccination programs, cohorting patients, hand hygiene, HCW education and training, monitoring and identification of infectious agents, and surface disinfection procedures as well as engineering controls for air cleanliness. The various time-tested protocols and procedures for controlling infections need to be adhered to in order for them to work effectively, and air or surface disinfection techniques cannot be a complete solution without full implementation of standard precautions or other techniques for reducing infection risks. Figure 1.4 illustrates the major components of an effective program for controlling airborne nosocomial infections.

This book examines the etiology of airborne nosocomial infections and addresses the various scenarios in which such infections occur. The subjects of treatment, vaccination, and prophylaxis are not treated in this text, and there is only limited discussion of procedural controls because these

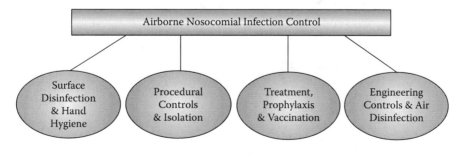

FIGURE 1.4
The major components of an effective airborne infection control program.

topics are general and are treated accurately and completely in many other texts. The subjects of air and surface contamination are addressed in detail along with the identification of the specific microbial hazards for each scenario. The methods for dealing with contamination problems are covered and these include hospital surface disinfection, hand hygiene, isolation, and air disinfection. The primary air disinfection methods addressed include ventilation dilution, air filtration, and ultraviolet germicidal irradiation (UVGI). Ultraviolet (UV) disinfection is also addressed separately as a surface disinfection technology. A variety of alternative technologies for air and surface disinfection are also discussed that may be relevant to particular or general applications.

References

ASHRAE (2009). *ASHRAE Position Document on Airborne Infectious Diseases.* American Society of Heating, Refrigerating and Air-Conditioning Engineers, Atlanta, GA.

Baron, S. (1996). *Medical Microbiology.* University of Texas Medical Branch, Galveston, TX.

Bennett, J. V., Jarvis, W. R., and Brachman, P. S. (2007). *Bennett & Brachman's Hospital Infections.* Lippincott Williams & Wilkins, Philadelphia.

Best, E. L., Fawley, W. N., Parnell, P., and Wilcox, M. H. (2010). The potential for airborne dispersal of *Clostridium difficile* from symptomatic patients. *Clin Inf Dis* 50, 1450–1457.

Bonten, M. J. M., and Bergmans, D. C. J. J. (1999). Nosocomial pneumonia; in *Hospital Epidemiology and Infection Control,* C. G. Mayhall, ed., Lippincott Williams & Wilkins, Philadelphia, 211–238.

Castle, M., and Ajemian, E. (1987). *Hospital Infection Control.* John Wiley & Sons, New York.

Caul, E. O. (1994). Small, round structured viruses: Airborne transmission and hospital control. *Lancet* 343, 1240–1242.

CDC (1996). National Nosocomial Infections Surveillance (NNIS) Report, data summary from October 1986–April 1996, Issued May 1996. *AJIC* 24(5), 380–388.

_____. (1999). *Guideline for Prevention of Surgical Site Infection.* Centers for Disease Control, Atlanta, GA.

_____. (2005). *Guidelines for Preventing the Transmission of Mycobacterium tuberculosis in Health-Care Facilities.* Centers for Disease Control, Atlanta, GA.

Chadwick, P. R., and McCann, R. (1994). Transmission of a small round structured virus by vomiting during a hospital outbreak of gastroenteritis. *J Hosp Infect* 26, 251–259.

Charney, W., and Fragala, G. (1999). *The Epidemic of Health Care Worker Injury: An Epidemiology.* CRC Press, Boca Raton, FL.

Cordell, R. L., and Solomon, S. L. (1999). Infections acquired in child care centers; in *Hospital Epidemiology and Infection Control,* C. G. Mayhall, ed., Lippincott Williams & Wilkins, Philadelphia, 695–716.

Davis, J. M., and Shires, G. T. (1991). *Principles and Management of Surgical Infections.* J.B. Lippincott, New York.

Durmaz, G., Kiremitci, A., Akgun, Y., Oz, Y., Kasifoglu, N., Aybey, A., and Kiraz, N. (2005). The relationship between airborne colonization and nosocomial infections in intensive care units. *Mikrobiyol Bul* 39(4), 465–471.

Eickhoff, T. C. (1994). Airborne nosocomial infection: A contemporary perspective. *Infect Control Hosp Epidemiol* 15(10), 663–672.

Farr, B. M. (1999). Nosocomial infections related to use of intravascular devices inserted for short-term vascular access; in *Hospital Epidemiology and Infection Control,* C. G. Mayhall, ed., Lippincott Williams & Wilkins, Philadelphia, 157–164.

Fisk, W., and Rosenfeld, A. (1997). Improved productivity and health from better indoor environments. *Center for Building Science News* Summer, 5.

Freeman, J. (1999). Modern quantitative epidemiology in the hospital in *Hospital Epidemiology and Infection Control,* C. G. Mayhall, ed., Lippincott Williams & Wilkins, Philadelphia, 15–48.

Gellert, G. A., Waterman, S. H., and Ewert, D. (1990). An oubreak of acute gastroenteritis caused by a small round structured virus in a geriatric convalescent facility. *Inf Contr Hosp Epidemiol* 11, 459–464.

Grieble, H., Bird, T., Nidea, H., and Miller, C. (1974). Chute-hydropulping waste disposal system: A reservoir of enteric bacilli and *Pseudomonas* in a modern hospital. *J Infect Dis* 130, 602.

Hall, C. B., and Douglas, R. G. Jr. (1981). Modes of transmission of respiratory syncytial virus. *J Pediatr* 99, 100–103.

Hall, C. B., Douglas, R. G., and Geiman, J. M. (1980). Possible transmission by fomites of respiratory syncytial virus. *J Infect Dis* 141(1), 98–102.

Heymann, D. L. (2008). *Control of Communicable Diseases Manual.* American Public Health Association, Washington, DC.

HICPAC (1998). *Guideline for Infection Control in Health Care Personnel.* Centers for Disease Control Atlanta, GA.

_____. (2003). Guidelines for environmental infection control in health-care facilities. *MMWR* 52(RR-10), 1–48.

_____. (2007). *Guideline for Isolation Precautions: Preventing Transmission of Infectious Agents in Healthcare Settings.* Centers for Disease Control, Atlanta, GA.

Hoffman, P. N., Bennett, A. M., and Scott, G. M. (1999). Controlling airborne infections. *J Hosp Infect* 43(Suppl), 203–210.

Kelsen, S. G., and McGuckin, M. (1980). The role of airborne bacteria in the contamination of fine particle neutralizers and the development of nosocomial pneumonia. *Ann NY Acad Sci* 353, 218.

Klevens, R. M., Edwards, J. R., Richards, C. L., Horan, T. C., Gaynes, R. P., Pollock, D. A., and Cardo, D. M. (2007). Estimating health care-associated infections and deaths in U.S. hospitals, 2002. *Pub Health Rep* 122, 160–166.

Kowalski, W. J. (2006). *Aerobiological Engineering Handbook: A Guide to Airborne Disease Control Technologies.* McGraw-Hill, New York.

_____. (2007). Airborne superbugs: Can hospital-acquired infections cause community epidemics? *Consulting Specifying Engineer* 41(3), 28–36, 69.

Lowbury, E. J. L., and Fox, J. (1954). The epidemiology of infection with *Pseudomonas pyocyanea* in a burn unit. *J Hyg* 52, 403–416.

McDowell, N. (2003). Air ionizers wipe out hospital infections. NewScientist.com news service. www.newscientist.com/article.ns?id=dn3228

NCIRD (2011). Mumps; in *Epidemiology and Prevention of Vaccine-Preventable Diseases, 12th ed.* Centers for Disease Control and Prevention (CDC), Atlanta, GA, 189–198.

Nielsen, P. (2008). *Clostridium difficile* aerobiology and nosocomial transmission. Northwick Park Hospital Harrow, Middlesex, UK.

NRC (1998). *Department of the Air Force Operating Room Ventilation Update.* National Resource Center Brooks Air Force Base, TX.

Ransjo, U. (1979). Attempts to control clothes-borne infection in a burn unit. *J Hyg* 82, 369–384.

Ryan, K. J. (1994). *Sherris Medical Microbiology.* Appleton & Lange, Norwalk.

Sawyer, L. A., Murphy, J. J., Kaplan, J. E., Pinsky, P. F., Chacon, D., Walmsley, S., Schonberger, L. B., Phillips, A., Forward, K., Goldman, C., Brunton, J., Fralick, R. A., Carter, A. O., Gary, W. G., Glass, R. I., and Low, D. E. (1988). 25- to 30-nm virus particle associated with a hospital outbreak of acute gastroenteritis with evidence for airborne transmission. *Am J Epidemiol* 127, 1261–1271.

Scott, R. D. (2009). *The Direct Medical Costs of Healthcare-Associated Infections in U.S. Hospitals and the Benefits of Prevention.* Centers for Disease Control, Atlanta, GA.

Snelling, A. M., Beggs, C. B., Kerr, K. G., and Sheperd, S. J. (2011). Spores of *Clostridium difficile* in hospital air. *Clin Infect Dis* 51, 1104–1105.

Wilcox, M. H., Bennett, A., Best, E. L., Fawley, W. N., and Parnell, P. (2011). Reply: Spores of *Clostridium difficile* in hospital air. *Clin Infect Dis* 51, 1105.

Wong, S. (1999). Surgical site infections; in *Hospital Epidemiology and Infection Control,* C. G. Mayhall, ed., Lippincott Williams & Wilkins, Philadelphia, 189–210.

2

Airborne Nosocomial Epidemiology

Introduction

Epidemiology is the study of disease transmission through the use of statistical and mathematical models. In this chapter the focus is on the epidemiology and dosimetry of airborne nosocomial infections. Relevant epidemiological statistics were summarized in the previous chapter and will not be revisited here. This chapter reviews transmission routes and summarizes the mathematics of airborne epidemiology and dosimetry, adapted for the hospital environment. These modeling tools will enable designers and infection control personnel to evaluate outbreaks or epidemics involving airborne transmission as well as surface-borne outbreaks. It must be reiterated here that the subject of airborne transmission is inextricably tied to surface transmission and that these subjects must be treated together in most cases. Obligate airborne transmission is thought to be a property of certain pathogens such as TB or influenza, but even these microbes may be transmitted via direct contact or contact with fomites. It will be shown herein that the infectious dose of any pathogen received at any entry point, whether the respiratory tract or an open wound, is the cumulative dose received by all routes, be they airborne, direct contact, or even endogenous migration. For each pathogenic species or each infection, one route may predominate but no routes should be neglected.

It is also important to qualify certain useful concepts and distinctions such as airborne transport and droplet nuclei. *Airborne transmission* can be defined as the process in which an airborne microorganism produces an infection, and this may involve inhalation or settling of pathogens on the site of infection. *Airborne transport* is simply defined as the movement of microbes in the air from one point to another, as occurs when spores drift on air currents through building spaces. Airborne transport of pathogens can result in inhalation or in contamination of surfaces with fomites.

Droplet nuclei are defined as the residue of evaporated droplets and these nuclei are regarded as being less than about 5 microns in size. This size range gives them the ability to become suspended in air. Droplet nuclei may contain as little as a single microbe or many thousands of microbes. *Droplet spray* can be defined as droplets larger than 5 microns that do not quickly evaporate but are projected by the expulsion force of coughing or sneezing out to a distance of 6 feet or more. Droplets in droplet spray are not truly airborne in the sense that they are suspended in air but are transported through the air until they arrive at a point of infection or at some surface where they may become fomites. Many sources consider droplet spray to be a form of direct contact, which cannot be argued, but this distinction does not add any parameters to the discussion that are not already being considered. Suffice it to say that droplet spray produces fomites whether they directly contact mucosal surfaces or produce surface contamination. *Fomites* are any particles, droplets, droplet nuclei, dust particles, or skin squames that contain infectious microbes that remain viable for periods of time, which depends on the ability of each pathogen to survive outside a host. *Aerosol clouds* are produced whenever droplets, droplet nuclei, or skin squames become suspended in the air and become subject to airborne transport (Liu et al. 2000). *Shedding* can result in particles that fall directly downward or in aerosols that may remain suspended for some time.

Airborne Transmission Routes

Nosocomial infections that involve airborne transmission routes may include pathogens that are inhaled, pathogens that settle from the air, or pathogens that contaminate equipment and hands. These distinctions are not absolute because multiple routes may be involved in any infection and, in fact, surface contamination probably plays a major role in most airborne infections.

Airborne respiratory infections such as upper respiratory infection (URI) can be treated and evaluated separately from other forms of airborne nosocomial infections. The same is true for lower respiratory infections (i.e., pneumonia) and many other upper airway infections like nasopharyngeal infections. An exception is ventilator-assisted pneumonia (VAP), which is almost invariably the result of contaminated equipment. However, part of the equipment contamination problem in VAP surely results from airborne transport of pathogens from environmental sources, such as when *Aspergillus* contamination occurs. Many respiratory infections may be the result of direct contact with fomites, but as mentioned earlier, the problem of surface-borne infections is interwoven with the problem of airborne infections, and this is especially true in the case of respiratory infections that produce droplets.

Surgical site infections (SSIs) involve both direct contact and airborne transmission, as has been fairly well established in numerous studies (see Chapters 4 and 12). No doubt the majority of SSIs involve endogenous microbes that either migrate to or otherwise contaminate the open wound before, during, or after surgery, but the evidence also indicates that some airborne microbes inevitably settle on wounds during surgery. Endogenous microbes shed from the operating team or ambient environmental contaminants that enter the operating room (OR) contribute to the contamination of surgical sites. Shedding of endogenous microbes like *Staphylococcus aureus* or skin squames containing bacteria is akin to droplet spray in that these particles may fall through the air without actually becoming suspended in air. It has been shown, however, that air currents in the OR can either drive such particles away or toward the surgical site. SSIs are a category of infections that are not directly governed by the epidemiological principles presented in this chapter because they are not contagious and do not spread in epidemic fashion. Outbreaks of SSIs, however, might be considered point source epidemics and modeled as such.

Nonrespiratory infections may sometimes be transported by the airborne route even though the actual infection results from ingestion or other type of direct contact. Gastrointestinal infections such as *Clostridium difficile* and noroviruses fall into this category.

Burn wound infections are similar to surgical wounds in that infecting microbes may arrive via settling from the air or may result from equipment that has been contaminated by settled microbes. Burn wounds are also subject to opportunistic infections because the host's natural defenses in the skin are compromised.

Immunocompromised infections comprise a loosely separate category because they may involve respiratory infections with opportunistic endogenous and environmental microbes that normally pose no serious threat to healthy individuals. Immunocompromised patients are also subject to skin and other infections from microbes that may settle from the air or come from contaminated surfaces and equipment.

The epidemiological models presented herein apply mainly to respiratory and nonrespiratory categories of contagious infections that may be spread in hospitals. The models of microbial survival and dosimetry presented here apply to all categories.

Epidemiological Principles

Diseases or infections must be treated as either communicable or noncontagious (noncommunicable) for purposes of evaluating the epidemiology. The degree to which a hospital population is affected by an outbreak of a noncommunicable disease depends primarily upon the virulence of the microbes, the dose received by individuals, and the natural resistance or immunity of the individuals. All of these factors tend to behave probabilistically and can be described by a Gaussian or normally distributed curve.

The basic epidemiological equation, called the Soper equation, is as follows (Wilson and Worcester 1944):

$$C = rIS \qquad (2.1)$$

where
C = number of new infections
r = average contact rate, fractional
I = number of infected disseminators
S = number of susceptible individuals

The production of secondary or new infections in a susceptible population of patients or HCWs is termed a generation. In an epidemic, several generations may occur as the infection is constantly retransmitted to new susceptible patients. Once the supply of susceptibles has been exhausted the epidemic will end. Normally, not everyone gets infected because there will be insufficient population density to propagate the epidemic as it winds down. That is, the density of susceptibles becomes so low that no new transmissions occur. Mathematically, if $C/I > 1.0$ it denotes a propagating epidemic. If $C/I < 1.0$ the epidemic will rapidly diminish.

Figure 2.1 shows an example of an outbreak of a contagious infection when the average contact rate $r = 0.4$. The number of susceptibles decreases in sigmoid fashion while the number of infection cases forms a bell curve.

Equation (2.1) is independent of the mode of transmission and, with appropriate adjustments, can be applied to direct contact, airborne, or any combination of transmission modes. It can also be applied to outbreaks of surgical site infections. The model can be adapted to other noncontagious infections that hail from a common source, such as *Legionella* outbreaks.

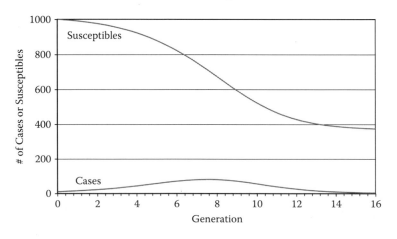

FIGURE 2.1
Example of contagious disease transmission for general outbreaks.

In a rearrangement of Equation (2.1),

$$\frac{C}{I} = rS \qquad (2.2)$$

The value rS determines the value of C/I and is called the contagious potential. The rate r can only be determined by epidemiological data, but a value of between 0.1 and 0.2 would be typical and conservative for most respiratory infections. Other computational methods are available to model epidemics, but they produce results virtually identical to Figure 2.1 (Ackerman, Elveback, and Fox 1984; Daley and Gani 1999; Frauenthal 1980).

The epidemic spread of respiratory infections in hospitals can have complex transmission factors that include inhalation, direct contact, and fomite spread. Riley (1980) introduced a new definition of the contact rate in terms of the release rate of the agent, the breathing rate, the exposure time, and the dilution rate of the room air. Using these concepts, the previous equations can be adapted to model infections spread by ventilation systems by defining these terms:

d = quanta of infection produced by each infective individual
Q = volumetric flow of fresh ventilation air (m^3/min)
p = volume of air breathed by each susceptible individual (m^3/min)
x = building characteristic constant

The ventilation system model is then defined by the equation

$$C = \left(\frac{xpd}{Q} \right) IS \qquad (2.3)$$

Comparison with Equation (2.1) makes it clear that the contact rate will be

$$r = \frac{xpd}{Q} \qquad (2.4)$$

For a patient or HCW breathing at rest the value of p is approximately 0.015 m^3/min. The value of d depends on the particular pathogen. For measles the value has been found to be about 9.1 based on interpretation of epidemiological studies (Remington et al. 1985). The factor "x" accounts for the effect of the air exchange rate of each hospital building or separate ventilation zone within the building. Figure 2.2 provides an application of the ventilation model to the epidemiological data from the 2003 Hong Kong SARS outbreak, which spread in hospital settings (HWFB 2003). Due to the large number of

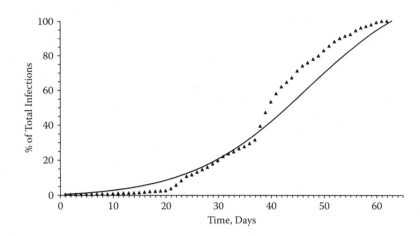

FIGURE 2.2
Epidemic model applied to SARS outbreak in Hong Kong. Line represents ventilation model.
Dots represent new infection cases.

infections, 1348, the data remain relatively well behaved. Smaller data sets may not produce such good conformity to the model.

These epidemiological models can be used to assess the effectiveness of intervention techniques such as vaccines as well as air disinfection technologies. Vaccines will decrease the population of susceptibles in Equations (2.1)–(2.4). Increased airflow rate (Q) will decrease the infections in Equation (2.3), as will the air exchange rate and any equivalent air exchange rate due to air disinfection technologies.

Dosimetry of Airborne Disease

It may be useful to estimate the infections that may occur from exposure to disease agents. The dose of any pathogen depends on the route of exposure. There are three possible routes of exposure: inhalation, skin exposure, and ingestion. The main factor relevant to airborne nosocomial dosimetry is the mean infectious dose, defined as follows:

> ID_{50}: Mean Infectious Dose. The dose or number of pathogens that will cause infections in 50% of an exposed population. Units are in terms of the number of colony-forming units per cubic meter (cfu/m^3). Similarly, the proper term for describing viable viruses in culture is "pfu" or plaque-forming units.

For viruses, the infectious dose is often expressed in terms of the median tissue culture infectious dose ($TCID_{50}$), which is the amount of a virus that

TABLE 2.1

Doses for Airborne Nosocomial Pathogens

Pathogen	Microbe	Infectious Dose ID_{50}
Adenovirus	Virus	150
Blastomyces dermatitidis	Fungal spore	11,000
Bordetella pertussis	Bacteria	(4)
Clostridium perfringens	Bacteria	10 per g of food
Coccidioides immitis	Fungal spore	100–1350
Coxsackievirus	Virus	67
Cryptococcus neoformans	Fungal yeast	1000
Histoplasma capsulatum	Fungal spore	10
Influenza A virus	Virus	20
Legionella pneumophila	Bacteria	<129
Mycobacterium tuberculosis	Bacteria	1–10
Mycoplasma pneumoniae	Bacteria	100
Neisseria meningitidis	Bacteria	110
Paracoccidioides	Fungal spore	8,000,000
Parvovirus B19	Virus	0.5 mL of serum
Respiratory syncytial virus	Virus	100–640
Rhinovirus	Virus	100
Rubella virus	Virus	10–60

will produce pathological change in 50% of inoculated cell cultures and is expressed as $TCID_{50}/mL$ (Bischoff 2010). There is also a lethal infectious dose, LD_{50}, but it is less well understood. An infectious dose can cause infections in individuals but not necessarily any fatalities. The lethal dose will always be higher than the infectious dose, although not necessarily by much. Many of the known infectious doses and lethal doses for microorganisms are not known with certainty, or are based on animal studies. Table 2.1 summarizes some of the doses that have been quantified (see Kowalski 2006 for specific references for microbial doses).

The dose-response curve for nosocomial infections produces a normal distribution (a bell curve). Some patients will acquire infections at very low doses while others require a large dose to become infected. If the concentration of airborne microorganisms is approximately constant, then the acquired dose is a linear function of exposure time. Defining E_t as exposure time and C_a as airborne concentration, we can write this equation as follows:

$$Dose = E_t C_a \tag{2.5}$$

Equation (2.5) ignores doses from other routes, such as direct contact, and illustrates the principle for airborne inhalation only. If the airborne concentration is such that the ID_{50} is achieved at 4 hours, then 50% of the exposed

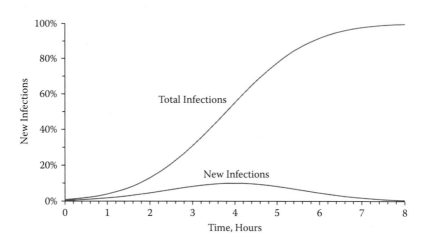

FIGURE 2.3
New infections and total infections over time under exposure to a constant concentration of airborne pathogens.

population will have been infected at that time. Figure 2.3 shows a plot of both new infections and total infections over an 8-hour exposure period, after which 100% infections are reached. The dose-response curve is virtually identical to the epidemiological curve presented in Figure 2.1.

The mathematical relation for predicting the total infections can be developed from the statistical definition of a normal bell curve. If y represents the number of new cases and x represents the dose, the normal distribution curve is given by

$$y = \frac{1}{\sigma\sqrt{2\pi}} e^{-0.5\left(\frac{x-ID_{50}}{\sigma}\right)^2} \tag{2.6}$$

where σ represents the standard deviation.

The standard deviation may not be the same for all pathogens, but it could be any value between a small fraction of the mean and some multiple thereof. Comparison with epidemiological data suggest that the standard deviation must be about 0.25–0.5 of the mean to provide reasonable and elastic results. Some data are available from published sources that have tabulated dose-response data for several pathogens (Haas 1983; Haas, Rose, and Gerba 1999). The standard deviations ranged from 0.028 to 2 times the mean. Based on Kowalski (2006), the value of 0.5 is a representative value and Equation (2.6) is rewritten as follows:

$$y = \frac{2}{ID_{50}\sqrt{2\pi}} e^{-2\left(\frac{x-ID_{50}}{ID_{50}}\right)^2} \tag{2.7}$$

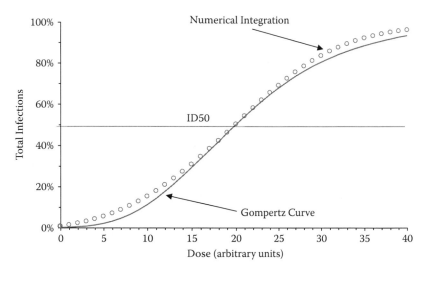

FIGURE 2.4
Comparison of fitted Gompertz curve with total infections computed by numerical integration.

Equation (2.7) can be numerically integrated to predict new infections by summing its results with software or on a spreadsheet. It can be convenient, however, to have a closed form of the equation that will directly predict the total infections for any given dose. The simplest form of Equation (2.7) that will satisfy this purpose is known as a Gompertz curve (Boyce and DiPrima 1997; Whiting 1993). Per Kowalski (2006), the following approximation simulates the numerical solution to Equation (2.7) and gives the sum of y in closed form:

$$y_{tot} = 0.5^{0.1^{\left(\frac{x-ID_{50}}{ID_{50}}\right)}}$$

(2.8)

Figure 2.4 compares Equation (2.8) with the numerical integration of Equation (2.7). Although the curve is not a perfect fit, the range of poorer fit is an area of uncertainty while the area of interest near the ID_{50} value is a good fit.

Equation (2.8) is not intended to supplant the classic epidemiological model, but it serves as a convenient means of estimating infections. It compares reasonably well with dose-response data on inhalation anthrax from other sources (Druett et al. 1953; Hass 2002). It can be used, for example, to evaluate the theoretical performance of air disinfection systems in terms of reduced infections (Kowalski 2003).

Having quantified the epidemiology of airborne disease transmission in the previous section, and having examples of dosimetry from Table 2.1, we need only define the release rates and we have a complete model of hospital contagious disease transmission. These models can be used to simulate epidemic outbreaks of respiratory and nonrespiratory diseases in nosocomial settings.

The actual number of airborne microbes generated by an infectious patient is not known with any certainty and varies greatly between species. The rate of release of TB bacilli from an infected individual is about 1–250 quanta per hour, where the quanta could be as little as a single TB bacilli (Nardell et al. 1991). In one school outbreak, measles was produced at a rate of 5480 quanta per hour (Wheeler 1993). In the case of viruses like measles, each quanta may represent thousands of virions. Remington et al. (1985) reports on a measles case in which an index patient was producing 8640 quanta per hour. As scant as these data are, they are sufficient to roughly define the bounds of the problem. In fact, if an epidemiological model is used to evaluate the effectiveness of an air treatment system, the dose can be assumed to be unitary to render the model generic, and the efficiency of the air disinfection process alone will determine the reduction in infections for any particular pathogen (Kowalski 2003).

Survival of Microbes Outside the Host

All microbes eventually die a natural death and will not survive outside a host indefinitely even under the most favorable conditions (Mitscherlich and Marth 1984). Fungal and bacterial spores are an exception, because life functions are suspended during their dormancy. Freezing some microbes like viruses under controlled conditions may also preserve them indefinitely. Under all other conditions the microbe will either age and die naturally or die prematurely. Health care facilities already provide favorable conditions for most microbes to survive outside a host by virtue of providing a comfortable environment for humans. Warmth, moisture, humidity, shade, material substrates (i.e., carpets, furnishings), and the presence of food or nutrients ensures that many bacteria, fungi, and viruses can survive indoors for prolonged periods.

Each microbe has a natural decay rate that may be accelerated by various factors including sunlight, dehydration or desiccation, heat, freezing, oxygenation, and the effects of pollutants or disinfectants (Henis 1987; de Mik and de Groot 1977). Dehydration renders microorganisms more susceptible to oxygenation (Cox, Baxter, and Maidment 1973). Most pathogens are mesophiles and prefer temperatures between ambient and body temperature.

Although each individual microbe may age and die from natural or environmental causes, populations of microbes may survive indefinitely if they have nutrients and favorable conditions for growth and multiplication. This applies strictly to bacteria and fungi, but not to viruses, which cannot replicate without a host.

Disinfectants such as those used on floors, furnishings, and equipment will certainly decrease survival on surfaces. The type of surface may impact the

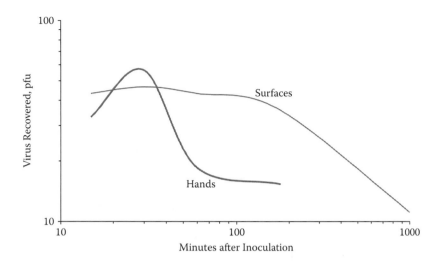

FIGURE 2.5
Survival of Rhinoviruses on skin and surfaces over time. Based on data from Reed (1975).

viability of microbes. Copper and silver surfaces can generate ions that can be lethal to microbes (Thurman and Gerba 1989). Any microbe that does not die instantly on some surfaces may remain as a fomite. Surfaces can include the hands of health care workers when the microbe is a transient colonizer (as opposed to normal microflora or human commensals). Microbial survival times on surfaces or in the environment are provided in Chapter 4 for specific microorganisms where the information is available.

Some microbes survive well on surfaces while others die out quickly. Figure 2.5 shows an example of Rhinovirus survival on skin and surfaces. The warmth or moisture of human hands provides bacteria a more favorable environment than inanimate surfaces, but this is not always true for viruses. Rhinovirus can survive 3–4 hours on hands while parainfluenza and RSV last less than 1 hour on hands (Hurst 1991). Figure 2.6 shows the results of surface survival for three common nosocomial pathogens.

Microbes cannot grow or multiply in air; they can only survive temporarily in air as they are transported to new hosts or more favorable environments. The ultimate fate of aerosolized microbes is to settle out of the air. Larger droplets settle first and the smallest droplets may remain suspended for many hours before they settle. Microbes that are inhaled may find a suitable environment to grow and cause respiratory infections. Particles that settle out on surfaces or equipment (fomites) may remain viable for hours or even days if conditions are right.

Environmental conditions inside hospitals can affect the survival of airborne pathogens. Lowen et al. (2007) demonstrated that influenza virus transmits through the air most readily in cold, dry conditions. High temperatures of 30°C (86°F) were found to block aerosol transmission but not contact

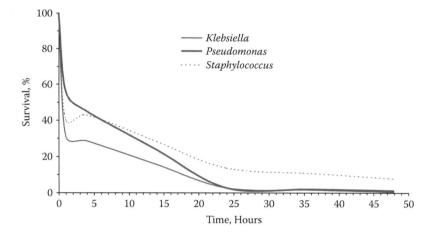

FIGURE 2.6
Survival of nosocomial microbes on laminate surfaces. Based on data from Scott and Bloomfield (1990).

transmission of influenza (Lowen et al. 2008). In general, viruses with lipid envelopes (influenza, coronavirus, RSV, parainfluenza, measles, rubella, VZV) will survive longer at a lower relative humidity (RH) of about 20–30% (Tang 2009). Nonlipid enveloped viruses (adenovirus, rhinovirus) survive longer in high RH (70–90%). Influenza survival is lowest at 40–60% RH. The survival of Gram-negative bacteria, including *Pseudomonas, Enterobacter,* and *Klebsiella,* is highest under high RH conditions. Temperatures above about 24°C (75°F) tend to decrease the survival of most bacteria including *Pseudomonas, Serratia, Bordetella,* and *Mycoplasma.* Studies on bacteria have found that intermediate to high RH of 50–90% produced the lowest airborne survival rates for *Pseudomonas, Proteus, Staphylococcus albus,* and *Streptococcus pneumoniae. Legionella* is most stable at 65% RH and least stable at 55–60% RH. *Mycoplasma* showed increased stability at both low RH (<25%) and high RH (>80%).

The aerosolization process affects the viability of airborne pathogens (Cox 1989). When microbes are disseminated in the wet state, as in sneezing and coughing, they tend to desiccate in the air. When microbes are disseminated in the dry state (i.e., from reaerosolized fomites) they tend to absorb moisture from the air. These processes can affect viability depending on the RH.

Settling of Microbes in Air

Airborne droplets are either evaporated droplets or droplet nuclei smaller than 5 μm in diameter (CDC 2003). Airborne droplets containing viable viruses or bacteria tend to evaporate rapidly in air and condense to droplet nuclei (Nardell 1990). Droplet nuclei are evaporated residues and are of such a small diameter, being typically less than 5 μm, that they can remain airborne for extended periods of time. Studies have shown that the larger droplets are less likely to cause infections because the larger the droplet nuclei, the more likely it is to settle out in the nasopharynx (Druett et al. 1953). Single micro-organisms are more likely to penetrate to the lower respiratory tract and are more infectious in this form (Druett et al. 1956). Particles smaller than 3–5 μm in size have the potential to reach the alveoli and remain there, while particles smaller than 0.25 μm may actually be exhaled sometimes (Gordon and Ingalls 1957).

Particles larger than about 0.3 μm in diameter will tend to settle out over time and these will include most bacteria, fungal spores, dust particles, and droplets or droplet nuclei, which may include clumps of viruses or bacteria. Settling causes particles to accumulate on the floor or the topside of horizontal surfaces. Many particles in the size range of 1–5 microns have settling velocities on the order of 3 feet per hour. Spores like *Aspergillus* often have natural rough surfaces that enhance their ability to float in air. Data showing the settling time for droplets in the size range of about 1–20 μm was presented by Duguid (1945) and is shown plotted in Figure 2.7. Obviously the smaller particles in the 1–2 μm size range can remain airborne for extended periods of time—long enough to diffuse throughout a room or be recirculated by ventilation systems.

Particles smaller than about 0.3 μm (mostly viruses and small bacteria) will tend to remain suspended in air and are subject to the effects of turbulence and diffusion, which may cause them to attach to surfaces. These latter

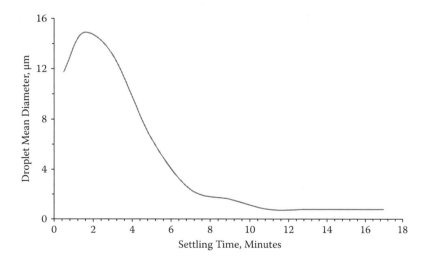

FIGURE 2.7
Settling time versus droplet size.

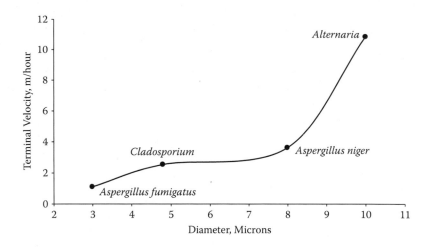

FIGURE 2.8
Terminal falling velocities of some common fungal spores. Based on data from Gregory (1973).

particles may attach to vertical surfaces and the underside of horizontal surfaces. Gregory (1973) studied the terminal velocities of several fungal spores and showed a relationship between falling velocity and size. These results are shown in Figure 2.8.

In the OR, airborne particles originating from the surgeon end up settling on open wounds during surgery (Alexakis et al. 1976). More bacteria and particles are released from skin during activity than during rest, and activity in the OR can directly impact both airborne and wound contamination (Mackintosh et al. 1978; Quraishi et al 1983). ORs often have higher air velocities due to the high volumes of air used (typically 12–15 ACH) and these currents can result in particles greater than 5 microns in size being suspended briefly or carried short distances.

References

Ackerman, E., Elveback, L. R., and Fox, J. P. (1984). *Simulation of Infectious Disease Epidemics*. Charles C. Thomas, Springfield, IL.

Alexakis, P. G., Feldon, P. G., Wellisch, M., Richter, R. E., and Finegold, S. M. (1976). Airborne bacterial contamination of operative wounds. *West J Med* 124, 361–369.

Bischoff, W. E. (2010). Transmission route of rhinovirus type 39 in a monodispersed airborne aerosol. *Inf Contr Hosp Epidemiol* 31(8), 857–859.

Boyce, W. E., and DiPrima, R. C. (1997). *Elementary Differential Equations and Boundary Value Problems*. John Wiley & Sons, New York.

CDC (2003). Guidelines for environmental infection control in health-care facilities. *MMWR* 52(RR-10), 1–48.

Cox, C. S. (1989). Airborne bacteria and viruses. *Sci Prog* 73, 469–499.

Cox, C. S., Baxter, J., Maidment, B. J. (1973). A mathematical expression for oxygen-induced death in dehydrated bacteria. *J Gen Microbiol* 75, 179–185.

Daley, D. J., and Gani, J. (1999). *Epidemic Modelling: An Introduction.* Cambridge University Press, New York.

de Mik, G., and de Groot, I. (1977). The germicidal effect of the open air in different parts of the Netherlands. *J Hygiene* 78, 175–187.

Druett, H. A., Henderson, D. W., Packman, L., and Peacock, S. (1953). The influence of particle size on respiratory infection with anthrax spores. *J Hygiene* 51, 359.

Druett, H. A., Robinson, J. M., Henderson, D. W., Packman, L., and Peacock, S. (1956). Studies on respiratory infection, II & III. *J Hygiene* 54, 37–57.

Duguid, J. P. (1945). The size and the duration of air-carriage of respiratory droplets and droplet-nuclei. *J Hygiene* 54, 471–479.

Frauenthal, J. C. (1980). *Mathematical Modeling in Epidemiology.* Springer-Verlag, New York.

Gordon, J. E., and Ingalls, T. H. (1957). Preventive medicine and epidemiology. *Prog of Med Science* 233, 334–357.

Gregory, P. H. (1973). *Microbiology of the Atmosphere.* Leonard Hill, Plymouth.

Haas, C. N. (1983). Estimation of risk due to low doses of microorganisms. *Am J Epidem* 118(4), 573–582.

_____. (2002). On the risk of mortality to primates exposed to anthrax spores. *Risk Analysis* April.

Haas, C. N., Rose, J. B., and Gerba, C. P. (1999). *Quantitative Microbial Risk Assessment.* John Wiley & Sons, New York.

Henis, Y. (1987). *Survival and Dormancy of Microorganisms.* Wiley-Interscience, New York.

Hurst, C. J. (1991). *Modelling the Environmental Fate of Microorganisms.* American Society for Microbiology, Washington.

HWFB (2003). *SARS Bulletin (24 April 2003).* Health, Welfare, and Food Bureau, Government of the Hong Kong Special Administrative Region. Hong Kong.

Kowalski, W. J. (2003). *Immune Building Systems Technology.* McGraw-Hill, New York.

_____. (2006). *Aerobiological Engineering Handbook: A Guide to Airborne Disease Control Technologies.* McGraw-Hill, New York.

Liu, H. (2000). *Science and Engineering of Droplets: Fundamentals and Applications.* William Andrew, Norwich, NY.

Lowen, A. C., Mubarcka, S., Steel, J., and Palese, P. (2007). Influenza virus transmission is dependent on relative humidity and temperature. *PLoS Pathog* 3, 1470–1476.

Lowen, A. C., Steel, J., Mubarcka, S., and Palese, P. (2008). High temperature (30°C) blocks aerosol but not contact transmission of influenza virus. *J Virol* 82, 5650–5652.

Mackintosh, C. A., Lidwell, O. M., Towers, A. G., and Marples, R. R. (1978). The dimensions of skin fragments dispersed into the air during activity. *J Hyg (London)* 81, 471–479.

Mitscherlich, E., and Marth, E. H. (1984). *Microbial Survival in the Environment.* Springer-Verlag, Berlin.

Nardell, E. A. (1990). Dodging droplet nuclei. *Am Rev Resp Dis* 142, 501–503.

Nardell, E. A., Keegan, J., Cheney, S. A., and Etkind, S. C. (1991). Airborne infection: Theoretical limits of protection achievable by building ventilation. *Am Rev Resp Dis* 144, 302–306.

Quraishi, Z. A., Blais, F. X., Sottile, W. S., and Adler, L. M. (1983). Movement of personnel and wound contamination. *AORN J* 38, 146–147, 150–156.

Reed, S. E. (1975). An investigation of the possible transmission of rhinovirus colds through indirect contact. *J Hygiene* 75, 249–258.

Remington, P. L., Hall, W. N., Davis, I. H., Herald, A., and Gunn, R. A. (1985). Airborne transmission of measles in a physician's office. *JAMA* 253(11), 1574–1577.

Riley, R. L. (1980). The role of ventilation in the spread of measles in an elementary school *Ann NY Acad Sci: Airborne Contagion* 353, 25–34.

Scott, E., and Bloomfield, S. F. (1990). The survival and transfer of microbial contamination via cloths, hands and utensils. *J Appl Bacteriol* 68(3), 271–278.

Tang, J. W. (2009). The effect of environmental parameters on the survival of airborne infectious agents. *J. R. Soc. Interface* 6, S737–S746.

Thurman, R., and Gerba, C. (1989). The molecular mechanisms of copper and silver ion disinfection of bacteria and viruses. *CRC Crit Rev Environ Control* 18, 295–315.

Wheeler, A. E. (1993). Better filtration: A prescription for healthier buildings; in *IAQ 93: Operating and Maintaining Buildings for Health, Comfort, and Productivity*, K. Y. Teichman, ed., American Society of Heating, Refrigerating and Air-Conditioning Engineers, Inc., Atlanta, GA, 201–207.

Whiting, R. C. (1993). Modeling bacterial survival in unfavorable environments. *J Ind Microb* 12, 240–246.

Wilson, E. B., and Worcester, J. (1944). The law of mass action in epidemiology. *Proc Nat Acad Sci* 31, 24–34.

3

Hospital Aerobiology

Introduction

The aerobiology of the hospital environment is a mirror of the hygienic state of the hospital. Hospital air contains bacteria, viruses, fungal spores, pollen, and other contaminants. Bacteria and fungi are commonly sampled and isolated as contaminants of hospital air. Viruses are rarely isolated in hospital air, although new DNA and RNA identification techniques are beginning to show evidence that some viruses like VZV, RSV, and measles are commonly airborne in nosocomial settings. Wherever airborne pathogens are found there is also likely to be surface contamination, and surface contamination may also be considered an indicator of hospital hygiene and of aerobiological quality whenever the microbe detected is an airborne pathogen.

Although no standards have been set in the United States for limits on airborne microbial levels, many other countries are adopting or already mandating certain limits (see Chapter 18). Even without standards, the measured levels of airborne microbes in hospital air provide a basis for understanding the epidemiology and etiology of airborne nosocomial disease transmission. It has been said that for pathogens no airborne concentration can be considered safe for human occupancy, but completely sterile air may be an almost unachievable ideal whenever human occupants are present. At best we can only hope to reduce airborne concentrations of pathogens to levels that will reduce the risk of infection. In order to know whether disinfection techniques are effective we first must know what the normal levels of airborne microbes are for any given hospital environment. The following sections describe the results of airborne sampling in various hospital environments. It should be noted, and will be demonstrated in later chapters, that the normal levels of requisite air filtration should completely eliminate virtually all fungal spores and environmental bacteria, and therefore the presence of these environmental microbes in the indoor air of hospitals indicates either a failure of the filters to perform or, more likely, the fact that there are alternate pathways by which such microbes may enter the hospital.

Airborne Levels of Bacteria

Bacteria that occur in hospitals hail from one of two major sources, the occupants or the environment. Endogenous bacteria, such as *Staphylococcus aureus*, are present on patients and HCWs, and these may include opportunistic pathogens that transiently colonize healthy individuals. Staphylococci routinely inhabit the air in any type of building and we inhale these bacteria every day. Environmental bacteria are largely innocuous to healthy individuals but can present severe risks to immunocompromised patients or to those undergoing surgery or who have burn wounds. Environmental bacteria typically enter from the outdoors or from other external sources, but sometimes they may obtain a niche in buildings and can multiply internally—the building becomes an amplifier.

Table 3.1 summarizes some of the studies on hospital airborne concentrations of microorganisms. Considering that the World Health Organization (WHO) recommends not more than 50 cfu/m³ of fungi in hospital air, it would seem that over half of the facilities tested exceeded this limit (Ross et al. 2004). For bacteria, WHO recommends a limit of 100 cfu/m³ and here we see about 30% of facilities beyond this limit. For operating rooms (ORs) the suggested limit of 10 cfu/m³ is exceeded in most cases.

The microbial composition of the air in hospitals varies between wards, and often the highest number of isolates is found in corridors, followed by operating rooms. In one study of airborne microbial contamination in the operating room and intensive care units (ICUs) of a surgery clinic, Holcatova, Bencko, and Binek (1993) measured bacterial concentrations of 150–250 cfu/m³. The most frequently isolated microorganisms included *Staphylococcus epidermis*, *Staphylococcus haemolyticus*, *Enterococcus* spp., *Enterobacter*, *Pseudomonas* spp., *Micrococcus*, *Corynebacteria*, and *Streptococcus faecalis*. The microbes most frequently cultured in the air of operating rooms include *S. epidermis* and *S. aureus*.

Airborne MRSA plays a role in the colonization of nasal cavities and in respiratory tract MRSA infections. In a study by Shiomori, Miyamoto, and Makishima (2001), MRSA was found in air samples collected in single-patient rooms during both rest periods and during bed sheet changes. About 20% of the MRSA were less than 4 μm in size. MRSA was also isolated from inanimate environments, such as sinks, floors, and bed sheets, as well as from the patients' hands. The clinical isolates of MRSA were of one origin and were identical to the MRSA strains that infected or colonized new patients. MRSA was recirculated among the patients, the air, and the local room environments, especially during movement in the rooms.

Patients may bring airborne infections into waiting areas. Remington et al. (1985) reported on an unusual outbreak of measles in a pediatrician's office in which three children developed measles, after arriving about an hour after

TABLE 3.1

Typical Airborne Concentrations of Bacteria in Hospitals

Area	Mean Level (cfu/m³)	Reference
General areas	55	Ross et al. 2004
General areas	80	Andrade and Brown 2003
General areas	207	Tighe and Warden 1995
General wards	31	Ekhaise, Ighosewe, and Ajakpovi 2008
Hospital room	1224	Solberg et al. 1971
ICU and critical care	83	Tighe and Warden 1995
Isolation room	314	Solberg et al. 1971
NICU	36	Kowalski and Bahnfleth 2002
Nurse's stations	52	Tighe and Warden 1995
Patient rooms	104	Tighe and Warden 1995
Ultraclean/laminar OR	1.5	Solberg et al. 1971
Ultraclean/laminar OR	7	Ritter et al. 1975
Ultraclean/laminar OR	7.7	Berg, Bergman, and Hoborn 1991
Ultraclean/laminar OR	19	Luciano 1984
Ultraclean/laminar OR	22	Friberg and Friberg 2005
Conventional OR	23	Bergeron et al. 2007
Conventional OR	24	Berg, Bergman, and Hoborn 1989
Conventional OR	28	Nelson 1978
Ultraclean/laminar OR	29	Brown et al. 1996
Conventional OR	35	Lidwell 1994
Conventional OR	65	Lowbury and Lidwell 1978
Conventional OR	74	Hambraeus, Bengtsson, and Laurell 1977
Conventional OR	74	Tighe and Warden 1995
Mean level for operating rooms	**34**	—
Mean level for ICUs, NICUs, isolation	**144**	—
Mean level for general areas	**250**	—

an infectious child had departed. Based on an airborne transmission model, it was estimated that the index patient was producing 144 units of infection (quanta) per minute while in the office. Characteristics such as coughing, increased warm air recirculation, and low relative humidity may have increased the likelihood of transmission.

The aerobiology of operating rooms is primarily dependent on the microbial flora of the occupants, with common skin microbes like *Staphylococcus* and *Streptococcus* and some intestinal flora like *Enterobacter* contributing to air and surface contamination. Environmental contaminants like *Pseudomonas aeruginosa* and *Bacillus subtilis* can also make their way into ORs.

Airborne Levels of Fungal Spores

Fungal spores hail from the environment but should not be able to significantly penetrate the standard recommended filters for hospitals, these being the MERV 8 prefilter combined with a MERV 15 final filter (AIA 2006; ASHRAE 2003). Therefore, unless the hospital has open windows, the only pathway for spores to enter is through the doors and with the traffic. In some cases, microbial growth inside a building can also lead to airborne and surface contamination throughout a building. In such cases the building itself becomes a vector for disease.

Table 3.2 summarizes the results of various studies on the concentration of airborne fungi in hospital environments, identifying the specific areas in the hospital and the overall averages. These levels can be considered representative of hospitals and are, in fact, typical for many commercial ventilated buildings.

In a study of mold spores in the air of a hospital ward, Tormo et al. (2002) found 22 different types of spores, with total concentrations of 175–1396 spores/m³. The most frequently isolated were *Cladosporium, Ustilago,* and various basidiospores. For *Aspergillus-Penicillium* spores, the concentration was higher indoors than outdoors, although for most spores lower levels were found indoors, with a mean indoor/outdoor ratio of 1:4.

TABLE 3.2

Typical Airborne Concentrations of Fungi in Hospitals

Area	Mean Level (cfu/m³)	Reference
ICU	9	Centeno and Machado 2004
NICU	15	Kowalski 2003
Wards	17	Streifel and Rhame 1993
ICU and critical care	23	Tighe and Warden 1995
Nurse's stations	23	Tighe and Warden 1995
Wards	32	Ekhaise, Ighosewe, and Ajakpovi 2008
Wards	43	Tighe and Warden 1995
General lobby	58	Streifel and Rhame 1993
General areas	84	Tighe and Warden 1995
Medical compressed air	140	Andrade and Brown 2003
General areas	194	Ross et al 2004
General areas	100	Andrade and Brown 2003
General areas	786	Tormo et al. 2002
OR	52	Tighe and Warden 1995
Mean level for ICUs, NICUs, isolation	**16**	—
Mean level for general areas	**149**	—

In one of the rare studies of pollen and spores in the air of a hospital ward, Tormo et al. (2002) conducted aerobiological studies and found 20 types of pollen grains whose concentrations ranged from 2.7 to 25.1 grains/m^3. The most frequently isolated were, in order, grasses, evergreen, oak, water plantain, and olive. Comparison with outdoor levels showed that the three most abundant pollen types had an indoor/outdoor ratio of 30:1.

Anderson et al. (1996) detected airborne *Aspergillus* spores in the air of a pediatric ward and found that they mirrored levels in the outdoor air, or about 0–6 cfu/m^3. It was also found that levels of spores near the vacuum cleaners increased from 24 to 62 cfu/m^3 when the vacuum cleaners were turned on. The most frequently isolated fungi in the air of hospitals, according to a study by Hong et al. (1999), were *Cladosporium, Penicillium, Aspergillus,* and *Alternaria,* in that order.

Spores, bacterial or fungal, are hardier and more likely to see greater distribution from the reservoir or source, whether it be a single infected patient or an entire ward. Entryways could be focused on as a design component of a spore control program, but a more feasible approach would be for visitors to disinfect themselves before every visit. Lobbies that are anterooms for the entire building are one option, or isolating and pressurizing the entire first floor (Kowalski 2003). Because the fungal spores that disseminate around hospital wards are likely undergoing spurts of airborne transport along with local activity or air currents, they will drift and diffuse in all possible directions. The distance they travel will be related to their natural life span or their ability to resist the environment. In hospital areas that are routinely cleaned, survival in the environment will normally be difficult. For this reason, bacteria will tend to stay very local, usually directly in the room of a patient source, while spores will tend to be transported farther from the source.

Airborne Virus Levels

Viruses have been rarely detected in hospital air due mainly to the many difficulties associated with virus air sampling. Recent advances in DNA detection using polymerase chain reaction (PCR) technology are likely to provide considerable data in the future. Insufficient data are available to present any representative examples of virus levels in hospital areas. Unlike endogenous or environmental bacteria, viruses are likely to appear only sporadically during outbreaks.

Airborne measles virus and RSV have been detected in hospital wards housing infected patients (Agranovski et al. 2008). Blachere et al. (2009) detected airborne particles containing influenza in the air of a hospital emergency ward in some 84% of air samples. Over one-half of the viral particles detected were less than 4 microns in size and therefore were in the respirable particle size range. VZV has been detected in air samples from a hospital (Sawyer et al. 1994). More data on airborne virus levels in hospitals is likely to accrue as new PCR assays become widely available for specific nosocomial pathogens.

References

Agranovski, I. E., Safarov, A. S., Agafonov, A. P., Pyankov, O. V., and Sergeev, A. N. (2008). Monitoring of airborne mumps and measles viruses in a hospital. *Clean Soil, Air, Water* 36(10-11), 845–849.

AIA (2006). *Guidelines for Construction and Equipment of Hospital and Medical Facilities.* Mechanical Standards American Institute of Architects, ed., Washington.

Anderson, K., Morris, G., Kennedy, H., Croall, J., Michie, J., Richardson, M. D., and Gibson, B. (1996). Aspergillosis in immunocompromised paediatric patients: Associations with building hygiene, design, and indoor air. *Thorax* 51, 256–261.

Andrade, C. M., and Brown, T. (2003). Microbial contamination of central supply systems for medical air. *Brazilian J Microbiol* 34(Suppl), 29–32.

ASHRAE. (2003). *HVAC Design Manual for Hospitals and Clinics.* American Society of Heating, Ventilating, and Air Conditioning Engineers, Atlanta.

Berg, M., Bergman, B. R., and Hoborn, J. (1989). Shortwave ultraviolet radiation in operating rooms. *J Bone Joint Surg* 71-B(3), 483–485.

———. (1991). Ultraviolet radiation compared to an ultra-clean air enclosure. Comparison of air bacteria counts in operating rooms. *JBJS* 73(5), 811–815.

Bergeron, V., Reboux, G., Poirot, J. L., and Laudinet, N. (2007). Decreasing airborne contamination levels in high-risk hospital areas using a novel mobile air-treatment unit. *Inf Contr Hosp Epidemiol* 28(10), 1181–1186.

Blachere, F. M., Lindsley, W. G., Pearce, T. A., Anderson, S. E., Fisher, M., Khakoo, R., Meade, B. J., Lander, O., Davis, S., Thewlis, R. E., Celik, I., Chen, B. T., and Beezhold, D. H. (2009). Measurement of airborne influenza virus in a hospital emergency ward. *Clin Infect Dis* 48, 438–440.

Brown, I. W., Moor, G. F., Hummel, B. W., Marshall, W. G., and Collins, J. P. (1996). Toward further reducing wound infections in cardiac operations. *Ann Thorac Surg* 62(6), 1783–1789.

Centeno, S., and Machado, S. (2004). Assessment of airborne mycoflora in critical areas of the Principal Hospital of Cumana, state of Sucre, Venezuela. *Invest Clin* 45(2), 137–144.

Ekhaise, F. O., Ighosewe, O. U., and Ajakpovi, O. D. (2008). Hospital indoor airborne microflora in private and government owned hospitals in Benin City, Nigeria. *World J Med Sci* 3(1), 19–23.

Friberg, B., and Friberg, S. (2005). Aerobiology in the operating room and its implications for working standard. *Proc Inst Mech Eng* 219(2), 153–160.

Hambraeus, A., Bengtsson, S., and Laurell, G. (1977). Bacterial contamination in a modern operating suite. *J Hyg* 79, 121–132.

Holcatova, I., Bencko, V., and Binek, B. (1993). Indoor air microbial contamination in the operating theatre and intensive care units of the surgery clinic. *Indoor Air* 93, 375–378.

Hong, W. P., Shin, J. H., Shin, D. H., Sul, Y. A., Lee, C. J., Suh, S. P., and Ryang, D. W. (1999). Filamentous fungi isolated from hospital air and from clinical specimens. *Korean J Nosoc Inf Contr* 4(1), 17–25.

Kowalski, W. J. (2003). *Immune Building Systems Technology*. McGraw-Hill, New York.

Kowalski, W. J., and Bahnfleth, W. P. (2002). Innovative strategies to protect hospitalized premature infants against airborne pathogens and toxins. Hershey Medical Center NICU, Hershey, PA.

Lidwell, O. M. (1994). Ultraviolet radiation and the control of airborne contamination in the operating room. *J Hosp Infect* 28, 245–248.

Lowbury, E. J., and Lidwell, O. M. (1978). Multi-hospital trial on the use of ultraclean air systems in orthopaedic operating rooms to reduce infection: Preliminary communication. *J Royal Soc Med* 71, 800–806.

Luciano, J. R. (1984). New concept in French hospital operating room HVAC systems. *ASHRAE Journal* Feb., 30–34.

Nelson, P. J. (1978). Clinical use of facilities with special air handling equipment. *Hosp Topics* 57(5), 32–39.

Remington, P. L., Hall, W. N., Davis, I. H., Herald, A., and Gunn, R. A. (1985). Airborne transmission of measles in a physician's office. *JAMA* 253(11), 1574–1577.

Ritter, M. A., Eitzen, H. E., French, M. L. V., and Hart, J. B. (1975). The operating room environment as affected by people and the surgical face mask. *Clin Ortho* 111, 147–150.

Ross, C., deMenezes, J. R., Svidzinski, T. I. E., Albino, U., and Andrade, G. (2004). Studies on fungal and bacterial population of air-conditioned environments. *Brazilian Arch Biology Technol* 47(5), 827–835.

Sawyer, M. H., Chamberlin, C. J., Wu, Y. N., Aintablian, N., and Wallace, M. R. (1994). Detection of varicella-zoster virus DNA in air samples from hospital rooms. *J Infect Dis* 169, 91–94.

Shiomori, T., Miyamoto, H., and Makishima, K. (2001). Significance of airborne transmission of methicillin-resistant *Staphylococcus aureus* in an otolaryngology-head and neck surgery unit. *Arch Otolaryngol Head Neck Surg* 127(6), 644–648.

Solberg, C. O., Matsen, J. M., Vesley, D., Wheeler, D. J., Good, R. A., and Meuwissen, H. J. (1971). Laminar airflow protection in bone marrow transplantation. *Am Soc Microbiol* 21, 209–216.

Streifel, A. J., and Rhame, F. S. (1993). Hospital air filamentous fungal spore and particle counts in a specially designed hospital. *Indoor Air* 93, 161–165.

Tighe, S. W., and Warden, P. S. (1995). An investigation of microbials in hospital air environments. *Indoor Air Rev* May.

Tormo, M. R., Gonzalo, G. M. A., Munoz, R. A. F., and Silva, P. I. (2002). Pollen and spores in the air of a hospital out-patient ward. *Allergol Immunopathol* 30(4), 232–238.

4

Airborne Nosocomial Microorganisms

Introduction

Nosocomial microorganisms comprise a distinct class of pathogens that can be categorized by their prevalence, their taxonomy, the type of infections they cause, and the type of hospital facility in which they occur. Airborne nosocomial microorganisms are a subset of this general class and share certain characteristics, such as the ability to survive airborne transport and to survive on dry surfaces. It is these shared characteristics that allow them to be eliminated from air and surfaces through the use of disinfection technologies such as surface disinfectants and air cleaning technologies. This chapter summarizes all of the common nosocomial pathogens and all relevant parameters that may describe their transmission potential.

There are a number of environmental and commensal microbes that are not normally pathogenic for healthy individuals but that can cause infections in the immunocompromised or can infect burn wounds or surgical sites. These are commonly known as opportunistic nosocomial pathogens. All of these opportunistic microbes are considered pathogens for our purposes. The only distinction of any relevance is that opportunistic pathogens present no major threat to healthy hospital workers, and this distinction may affect some procedural controls (where discrimination is an option) but will not affect any disinfection technologies (which do not discriminate).

Airborne Nosocomial Pathogens

Table 4.1 summarizes those airborne nosocomial pathogens that have been identified as partly or potentially airborne from an extensive review of the literature, along with size (aerodynamic logmean diameter), disease group, Source, Biosafety Level (BSL), Airborne Class, and Category. The Category refers to the seven categories previously identified for airborne nosocomial pathogens, as follows:

TABLE 4.1

Airborne Nosocomial Pathogens

Microbe	Type	Size (µm)	Source	BSL	Category 1	2	3	4	5	6	7	Airborne Class
Acinetobacter	Bacteria	1.225	E	RG 2	X			X		X	X	2
Adenovirus	Virus	0.079	H	RG 2	X	X			X	X		2
Alcaligenes	Bacteria	0.775	HE	RG 2	X							2
Alternaria alternata	Fungi	11.225	E	RG 1				X	X			2
Aspergillus	Fungi	3.354	E	RG 2	X	X	X	X	X			1
Blastomyces dermatitidis	Fungi	12.649	E	RG 2	X				X			2
Bordetella pertussis	Bacteria	0.245	H	RG 2	X						X	1
Clostridium difficile	Bacteria	2	H	RG 2		X			X	X	X	1
Clostridium perfringens	Bacteria	5	HE	RG 2		X					X	2
Coccidioides immitis	Fungi	3.464	E	RG 3	X				X			2
Coronavirus (SARS)	Virus	0.11	H	RG 2	X						X	1
Corynebacterium diphtheriae	Bacteria	0.698	H	RG 2	X					X		2
Coxsackievirus	Virus	0.027	H	RG 2	X						X	2
Cryptococcus neoformans	Fungi	4.899	E	RG 2	X				X			2
Enterobacter	Bacteria	1.414	HE	RG 1	X		X	X	X	X		2
Enterococcus (VRE)	Bacteria	1.414	H	RG 1-2	X		X	X	X	X	X	2
Fugomyces cyanescens	Fungi	2.12	E	RG 1					X			2
Fusarium	Fungi	11.225	E	RG 1			X	X				2
Haemophilus influenzae	Bacteria	0.285	H	RG 2	X					X	X	2
Haemophilus parainfluenzae	Bacteria	1.732	H	RG 2	X							2
Histoplasma capsulatum	Fungi	2.236	E	RG 3	X	X			X			1
Influenza A virus	Virus	0.098	H	RG 2	X				X	X	X	1
Klebsiella pneumoniae	Bacteria	0.671	HE	RG 2	X		X	X	X	X		2
Legionella pneumophila	Bacteria	0.52	E	RG 2	X						X	1
Measles virus	Virus	0.158	H	RG 2	X					X		1
Mucor	Fungi	7.071	E	RG 1	X			X				2
Mumps virus	Virus	0.164	H	RG 2	X							1
Mycobacterium avium	Bacteria	1.118	E	RG 2					X			2

TABLE 4.1 (*Continued*)

Airborne Nosocomial Pathogens

Microbe	Type	Size (µm)	Source	BSL	Category 1	2	3	4	5	6	7	Airborne Class
Mycobacterium tuberculosis	Bacteria	0.637	H	RG 3	X	X			X	X	X	1
Mycoplasma	Bacteria	0.177	H	RG 2	X					X	X	2
Neisseria meningitidis	Bacteria	0.775	H	RG 2						X		2
Nocardia asteroides	Bacteria	1.118	E	RG 2	X							2
Norwalk virus	Virus	0.029	E	RG 2		X					X	1
Parainfluenza virus	Virus	0.194	H	RG 2	X					X	X	2
Parvovirus B19	Virus	0.022	H	RG 2						X		2
Penicillium	Fungi	3.262	E	RG 2					X			2
Pneumocystis jirovecii	Fungi	2	HE	RG 1	X				X			2
Proteus mirabilis	Bacteria	0.494	H	RG 2	X		X	X		X		2
Pseudallescheria boydii	Fungi	3.162	E	RG 1					X			2
Pseudomonas aeruginosa	Bacteria	0.494	E	RG 1	X	X	X	X	X	X		1
Reovirus	Virus	0.075	H	RG 2						X		2
Respiratory syncytial virus (RSV)	Virus	0.19	H	RG 2	X				X	X	X	1
Rhinovirus	Virus	0.023	H	RG 2	X						X	2
Rhizopus	Fungi	6.928	E	RG 2	X		X					2
Rotavirus	Virus	0.073	H	RG 2					X	X	X	2
Rubella virus	Virus	0.061	H	RG 2	X							1
Scedosporium	Fungi	3.162	E	RG 1					X			2
Serratia marcescens	Bacteria	0.632	E	RG 1	X		X	X	X	X		2
Staphylococcus aureus (MRSA)	Bacteria	0.866	H	RG 2	X	X	X	X	X	X	X	1
Staphylococcus epidermis	Bacteria	0.866	H	RG 1		X	X		X			2
Streptococcus pneumoniae	Bacteria	0.707	H	RG 2	X	X			X	X	X	2
Streptococcus pyogenes	Bacteria	0.894	H	RG 2	X	X		X			X	1
Trichosporon	Fungi	8.775	E	RG 3					X			2
Varicella-zoster virus (VZV)	Virus	0.173	H	RG 2	X				X	X		1

- Category 1: Respiratory Infections
- Category 2: Nonrespiratory
- Category 3: Surgical Site Infections
- Category 4: Burn Wound Infections
- Category 5: Immunocompromised Infections
- Category 6: Pediatric Infections
- Category 7: Nursing Home Infections

The categories given in Table 4.1 only apply to those pathogens that have caused infections in nosocomial settings. Microbes that have not appeared in health care facilities will not be listed. Some respiratory pathogens, for example, that may be transmitted in the community are not listed if they have not yet presented themselves as a nosocomial problem.

Airborne Class refers to the evidence for classifying the pathogen as airborne, as stated in Chapter 1, with Airborne Class 1 being those pathogens that have been officially recognized as airborne in nosocomial settings or for which there is compelling evidence in the literature that the microbe has been transmitted via the airborne route and caused infections in nosocomial settings. Airborne Class 2 refers to those pathogens for which airborne transmission or transport is plausible or suspected but not yet demonstrated or proven in nosocomial settings. The latter class includes pathogens that may be airborne in the community but have not yet posed a significant problem in health care facilities. The pathogens listed in Table 4.1 will be revisited in subsequent chapters, including the chapters on filtration and UVGI, in which the ability of air cleaning systems to remove or disinfect these microbes will be demonstrated analytically. A summary of these pathogens is provided in the Appendix.

There are perhaps another 100 potential airborne pathogens that are not addressed here for the simple reason that they have not yet been noted as airborne in hospital settings or because they are simply very rare in any setting. Well-known agents like smallpox and plague (*Yersinia pestis*), many zoonotic diseases, and common allergens are specifically not included in this database. This database, however, covers such a broad range of pathogens that virtually any existing pathogen, airborne or otherwise, is represented either in terms of similar species or in terms of similar aerodynamic size or UV susceptibility. This array of pathogens therefore can be considered representative of a much broader range of pathogens for analytical purposes.

In addition to the microbes in Table 4.1, there are a number of other bacteria and fungi that present a risk to patients with immunodeficiency. Table 4.2 lists such opportunistic microbes summarized from the literature (Flannigan, McCabe, and McGarry 1991; Kowalski 2006; Mayhall 1999a) but excluding any genus previously identified or any associated fungal teleomorphs. Where more than one species occurs, only the genus is given. The

TABLE 4.2

Additional Microbes Presenting Opportunistic Risks

Bacteria	
Actinomyces israelii	*Moraxella*
Aeromonas	*Mycobacterium abcessus*
Bacillus cereus	*Mycobacterium fortuitum*
Bacteroides fragilis	*Mycobacterium kansasii*
Burkholderia cenocepacia	*Mycobacterium marinum*
Burkholderia mallei	*Mycobacterium ulcerans*
Burkholderia pseudomallei	*Nocardia brasiliensis*
Cardiobacterium	*Saccharopolyspora rectivirgula*
Chlamydia pneumoniae	*Thermoactinomyces sacchari*
E. coli O157:H7	*Thermoactinomyces vulgaris*
Micromonospora faeni	*Thermomonospora viridis*
Fungi	
Absidia	*Geotrichum*
Acremonium	*Graphium eumophum*
Arthrinium phaeospermum	*Helminthosporium*
Aureobasidium pullulans	*Malassezia*
Botrytis cinerea	*Paecilomyces variotii*
Candida albicans	*Paracoccidioides brasiliensis*
Chaetomium globosum	*Phialophora*
Cladosporium	*Phoma*
Cryptostroma corticale	*Rhizomucor pusillus*
Cunninghamella bertholletiae	*Rhodoturula*
Curvularia lunata	*Scopulariopsis*
Drechslera	*Trichoderma*
Emericella nidulans	*Trichophyton*
Epicoccum purpurascens	*Ulocladium*
Epidermophyton floccosum	*Ustilago*
Eurotium	*Verticillium*
Exophiala	*Wallemia sebi*

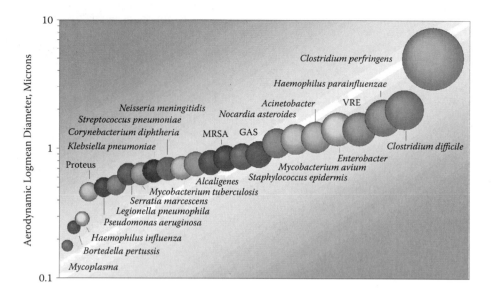

FIGURE 4.1
Relative size of airborne nosocomial bacteria based on aerodynamic logmean diameter. Volume of spheres are proportional to diameters.

microbes in Table 4.2 will not be addressed further as Table 4.1 provides adequate representation of airborne nosocomial pathogens throughout a wide range of characteristics.

Airborne Nosocomial Bacteria

Bacteria are the largest single cause of nosocomial infections. All those bacteria previously identified as Airborne Class 1 or 2, a total of 24 bacteria, are included and addressed individually in the Database section. These bacteria can be Communicable, Endogenous, or Noncommunicable. Figure 4.1 graphically illustrates the array of airborne nosocomial bacteria by aerodynamic logmean diameter shown in relative size, as indicated. Almost all of these bacterial cells are less then 5 microns in size and, if aerosolized, may remain suspended in air for prolonged periods. This array of bacteria consists of both communicable and noncontagious bacteria, many of which are endogenous commensals and opportunistic pathogens. A few come from the outdoor environment, such as the only bacterial spores, *Clostridium difficile*, *Clostridium perfringens*, and *Nocardia*.

Endogenous bacteria often live as commensal human microflora and rarely present a hazard to healthy individuals. In the immunocompromised and those undergoing surgery, endogenous microbes become pathogenic. This is often a case of the right microbe being in the wrong place, such as skin micro-flora entering tissues or the bloodstream. Endogenous microbes may come from a patient and infect the same or another patient. Endogenous microbes may also hail from hospital personnel and infect the immunocompromised.

Some endogenous microbes are excluded, such as *Neisseria meningitidis,* which is rare in hospital environments, although transmission may occur by contact with large droplets from the nose and throat of colonized or infected carriers (Simmons and Gelfand 1999). Noncommunicable opportunistic bacteria primarily hail from the environment or animal sources and are often pathogenic to healthy people, as well as being a constant threat to the immunocompromised. Other bacteria are excluded for obvious reasons or for their rarity in hospital settings, such as *Chlamydia, Yersinia, Clostridium tetani,* and most anaerobic bacteria.

Airborne Nosocomial Viruses

All viruses are pathogens and there are no truly endogenous viruses, but viruses may be communicable or noncommunicable, and humans are the ultimate reservoir for most of them. Figure 4.2 illustrates the distribution of viruses by aerodynamic logmean diameter and these are shown in relative

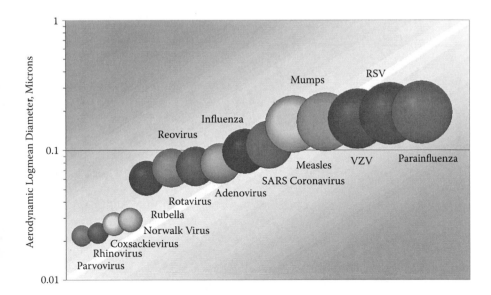

FIGURE 4.2
Relative size of airborne nosocomial viruses based on aerodynamic logmean diameter. Volume of spheres are proportional to logmean diameters.

size, as indicated. All virions in this size range may become aerosolized and remain suspended in air almost indefinitely. Human metapneumovirus, an emerging pathogen capable of causing severe respiratory infections but with an unknown etiology, is not included in this database—see Schlapbach et al. (2011) for more information on this virus.

Airborne Nosocomial Fungi

A wide variety of fungi are potentially airborne by virtue of their spores and are often present in outdoor and indoor air samples (Figure 4.3). Although most fungi are harmless to healthy humans, they can cause severe infections in the immunocompromised. All fungal spores or vegetative cells in this size range may become aerosolized and remain suspended in air for prolonged periods. Almost all fungi are considered noncommunicable, although they are ubiquitous in the environment and routinely contaminate homes, furnishings, and clothes. Only one fungi, *Pneumocystis jirovecii* (formerly *Pneumocystis carinii*) has been identified as communicable, and this has been primarily an immunocompromised infection. All of these fungi occur as spores except for the yeasts *Cryptococcus neoformans* and *Trichosporon*.

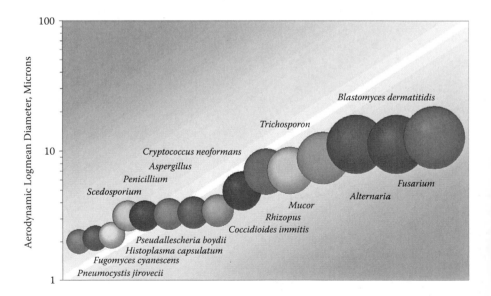

FIGURE 4.3
Relative size of airborne nosocomial fungi based on aerodynamic logmean diameter. Volume of spheres are proportional to logmean diameters.

Database of Airborne Nosocomial Pathogens

This section provides detailed information on the pathogens summarized in Table 4.1, including infection transmission modes and any references for evidence that the microbe may be airborne in nosocomial settings. Information is provided on disinfectants and microbial survival, and treatments are summarized in terms of vaccines and antibiotics. Precautions are noted where applicable (i.e., respiratory pathogens) and where authorities have specified them—these include Standard Precautions, Droplet Precautions, and Airborne Precautions. Pathogens are listed in alphabetical order.

Acinetobacter: Noncontagious Bacteria—Airborne Class 2

Infection can occur at any body site but the respiratory tract is the most frequent site of infection. *Acinetobacter* commonly inhabits soil, sewage, and water, and is frequently found in humans as part of the normal flora (Arnow and Flaherty 1999). Occurs mainly in the immunocompromised. Evidence suggests they are opportunistic pathogens that can cause meningitis and septic infections, mostly in immunocompromised hosts. Sometimes causes outbreaks in ICUs. Transient colonization of the pharynx occurs in 7% of healthy people, while cutaneous colonization occurs in 25%. Airborne transmission of *Acinetobacter* has been reported but not yet fully corroborated (Allen and Green 1987; Jawad et al. 1996). Air samples taken near colonized patients or even in other rooms near the index patient's room have yielded *Acinetobacter*, but airborne transmission has yet to be demonstrated conclusively (Crowe, Towner, and Humphreys 1995; Bernards et al. 1998; Brooks, Walczak, and Hameed 2000). Respiratory infections include bronchiolitis and tracheobronchitis in children. Some nosocomial infections are related to contaminated inhalation equipment.

Disease: opportunistic/septic infections, nosocomial infections, meningitis

Source: environmental, soil, sewage, indoor growth in potable water

Point of infection (POI): upper respiratory tract, skin

Treatment: combined use of beta-lactams and aminoglycosides

Survival: survives outdoors; can survive at least 2 weeks on various surfaces

Adenovirus: Communicable Virus—Airborne Class 2

Adenovirus causes acute respiratory infections of the lungs and sometimes the eyes. It is common in the adult population, and mild respiratory infections resemble the common cold. It can occur in epidemic form. Some types of this virus occur primarily in infants. The most likely route of transmission is fecal-oral between young children (Turner 1999). Evidence supports the spread of adenovirus by direct contact, indirect contact, droplets, and the fecal-oral route, but airborne transmission, though plausible, has not been demonstrated in any nosocomial outbreaks (Decker and Schaffner 1999). Airborne transmission has only been demonstrated experimentally (Couch et al. 1966). Some patients experience shedding for only one day while shedding may persist for years after infection. Contact and Droplet Precautions.

Disease: colds, fever, pharyngitis, acute respiratory disorder, pneumonia

Source: humans, sewage

POI: upper respiratory tract, eyes

Treatment: supportive therapy is the only treatment; no prophylaxis available

Vaccines: types 4 and 7

Survival outside host: 10 days on paper; 3–8 weeks on glass, steel, and tile (adenovirus type 2)

Alcaligenes: Noncontagious Bacteria—Airborne Class 2

May be an innocuous inhabitant of man, especially in the respiratory tract and gastrointestinal tract in hospitalized patients. It can infect immunocompromised hosts but is uncommon. Infection results when microbes are introduced into wounds or colonize immunosuppressed hosts. Most isolates of *Alcaligenes faecalis* from blood or respiratory secretions are related to contamination of hospital equipment or fluids. Some species affect other areas of the body. Nosocomial strains may be resistant to common antibiotics.

Disease: opportunistic infections of wounds or other parts of the body

Source: humans, soil, water

POI: upper respiratory tract, blood, urine, wounds

Treatment: aminoglycosides, beta-lactams, fluoroquinolones

Alternaria alternata: Noncontagious Fungal Spore—Airborne Class 2

A common indoor air contaminant; indoor levels can exceed outdoor levels. *Alternaria* can produce opportunistic skin infections in the immunocompromised. Can also occur on various foodstuffs and textiles. Standard Precautions.

Disease: allergic alveolitis, rhinitis, sinusitis, asthma, toxic reactions

Source: environmental, indoor growth on paint, dust, filters, and cooling coils

POI: upper respiratory tract

Treatment: antifungal therapy, amphotericin B

Survival outside host: survives outdoors

Aspergillus: Noncontagious Fungal Spore—Airborne Class 1

Aspergillus represents several related fungi that cause aspergillosis. This disease most often affects the external ear, but also affects the lungs. *Aspergillus* species are common in the soil, and spores become airborne in dry, windy weather. Spores can germinate in moist areas of buildings and ventilation systems. Sometimes associated with sick building syndrome. Can be fatal to those with immunodeficiency. Sometimes found on human body surfaces. *Aspergillus* infections are acquired by airborne transmission (Pennington 1980). Standard Precautions.

Disease: aspergillosis, alveolitis, asthma, allergic fungal sinusitis, organic dust toxicity syndrome (ODTS), toxic reactions, pneumonia possible

Source: environmental, indoor growth on insulation and coils

POI: upper respiratory tract

Treatment: amphotericin B, itraconazole, or voriconazole

Survival outside host: survives outdoors

Blastomyces dermatitidis: Noncontagious Fungal Spore—Airborne Class 2

A pathogenic fungi. Entry is through the upper respiratory tract, but can spread to other locations. Can resemble TB and spread beyond the lungs. Found mainly in north central and eastern United States. Males are more susceptible to this progressive disease than females. Like other fungi pathogenic for man, this fungi exhibits dimorphism, existing in one form in nature and another when causing infection. Standard Precautions.

Disease: blastomycosis, pneumonia possible

Source: environmental, nosocomial sources

POI: upper respiratory tract, skin

Treatment: amphotericin B, itraconazole, ketoconazole, hydroxystilbamidine; no prophylaxis, no vaccine

Survival outside host: survives outdoors

Bordetella pertussis: Communicable Bacteria—Airborne Class 1

Bordetella pertussis is the cause of whooping cough. It produces microbial toxins, which are primarily responsible for the disease symptoms. Occurring worldwide, this infection almost exclusively affects children. Almost two-thirds of cases are under 1 year of age. Asymptomatic cases are more frequent. It is highly contagious and transmits by fomites, by direct contact, and likely by aerosols. Droplet Precautions.

Disease: whooping cough, toxic reactions

Source: humans, nosocomial sources

POI: upper respiratory tract, trachea

Treatment: 14 days with erythromycin or trimethoprim-sulfamethoxazole (TMP-SMX), oxygenation, hydration, and electrolyte balance. Antibiotics are used for prophylaxis. Vaccine is available.

Survival outside host: 1 hour to 7 days.

Clostridium difficile: Noncontagious Bacterial Spore—Airborne Class 1

A spore-forming enteric bacterial pathogen that occurs in 2–4% of healthy adults. Can transmit between hospital personnel and is a ubiquitous nosocomial pathogen (Johnson and Gerding 1999). Generally affects adults but newborns are increasingly susceptible. *C. difficile* is acquired from endogenous sources, which may include other patients. Health care facilities are the primary reservoir, and asymptomatic carriage among HCWs is common. Newborn infants frequently carry *C. difficile* in high numbers with high levels of toxins in their stool, although disease is rare in this group. Environmental sources of *C. difficile* contamination are a potentially important source of nosocomial infections. Spores can survive for weeks, allowing them to spread throughout a hospital environment via airborne transport. The degree of contamination in hospitals is related to outbreaks, and the environment of infected patients is more frequently contaminated than other environments (Mulligan et al. 1979). Floors and bathrooms tend to be the most contaminated sites. Contaminated equipment can also serve as a reservoir for dissemination. Environmental disinfection can interrupt an outbreak (Kaatz et al. 1988). *C. difficile* has been transmitted to hospital workers and patients by way of fomites (Fekety et al. 1981). Evidence for airborne transmission has been presented by Best et al. (2010) and Snelling et al. (2011). *Clostridium difficile* infections were reduced in a hospital using air cleaning (Nielsen 2008). Standard and Contact Precautions.

Disease: diarrhea, pseudomembranous colitis (PMC), antimicrobic-associated diarrhea (AAD), toxicosis

Source: environmental, soil, nosocomial sources

POI: oral, enteric

Treatment: vancomycin, metronidazole

Survival outside host: can survive for months in hospital environments

Clostridium perfringens: Noncontagious Bacterial Spore—Airborne Class 2

Nonrespiratory but may settle on exposed foods. Not necessarily toxic if inhaled. Classified by the type of toxin produced, with Type A being the most important and common in the human colon and the soil. May grow on foods like meat. Often found in the intestines and in feces. Can cause contamination of wounds. Forms spores that are resistant to heat. May cause gas gangrene in infections. Standard Precautions.

Source: environmental, humans, animals, soil

POI: oral, intestinal

Treatment: penicillin; no prophylaxis, no vaccine

Survival outside host: survives outdoors

Coccidioides immitis: Noncontagious Fungal Spore—Airborne Class 2

The most dangerous fungal infection. Estimates are that 20–40 million people in the Southwest have had infections. Only about 40% of infections are symptomatic, and only 5% are clinically diagnosed. A self-limiting, nonprogressive form of the infection is commonly known as valley fever or desert rheumatism. Transmission occurs by inhalation. The natural reservoir is the soil. Standard Precautions.

Disease: coccidioidomycosis, valley fever, desert rheumatism, chronic pneumonia possible, potentially nosocomial and hazardous to the immunocompromised

Source: environmental, found in alkali soil in warm, dry regions, southwest United States, etc.

POI: upper respiratory tract

Treatment: amphotericin B, ketoconazole, itraconazole, fluoconazole for meningeal infections; no prophylaxis, no vaccine

Survival outside host: survives outdoors

Coronavirus: Communicable Virus—Airborne Class 1

Coronaviruses are one of the causes of the common cold. Accounts for about 10–30% of all colds with those of age 14–24 years most affected. Coronaviruses can infect other animals besides humans but strains are usually specific to one host. Occurs worldwide with predominance in late fall and early winter. Severe acute respiratory syndrome (SARS) can be fatal. Airborne transmission of SARS coronavirus has been implicated in hospital outbreaks (HWFB 2003). Standard Precautions for coronavirus. Droplet precautions for SARS.

Disease: colds, croup, severe acute respiratory syndrome (SARS)

Source: humans

POI: upper respiratory tract

Treatment: no specific treatment or antivirals; no prophylaxis, no vaccine

Survival outside host: up to 24 hours on metal

Corynebacterium diphtheriae: Contagious Bacteria—Airborne Class 2

Corynebacterium diphtheria is the causative agent of diphtheria, which was historically a disease of children. In modern times this disease is less prevalent, but increasingly afflicts those in older age groups. Small outbreaks occur periodically. Healthy carriers may harbor the bacteria in their throats and upper respiratory tracts asymptomatically for a lifetime. Diphtheria is spread by both respiratory droplets and by direct contact, and Droplet Precautions should be used (Decker and Schaffner 1999). Contact Precautions are appropriate for skin infections.

Disease: diphtheria, opportunistic infections

Source: humans

POI: upper respiratory tract

Treatment: Antitoxin is administered in conjunction with erythromycin and/or penicillin.

Prophylaxis: DTP. Vaccine is available.

Survival: 2.5 hours in air, less than 1 year in soil

Coxsackievirus: Communicable Virus—Airborne Class 2

A common cause of colds. Often subclinical. It can sometimes be found in feces and sewage, but is commonly isolated from the throat. This particular cold virus is prevalent in the summer and fall. Typically self-limiting. Occurs worldwide. Experimentally induced airborne infection was demonstrated by Couch et al. (1966). Standard Precautions.

Disease: colds, acute respiratory disorder (ARD)

Source: humans, feces, sewage

POI: upper respiratory tract

Treatment: no specific treatment; no antivirals available

Survival outside host: can survive 24 hours on paper and plastic; 2 weeks on glass, steel

Cryptococcus neoformans: Noncontagious Fungal Yeast—Airborne Class 2

Always occurs in yeast form. The cause of cryptococcosis, and can result in cryptococcus meningitis also. An opportunistic pathogen that can fatally infect those with impaired immune systems. Cells can enter the lungs, germinate, and produce mycelial growth. This is always in yeast form and can occur in dried pigeon excrement, from which it can become airborne. *Cryptococcus neoformans* is extremely common in the environment and most infections are subclinical and asymptomatic. It may become invasive in immunosuppressed patients and cause central nervous system infections. Sometimes found on human body surfaces. Standard Precautions.

Disease: cryptococcosis, cryptococcal meningitis, pneumonia possible

Source: environmental, indoor growth on floor dust

POI: upper respiratory tract, cutaneous infections possible

Treatment: combination therapy with amphotericin B and 5-fluorocyto-sine or ketoconazole; no prophylaxis, no vaccine

Survival outside host: survives in pigeon feces

Enterobacter cloacae: Noncontagious Bacteria—Airborne Class 2

Often found as a normal commensal in the intestines. Associated with a variety of infections, especially in nosocomial settings, including urinary, pulmonary, wound, bloodstream, and other opportunistic infections. Often found in mixed infections. Occurs worldwide in health care settings. Has caused a septicemia epidemic. Equipment contamination and the fecal-oral route are possible transmission mechanisms. Can resist some antibiotics. This species is representative of other species of *Enterobacter* where they occur throughout this text.

Disease: opportunistic infections, pneumonia possible from some other *Enterobacter* species

Source: humans, environmental, soil, and water

POI: wounds; infections of the lungs, blood, and urinary tract

Treatment: aminoglycosides, chloramphenicol, tetracyclines, TMP-SMX, nalidixic acid, nitrofurantoin. Prophylaxis is possible. No vaccine.

Survival outside host: 7–21 days in food

Enterococcus (VRE): Communicable Bacteria—Airborne Class 2

Enterococcus species including *E. faecalis*. *Enterococcus faecium* occasion-ally causes human disease. *Enterococcus durans* accounts for less than 2% of enterococci isolates. Nonrespiratory but may be airborne in nosocomial settings. Can cause opportunistic infections of the urinary tract and wounds. Normally commensals resident in the intestines and other areas of the body. *E. faecium* is found in the feces of about 25% of normal adults. These species are related to Group D streptococci and pneumococci, and cause similar clinical infections. Enterococci are hardy microbes and can survive in the environment and on the hands of HCWs (Chenoweth and Schaberg 1999). Enterococci have been recovered from environmental surroundings in hospitals with colonized patients. Vancomycin-resistant enterococci (VRE) have been found to contaminate gowns and linens, beds, bed rails, tables, commodes, and other equipment. Person-to-person spread is a significant mode of transmission of nosocomial enterococci. *E. faecalis* causes more diseases than other Group D streptococci. It is commonly present in the mouth of normal adults, and in small numbers throughout the intestines. Contact Precautions.

Disease: opportunistic infections, endocarditis, bacteremia

Source: humans

POI: URI, wounds, urinary tract and soft tissue infections, bacteremia

Treatment: ampicillin, combinations of penicillin and aminoglycoside. Often resistant to antibiotics, especially sulfanilomides. No vaccine.

Survival: survives well in the environment

Fugomyces cyanescens: Noncontagious Fungal Spore—Airborne Class 2

Formerly *Sporothrix schenckii*. Pulmonary lesions and pneumonia due to *Fugomyces* have been reported in organ transplant recipients and the immunocompromised. A hazard to the immunocompromised, but uncommon. Pulmonary sporotrichosis probably develops as the result of inhalation of spores. Skin infections may develop as the result of contamination of scratches and cuts. Outbreaks have occurred among children playing or working with hay. Often an occupational disease in agriculture and sporadic in nature. Epidemics have occurred. Worldwide distribution in environment. Laboratory-acquired infections have occurred. *F. cyanescens* may cause nosocomial pneumonia in the immunocompromised. Standard Precautions.

Disease: sporotrichosis, Rose gardener's disease, skin lesions, pulmonary lesions, pneumonia

Source: environmental, soil, decaying plant material

POI: upper respiratory tract, skin

Treatment: oral iodides or itraconazole, amphotericin B. No prophylaxis, no vaccine.

Survival outside host: months in vegetation

Fusarium: Noncontagious Fungal Spore—Airborne Class 2

Reportedly allergenic. Can grow on damp grains and a wide range of plants. Has been found growing in humidifiers. A common soil fungus. Several species in this genus can produce potent toxins. Produces vomitoxin on grains during damp growing conditions. Symptoms may occur either through ingestion of contaminated grains or possibly inhalation of spores. The genera can produce hemorrhagic syndrome in humans. Frequently involved in eye, skin, and nail infections. Standard Precautions.

Disease: allergic alveolitis, allergic fungal sinusitis, toxic reactions

Source: environmental, indoor growth on floor dust filters, and in humidifiers

POI: upper respiratory tract, skin, eyes

Treatment: no prophylaxis, no vaccine

Survival outside host: survives outdoors

Haemophilus influenzae: Communicable Bacteria—Airborne Class 2

A leading cause of meningitis before the development of a vaccine. Infants are main victims and it can be fatal under age 2. Can be pleomorphic in shape. In spite of the name this microbe is the cause of meningitis but not a major cause of the flu. It can occur as a secondary invader when influenza virus is present. Some species occur naturally as human oral flora. Sometimes found on human body surfaces. Use of Droplet Precautions is recommended to prevent the spread of the microbe to the elderly, the immunocompromised, and unimmunized children (Decker and Schaffner 1999).

Disease: meningitis, pneumonia, endocarditis, otitis media, flu, and opportunistic infections

Source: humans

POI: nasopharyngeal

Treatment: antibiotic therapy for 10–14 days using chloramphenicol, or cephalosporins

Prophylaxis: rifampin; vaccine available

Survival: 12 days in sputum

Haemophilus parainfluenzae: Communicable Bacteria—Airborne Class 2

Causes infections that are similar to or associated with *H. influenzae*, but is more common. A member of the normal flora in the upper respiratory tract (oral cavity and pharynx). Can be recovered in the throat of 10–25% of children. May cause pharyngitis, epiglottis, otitis media, conjunctivitis, pneumonia, meningitis, bacteremia, endocarditis, and other infections. Other respiratory tract infections may predispose patients to infections with *H. parainfluenzae.*

Disease: opportunistic infections, conjunctivitis, pneumonia, meningitis

Source: humans

POI: upper respiratory tract

Treatment: erythromycin, trimethoprim-sulfamethoxazole; no vaccine

Histoplasma capsulatum: Noncontagious Fungal Spore—Airborne Class 1

Histoplasma capsulatum causes histoplasmosis, an infection estimated to have afflicted 40 million Americans, mostly in the Southeast. It most often causes mild fever and malaise, but in 0.1–0.2% of cases the disease becomes progressive. The infection is inevitably airborne and enters through the lungs, from where it may spread to other areas. In the environment, it is most often found in pigeon roosts, bat caves, or old buildings. This infection can become fatal in some cases. Standard Precautions.

Disease: histoplasmosis, fever, malaise, pneumonia possible

Source: environmental

POI: upper respiratory tract

Treatment: amphotericin B; no prophylaxis, no vaccine

Survival outside host: survives outdoors

Influenza A Virus: Contagious Virus—Airborne Class 1

Causes periodic flu pandemics and can cause widespread fatalities, with sometimes many millions dead. Constant antigenic variations among the main types of influenza, Type A and Type B, ensure little chance of immunity developing. Pneumonia can result from secondary bacterial infections, usually *Staphylococcus* or *Streptococcus.* Current theory suggests that the virus passes to and from humans, pigs, and birds, in agricultural areas of Asia where their close association is common. Influenza is an important nosocomial pathogen especially for the elderly and those with other diseases. Influenza vaccination is often an effective strategy for controlling seasonal outbreaks. Influenza A and B viruses are among the most highly communicable diseases known and can produce explosive epidemics. Infected humans form the reservoir and person-to-person transmission is thought to be airborne (Valenti 1999). Evidence for airborne transmission of influenza was provided by McLean (1961) and Moser et al. (1979). Alford et al. (1966) demonstrated that inhalation of small-particle aerosols could cause human infections. Infection may also be transmitted by fomites on contaminated surfaces and hands. Small particle aerosols (less than 10 microns) of influenza are produced and disseminated by coughing and sneezing (Douglas 1975). Direct transmission involves inoculation of mucous membranes of the nose or eyes, and indirect transmission can occur when the host's hands are contaminated. The aerosol transmission mode may be responsible for the explosive nature of flu outbreaks (Graman and Hall 1989), and a single infective can expel a large number of droplets as well as aerosol clouds. The larger droplets can be sprayed by the force of expulsion directly onto a new host's mucous membranes, and such expulsion invariably generates aerosol clouds. Droplets can be sprayed up to 6 feet distant from the infected patient. Virus shedding from patients is exclusively from the upper respiratory tract and often lasts for only 1 week. Fabian et al. (2008) has demonstrated that influenza virus is emitted during tidal breathing, at a mean rate of about 11 virus particles per minute, and with most of the exhaled particles being less than 1 micron in diameter. Droplet Precautions.

Disease: flu, secondary pneumonia

Source: humans, birds, pigs

POI: upper respiratory tract

Treatment: No antibiotic treatments, fluids and rest. Prophylaxis available.

Survival outside host: 2–4 days on cloth, steel

Klebsiella pneumoniae: Communicable Bacteria—Airborne Class 2

These bacteria are members of the Enterobacteriaceae family and exist in the soil and in water as free-living microorganisms. They are also found in human intestines as commensal flora. It is only when they enter the upper respiratory tract that they become an infectious problem. Worldwide, two-thirds of *Klebsiella* infections are nosocomial. This bacterium causes 3% of cases of acute bacterial pneumonia, but the fatality rate is as high as 90% in untreated cases. The bacteria are primarily spread in the hospital from person to person via the hands of HCWs or from environmental reservoirs to patients (Bonten, Hariharan, and Weinstein 1999). Increases in bacteremia and respiratory tract infection have a demonstrated association with increases in air contamination with *Klebsiella* (Grieble et al. 1974; Kelsen and McGuckin 1980).

Disease: opportunistic infections, pneumonia, ozena, rhinoscleroma; nosocomial urinary and pulmonary infections, wound infections, secondary infections of lungs in cases of chronic pulmonary disease

Source: environmental, soil, humans, indoor growth in water

POI: upper respiratory tract

Treatment: aminoglycosides, cephalosporins, resists some other antibiotics. Prophylaxis possible. No vaccine.

Survival outside host: 4 hours to several days

Legionella pneumophila: Noncontagious Bacteria—Airborne Class 1

The cause of Legionnaires' disease, *Legionella pneumophila* exists in warm out-door ponds naturally and in indoor water supplies unnaturally. It becomes a problem when amplified by air conditioning equipment and aerosolized in ventilation systems. Extremely high concentrations of the bacteria can result in aerosolization by various means, including showerheads and sauna baths. Legionnaires' disease can be acquired by the inhalation of aerosols contain-ing *Legionella* or by aspiration of water or respiratory secretions contaminated with *Legionella* (Stout and Yu 1999). Approximately 20% of Legionnaires' dis-ease cases are nosocomial and these occur most frequently in immunocom-promised hosts. Aspiration of water contaminated with *Legionella* is probably the major mode of transmission in hospitals. A large fraction of hospitals have their water systems colonized with *Legionella*. Water disinfection sys-tems, including ultraviolet light systems, are an effective means of controlling waterborne contamination with *Legionella*. Standard Precautions.

> Disease: Legionnaires' disease, Pontiac fever, opportunistic infec-tions, pneumonia
>
> Source: environmental, growth in cooling tower water, spas, potable water
>
> POI: upper respiratory tract
>
> Treatment: erythromycin, rifampin, ciprofloxacin, oxygen and fluid replacement. Prophylaxis with antibiotics. No vaccine.
>
> Survival outside host: months in water

Measles Virus: Contagious Virus—Airborne Class 1

Mainly affects children in 2- to 3-year epidemic cycles. Morbillivirus causes the well-known children's disease measles, also called rubeola. Some 90% of adults carry immunity to this virus. It is contracted via the airborne route, and most often in schools, where recirculated air from ventilation systems has been directly implicated by epidemiology studies. Occurs primarily in winter and spring. Many measles outbreaks have occurred in health care settings, and it is commonly transmitted among patients in emergency departments, outpatient waiting areas, and physicians' offices (Wainwright et al. 1999; CDC 1983). Patients are the source for about 90% of measles cases acquired in health care, with HCWs the source for the remainder. Measles can be severe in immunocompromised hosts. Measles can survive for at least 2 hours in fine droplets, and airborne spread has been well documented (Remington et al. 1985; Bloch, Orenstein, and Ewing 1985; Ehresmann et al. 1995; Wells and Holla 1950; Sienko et al. 1987; Riley, Murphy, and Riley 1978; CDC 1983). Airborne measles virus has been detected in hospital infection wards (Agranovski et al. 2008). Airborne Precautions.

Disease: measles (rubeola), hard measles, red measles, morbilli

Source: humans

POI: upper respiratory tract

Treatment: No antibiotic treatment. Live vaccine available.

Survival outside host: 30 minutes to 2 hours as aerosol

Mucor: Noncontagious Fungal Spore—Airborne Class 2

An opportunistic pathogen that can infect the lungs or other locations. It can be fatal to those with impaired immune systems. Spores will enter and germinate to produce mycelial growth. Indoor levels can exceed outdoor levels and it can grow on dust and filters. The majority of patients with mucormycosis are seriously immunocompromised. Standard Precautions.

Disease: mucormycosis, rhinitis, pneumonia

Source: environmental, sewage, dead plant material, horse dung, fruits

POI: upper respiratory tract

Treatment: treatment with amphotericin B remains the only reliable therapy; no prophylaxis, no vaccine

Survival outside host: survives outdoors

Mumps Virus: Contagious Virus—Airborne Class 1

Paramyxovirus, or mumps virus, causes the common childhood disease in about 60% of children in spring and winter. Some 70% of infections are asymptomatic. It only affects humans and is seldom life threatening. Immunity runs at 60% in the adult population. Tends to be benign and self-limiting. Most cases of mumps are community acquired but transmission has occurred, though rarely, in hospital settings (Wainwright et al. 1999). Mumps is transmitted in saliva and respiratory secretions, and is spread by direct contact with infected droplet nuclei or saliva or through airborne transmission (NCIRD 2011; Habel 1945; Baron 1996). Bahlke, Silverman, and Ingraham (1949) used ultraviolet lights for air disinfection during a mumps epidemic in schools, but the effect was limited and no statistically significant reduction in infections resulted. One study by Agranovski et al. (2008) detected airborne mumps virus in hospital infection wards with patients suffering from mumps infections. Droplet Precautions.

Disease: mumps, viral encephalitis

Source: humans

POI: upper respiratory tract

Treatment: symptomatic and supportive treatment only. Prophylaxis possible. Vaccine available.

Mycobacterium avium: Contagious Bacteria—Airborne Class 2

A nontuberculous mycobacteria with TB-like symptoms. *Mycobacterium avium* and *Mycobacterium intracellulare* are nearly identical and are members of the atypical mycobacteria. They can be asymptomatic. They are often found in association with the tuberculosis bacilli (Garrett et al. 1999). Has been isolated from soil and water. Inhalation is believed to be the common route of infection. *M. avium* has occurred in nosocomial settings. Distributed worldwide. Nearly all cases of pulmonary infection occur in adults.

> Disease: cavitary pulmonary disease, opportunistic
>
> Source: environmental, water, dust, plants
>
> POI: upper respiratory tract
>
> Treatment: treatment with isoniazid, rifampin, and ethambutol; no prophylaxis, no vaccine
>
> Survival outside host: survives outdoors

Mycobacterium tuberculosis: Contagious Bacteria—Airborne Class 1

Tuberculosis infects over a third of the world's population. This bacterium causes TB, once called consumption because of the way it depleted a person to death. This disease poses one of the greatest modern health hazards due to the recent emergence of drug-resistant strains. It is highly contagious and a single bacilli is capable of causing an infection in lab animals. *M. tuberculosis* is carried in airborne droplet nuclei, which are produced when infected persons cough, speak, or sing (Garrett et al. 1999). Ambient air currents can keep them airborne for extended periods of time and carry them some distance. Airborne transmission is well established (CDC 2005; Ehrenkranz and Kicklighter 1972; Riley et al. 1957). The risk of infection is correlated with the airborne concentration of TB bacilli and the duration of exposure. Transmission by other routes is possible, usually involving needle stick injuries. *Mycobacterium bovis* can also cause infections via airborne transmission. TB is a serious concern for immunocompromised patients and very high attack rates can occur. Outbreaks of TB in nosocomial settings have included emergency departments, ICUs, a surgical suite, and other areas.

> Related spp.: *M. avium, M. intracellulare*. Infections with *Mycobacterium ulcerans* involve severe ulceration of the skin and subcutaneous tissue.
>
> Disease: tuberculosis (TB), pneumonia possible
>
> Source: humans, sewage (potential)
>
> POI: upper respiratory tract

Treatment: isoniazid, rifampin, streptomycin, ethambutol, pyrazin-amide. Prophylaxis possible. Vaccine available.

Survival outside host: 40–100 days

Mycoplasma pneumoniae: Communicable Bacteria—Airborne Class 2

Mycoplasma pneumoniae is a member of a class called Mollicutes, which are unlike other bacteria because they contain no cell wall. Weakly pathogenic for man and often found as commensals. Immune system disruption, typically caused by another disease, is required to produce an infection. It accounts for approximately 20% of all cases of pneumonia. Transmission is believed to occur by respiratory droplets and Droplet Precautions should be followed (Decker and Schaffner 1999). Endemic infections occur worldwide. Mainly affects those 5–15 years old.

Disease: pneumonia, pleuropneumonialike organisms (PPLO), walking pneumonia

Source: humans

POI: upper respiratory tract

Treatment: tetracyclines, gentamicin, doxycycline, macrolides

Prophylaxis: antibiotics; no vaccine

Survival outside host: 10–50 hours in air

Neisseria meningitides: Communicable Bacteria—Airborne Class 2

Infections are endemic throughout the world but epidemics occur also. The second leading cause of meningitis after *H. influenza*. Infection can result in asymptomatic colonization. Carriage of meningococci is common. Prolonged close contact is a risk factor, but transmission in hospital settings is rare. Transmission is by respiratory droplets, including contact with large droplets from the nose or mouth of an infected individual (Decker and Schaffner 1999). Droplet Precautions.

Disease: meningitis

Source: humans

POI: upper respiratory tract

Treatment: rifampin, ceftraixone, ciprofloxacin, penicillin; vaccine available

Prophylaxis: rifampin

Survival outside host: does not survive long

Nocardia asteroides: Noncontagious Bacterial Spore—Airborne Class 2

Considered pathogenic, this Gram-positive bacteria is classified as a pathogenic actinomycetes. This microorganism is a bacterium barely distinguishable from fungi. It can be found in some soils. It is an opportunistic pathogen and primarily affects patients who have been rendered susceptible by other diseases, especially those involving immunodeficiency.

Disease: nocardiosis, pneumonia

Source: environmental, soils, sewage

POI: upper respiratory tract

Treatment: surgical drainage, sulfanilamides (TMP-SMX, sulfisoxazole, sulfadiazine); no prophylaxis, no vaccine

Survival outside host: indefinitely in soil, water

Norwalk Virus: Contagious Virus—Airborne Class 1

Strictly an intestinal pathogen, it is responsible for explosive outbreaks of gastroenteritis in home, school, and community settings (Decker and Schaffner 1999). Almost half of outbreaks involve nursing homes and hospitals. Secondary attack rates are high. Aerosolization can occur during vomiting, or from outdoor air sprays of warm, contaminated seawater, such as may happen on cruise ships. Airborne transmission has been suspected in some hospitals, restaurants, and cruise ship outbreaks (Sawyer et al. 1988; Chadwick 1994; Caul 1994; Gellert, Waterman, and Ewert 1990; Marks et al. 2000). Projectile vomiting associated with Norwalk gastroenteritis may aerosolize infectious droplets. Has caused repeated outbreaks on some cruise ships in warm tropical waters (Ho et al. 1989). Occasionally causes foodborne outbreaks on land. The viruses are passed in the stool of infected persons. People get infected by swallowing stool-contaminated food or water. Contact and Droplet Precautions.

Disease: gastroenteritis

Source: environmental, warm ocean waters

POI: gastrointestinal

Treatment: none, rehydration therapy; no prophylaxis, no vaccine

Survival outside host: survives outside host

Parainfluenza Virus: Contagious Virus—Airborne Class 2

Parainfluenza occurs worldwide and infects children (at a rate of 75–80%) more than adults. This virus is very contagious and also causes croup. Outbreaks usually occur in the fall. There are four types of parainfluenza viruses: parainfluenza 1, 2, 3, and 4. Parainfluenza 1 and 2 can cause croup. Parainfluenza 3 is a major cause of respiratory disease in infants and children. Reinfections are common but are generally associated with mild upper respiratory illness (Turner 1999). Viral shedding from the upper respiratory tract occurs 1–4 days before onset and continues about 7–10 days. Some patients may shed for 3–4 weeks. The route of spread is by direct person-to-person contact or via large droplets (Baron 1996). Standard and Droplet Precautions.

Disease: flu, colds, croup, pneumonia

Source: humans

POI: upper respiratory tract, lower respiratory tract

Treatment: no specific treatment available; no prophylaxis, no vaccine

Survival outside host: 4–10 hours on steel, cloth

Parvovirus B19: Contagious Virus—Airborne Class 2

An uncommon cause of fever. Parvovirus B19 is similar to the adenoviruses. Mostly occurs in children. Some 25% of infections are asymptomatic. Symptoms resolve in 7–10 days. Severe complications are unusual but anemic patients may develop transient aplastic crisis. May cause severe anemia in the immunosuppressed. Worldwide outbreaks occur, mainly in schoolchildren in winter and spring. Some 60% of adults have been exposed. Parvovirus outbreaks occur worldwide in communities and in schools, although it is relatively uncommon in health care facilities, and the incidence in HCWs is about 1% (Wainwright et al. 1999). Transmission is believed to occur most frequently through contact with respiratory secretions from infected patients. Prolonged close contact, such as in households, is a primary determinant of infection transmission. Parvoviruses remain infectious in the environment for a long time. Whether the transmission primarily involves droplets, fomites, or direct contact is uncertain. Standard Precautions.

Disease: fifth disease, anemia, fever

Source: humans

POI: upper respiratory tract

Treatment: treatment of symptoms only; no prophylaxis, no vaccine

Survival outside host: survives frozen for years

Penicillium: Noncontagious Fungal Spore—Airborne Class 2

Penicillium notatum and some closely related species of *Penicillium* are occasional causes of infections in man. Pulmonary, or lung, infections are rare, but it can infect the ear and cornea. *P. notatum* can produce penicillin, and some people are highly allergic to this antibiotic. Indoor levels of spores can exceed outdoor levels. Sometimes found on human body surfaces. Standard Precautions.

Disease: alveolitis, rhinitis, asthma, allergic reactions, irritation, ODTS, toxic reactions

Source: environmental, indoor growth on paint, filters, coils, and humidifiers

POI: upper respiratory tract, ear, eyes

Treatment: amphotericin B, itraconazole; no prophylaxis, no vaccine

Survival outside host: survives outdoors

Pneumocystis jirovecii: Noncontagious Fungal Spore—Airborne Class 2

Pneumocystis jirovecii (formerly *P. carinii*) was previously classified as a protozoa but is now recognized as a complex fungi. It is opportunistic and dangerous mainly to those with forms of immunodeficiency. Occurs worldwide and is reportedly airborne. It is an important cause of pneumonia, especially in AIDS patients. *P. jirovecii* exists as a saprophyte in the lungs of humans and a variety of animal species. Most healthy children have been exposed at an early age. Direct contact is usually required for transmission (Decker and Schaffner 1999). Outbreaks of pneumocystis have occurred in hospitals and orphanages. *Pneumocystis* may not grow in the hospital environment, but it can be readily detected with PCR methods and has been recovered from the air (Bartlett et al. 1997). *Pneumocystis* has been detected in 57% of air samples from rooms of infected patients, and has been found in hospital areas where there were no infected patients (Bartlett et al. 1997). *Pneumocystis* has been detected in the exhaled breath of infected patients (Bartlett and Lee 2010). Although absolute proof of airborne transmission has not yet been shown, there is a wealth of data suggesting that this is indeed the case, and the CDC recommends not placing immunocompromised patients in the same room as an infected patient because the microbe is disseminated into the air surrounding patients (Choukri 2010). Standard Precautions.

Disease: pneumocystosis, pneumonia possible

Source: environmental, humans

POI: upper respiratory tract

Treatment: TMP-SMX; no prophylaxis, no vaccine

Survival outside host: 48 hours

Proteus mirabilis: Noncontagious Bacteria—Airborne Class 2

Proteus mirabilis is a nonrespiratory opportunistic pathogen that is often found as intestinal flora and has a tendency to colonize the urinary tract.

Up to 10% of urinary tract infections are caused by *P. mirabilis*. Can cause problems when health is compromised. *Proteus* cultures tend to swarm over the surface of media rather than remain confined to colonies. Other species of *Proteus* may be infectious.

Disease: opportunistic infections, pneumonia possible

Source: humans

POI: upper respiratory tract, burns, wounds

Treatment: beta-lactams, quinolones. May resist ampicillin and cephalosporin. No prophylaxis or vaccines.

Pseudallescheria boydii: Noncontagious Fungal Yeast—Airborne Class 2

P. boydii is the most common filamentous fungus in the lungs of cystic fibrosis patients. The asexual stage is represented by *Scedosporium* and *Graphium*. Standard Precautions.

Disease: cutaneous infections, sinusitis, keratitis, lymphadenitis, endophthalmitis, meningoencephalitis, brain abscess, endocarditis, pneumonia, lung abscess, and other infections, sometimes collectively called pseudallescheriasis

Source: often associated with polluted environments, soil, sewage, contaminated water, agricultural manure

POI: skin, lungs, sinus, eyes, etc.

Treatment: minconazole, itraconazole, voriconazole, posaconazole, ravuconazole, triazole (UR-9825), echinocandins are effective in vitro. A modified tetracycline, CMT-3, has also shown some effectiveness.

Survival outside host: survives outdoors

Pseudomonas aeruginosa: Noncontagious Bacteria—Airborne Class 1

The primary cause of nosocomial pseudomonal infections. Its infectivity is limited mostly to immunosuppressed patients or those who have their health compromised by other illnesses. It is considered to exist ubiquitously in the environment, but is amplified in hospitals. Infection sites include the lungs, burn wounds, and open wounds. Can become fatal in 80% of cases. It produces some minor toxins. The source of *Pseudomonas* infections after intra-abdominal surgery is generally considered to be the patients (Arnow & Flaherty 1999). *Pseudomonas* is one of the most common burn wound infections. Dispersal of *Pseudomonas*, presumably from colonized patients, has resulted in contamination of air (Lowbury and Fox 1954; Ransjo 1979) as well as local surfaces and equipment. Increases in bacteremia and respiratory tract infection have a demonstrated association with increases in air contamination with *P. aeruginosa* (Grieble et al. 1974; Kelsen and McGuckin 1980). Standard Precautions.

Disease: pneumonia, toxic reactions

Source: environmental, sewage, indoor growth in dust, water, humidifiers

POI: upper respiratory tract, burns, wounds

Treatment: antibiotic treatment; regularly resistant to penicillin, ampicillin, and other antibiotics

Survival outside host: survives outdoors

Reovirus: Contagious Virus—Airborne Class 2

A comparatively rare cause of fever and colds. Reovirus can cause mild forms of fever in infants and children. Several types exist that have variable symptoms. Adults can be infected but symptoms are often mild. Upper respiratory infections may cause fever, pharyngitis, rhinitis, and sometimes rashes. Isolated cases of encephalitis, pneumonia, and renal disease have been reported. Reovirus remains stable and viable in aerosols (Adams et al. 1982). Can survive in the dry state on surfaces (Buckland and Tyrrell 1962).

Disease: colds, fever, pneumonia, rhinorrhea

Source: humans

POI: upper respiratory tract

Treatment: no treatment available; no prophylaxis, no vaccine

Survival outside host: 12 hours or more

Respiratory Syncytial Virus (RSV): Contagious Virus—Airborne Class 1

A common cause of pneumonia (40%) and bronchiolitis (90%) in infants. Occurs within a few months of birth. This virus is unaffected by maternal antibody. Occurs worldwide. Most common cause of viral pneumonia in children under 5 years of age. RSV infections may appear as bronchiolitis in infants and as a common cold in their older caregivers (Hierholzer and Archibald 1999). Reinfection is common with a nearly universal attack rate of 40%. Outbreaks peak in March and February. Can cause severe illness in the elderly and the immunocompromised. Close contact with an infected infant can transmit the virus. The virus may be transmitted directly or by contact with fomites or by large particle aerosols or droplet spread (Hall, Douglas, and Geiman 1980; Baron 1996). Aerosol transmission with large droplets only has been demonstrated experimentally, but RSV has been transmitted to patients by way of fomites (Hall and Douglas 1981). The virus must reach the respiratory mucosa to become infectious, and inoculation of the virus into either the eye or the nose is equally efficient (Turner 1999). Shedding of RSV can last for a week or two. With the ability to survive for 8 hours on surfaces, it may be capable of spreading locally with the patient as a reservoir. RSV has been detected in some 71% of air samples taken from the air of an urgent medical clinic using an air sampler and a PCR assay (Lindsley et al. 2010). Standard and Contact Precautions.

Disease: pneumonia, bronchiolitis

Source: humans

POI: lower respiratory tract

Treatment: ribavirin is beneficial when delivered as a nasal spray. Prophylaxis possible. No vaccine.

Survival outside host: up to 8 hours

Rhinovirus: Contagious Virus—Airborne Class 2

One of the causes of the common cold. Humans are the only hosts for the human strains. Rhinovirus colds are predominant in adults, while other cold viruses may predominantly afflict children. Brief hand contact is thought to be a primary mechanism of transmission. In experiments, rhinovirus was shown to transmit most efficiently by direct person-to-person contact, but transmission by large particle aerosols has also been shown (Dick, Blumer, and Evans 1967; Gwaltney, Moskalksi, and Hendley 1978; Dick et al. 1987). Shedding of the virus from infected patients may continue for 2–3 weeks. Treatment of the hands with a virucidal compound can prevent transmission of rhinovirus infection. The airborne infectious dose has been shown to be 100 $TCID_{50}$/mL (Bischoff 2010). Experimentally induced airborne infection was demonstrated by Couch et al. (1966). Droplet Precautions.

Disease: colds

Source: humans

POI: upper respiratory tract

Treatment: no specific antivirals, but sensitive to alpha-2 interferon; no prophylaxis, no vaccine

Survival outside host: 1–7 days on surfaces

Rhizopus: Noncontagious Fungal Spore—Airborne Class 2

Can infect the lungs and other locations. Can be fatal to those with impaired immune systems. An opportunistic pathogen. Spores will germinate and mycelial growth will result. *Rhizopus arrhizus* is the most common agent of zygomycosis, with *Rhizopus microsporus* being the second most common. The majority of patients with mucormycosis are seriously immunocompromised. Hemoptysis may develop with continued tissue necrosis, and the end result may be fatal pulmonary hemorrhage. Indoor growth may occur on dust, filters, and ductwork. Standard Precautions.

Disease: zygomycosis, allergic reactions, pneumonia, mucormycosis

Source: environmental, decaying fruit and vegetables, compost

POI: upper respiratory tract, sinus, skin, eye

Treatment: amphotericin B remains the only reliable therapy; no prophylaxis, no vaccine

Survival outside host: survives outdoors

Rotavirus: Contagious Virus—Airborne Class 2

Rotavirus is the principal agent of infantile diarrhea and has been responsible for nosocomial outbreaks (Decker and Schaffner 1999). Rotavirus is spread by the fecal-oral route, mainly by the hands of HCWs. Rotavirus can survive on surfaces, but environmental contamination has not been identified as a contributing factor in outbreaks. Sufficient evidence exists to warrant Droplet Precautions as well as Contact Precautions. Airborne transmission is plausible but not yet proven. Airborne rotavirus has been detected in hospital air using PCR assays (Dennehy et al. 1998).

Disease: diarrhea

Source: humans

POI: intestinal

Treatment: none

Survival outside host: up to 1 hour on surfaces (Keswick et al. 1983)

Rubella Virus: Contagious Virus—Airborne Class 1

The common cause of German measles in children. Up to 80% of adults have immunity. It is a mild disease, and those infected develop lifelong immunity. Occurs worldwide with prevalence in winter and spring. Some 30–50% of infections are asymptomatic. Endemic in most communities. Congenital rubella syndrome may occur in infants born to women with rubella in first trimester. Virus is shed in oropharyngeal secretions and is highly transmissible. In communities where vaccination is rare, spring outbreaks typically occur every few years. Children represent the largest number of cases. Nosocomial rubella has involved both HCWs and patients (Wainwright et al. 1999). It is most contagious while the rash is erupting, and virus may be shed for a week before and a week after the rash develops. The primary portals for virus entry are the mucosa of the upper respiratory tract and nasopharygeal lymphoid tissue. Transmission occurs from person to person via droplets shed from the respiratory secretions of infected patients. Droplet Precautions.

Disease: rubella (German measles)

Source: humans

POI: upper respiratory tract

Treatment: no antibiotic treatment, no specific treatment; no prophylaxis, no vaccine

Survival outside host: for short periods

Scedosporium: Noncontagious Fungal Spore—Airborne Class 2

An emerging opportunistic fungus that can cause infections in both immunocompetent and other patients. *Scedosporium prolificans* infections can be fatal. *Scedosporium apiospermum* is the counterpart of the teleomorph of *P. boydii*. Standard Precautions.

Disease: opportunistic infections

Source: environmental

Treatment: Optimum treatment is unknown. Amphotericin B alone or in combination with flucytosine, fluconazole, or itraconazole has been used. Success has been achieved with voriconazole and terbinafine, but some strains are resistant. Posaconazole, miltefosine, and albaconazole may be helpful.

Survival outside host: survives outdoors

Serratia marcescens: Noncontagious Bacteria—Airborne Class 2

Normally benign but is capable of causing serious infections in some cases, especially as nosocomial infections. Causes opportunistic infections of the eyes, blood, wounds, urinary tract, and respiratory tract. Important cause of nosocomial outbreaks in nurseries, intensive care wards, and renal dialysis units. Responsible for 4% of nosocomial pneumonias. They are primarily spread in the hospital from person to person via the hands of HCWs or from environmental reservoirs and medical equipment to patients (Bonten, Hariharan, and Weinstein 1999). *Serratia* thrives in moist environments. Oropharyngeal colonization of patients is common.

Disease: opportunistic infections, bacteremia, endocarditis, pneumonia

Source: environmental, indoor growth in potable water

POI: upper respiratory tract, wounds, eyes, urinary tract; opportunistic infections of the lungs, eyes, and urinary tract.

Treatment: aminoglycosides, amikacin, resistant to penicillins.
  Prophylaxis possible. No vaccine.

Survival outside host: 35 days or more

Staphylococcus aureus (MRSA): Communicable Bacteria—Airborne Class 1

S. aureus is generally a commensal microorganism. Can cause opportunistic infections when host resistance is compromised, especially when a primary infection such as influenza is present. The case mortality rate is high. Also causes food intoxication and toxic shock syndrome. The major reservoir is the anterior nares. Once a patient is colonized, the particular strain of *S. aureus* may disseminate by person-to-person contact, especially by spread on the hands or the dispersion of bacteria carried on skin squames (John and Barg 1999). *S. aureus* spreads in this manner among hospitalized patients. Nasal carriage is common among HCWs but not often the cause of infection, and in fact HCWs are at risk of acquiring a strain (i.e., MRSA) in the hospital. *Staphylococcus* is efficiently transmitted by direct contact and is also transmitted by the airborne route, albeit much less efficiently (Williams 1966; Williams et al. 1966). A clear role for airborne transmission of *S. aureus* was shown by Mortimer et al. (1966) in a nursery. It is likely that *S. aureus* strains from patients with pneumonia or burn infections may spread by the airborne route (John and Barg 1999). Epidemics are likely sustained by human carriers. MRSA can spread quickly and displace nasal flora in patients and HCWs. Some 13% of male and 5% of females are shedders, who carry a heavy nasal inoculum and disperse large numbers of microbes from their lower extremities and perineum into the air around them (Hare and Thomas 1999). In a seminal experiment, a physician with a rhinovirus infection dispersed *S. aureus* up to 20 feet away (Eichenwald, Kotsevalov, and Fasso 1960; Sherertz et al. 1996). Such "cloud adults," like cloud babies who disperse *S. aureus* in the nursery, can continuously contaminate air and surfaces in their vicinity (Sherertz, Bassetti, and Bassetti-Wyss 2001). People routinely inhale staphylococci in indoor environments. One early study showed that HCWs who were shedders were associated with outbreaks in the operating room (Walter, Kundsin, and Brubaker 1963). Skin squames are about 15 microns in diameter and should fall in still air within seconds, but they may be carried on air currents or wafted off the floor to become briefly airborne again. MRSA can be isolated from the immediate environment of colonized patients and has been recovered from many hospital surfaces including floors, linens, air vents, furniture, and equipment. Airborne transmission is a consideration whenever a patient acquires staphylococcal pneumonia and may be an important factor in burn units as well (Thompson, Cabezudo, and Wenzel 1982; Rutala et al. 1983). The spread of MRSA in a burn unit could not be halted by methods that were effective in a neonatal unit, reportedly because the problem involved environmental and airborne routes of transmission (Farrington, Ling, and French 1990). ICUs have often been the site of outbreaks of MRSA, especially surgical ICUs and neonatal units. Environmental reservoirs and the airborne spread of MRSA appear to be more important in burn units than in other facilities (Hartstein and Mulligan 1999). Standard Precautions supplemented by Contact Precautions when environment is contaminated.

Disease: staphylococcal pneumonia, opportunistic infections (esp. MRSA)

Source: humans, sewage

POI: upper respiratory tract; deep infections include endocarditis, meningitis, pneumonia

Prophylaxis: none; no vaccine

Survival outside host: 7–60 days, 72 hours on steel

Staphylococcus epidermis: Communicable Bacteria—Airborne Class 2

A normal commensal of the skin and the most frequently isolated microbe clinically. Nonrespiratory. May represent up to 90% of all isolates from skin. Can contaminate medical equipment via contact or settling in air. A common cause of nosocomial urinary tract infections. Virtually all *S. epidermis* infections are hospital acquired. Some people, particularly some males, shed much more *S. epidermis* than others. *S. epidermis* is one of a number of *coagulase-negative staphylococci* (CONS) that have been recovered from humans (Boyce 1999). Others include *S. saprophyticus, S. haemolyticus, S. lugdunensis, S. warneri, S. hominis, S. schleiferi, S. simulans, S. cohnii, S. capitis, S. saccharolyticus, S. auricularis, S. caprae,* and *S. xylosus.* Nosocomial infections of these staphylococci may be caused by endogenous strains that colonize patients at the time of admission or by strains acquired in the hospital. Areas of the skin may be populated by 10–100,000 CFU/cm² of coag-negative staphylococci. Only a fraction of the coag-negative staphylococci that cause infections during implant surgery can be traced to the patient's skin. HCWs are a likely source of infections. Studies have demonstrated that ultraclean air systems, surgical isolation systems, and the use of body exhaust suits reduces intraoperative contamination of surgical sites and infection rates in implant-related infections. Staphylococci shed from the skin, and once environmental surfaces are contaminated, these surfaces may serve as a secondary reservoir for further spread within a hospital. Airborne transmission is a plausible route by which coag-negative staphylococci may spread from HCWs to patients.

Disease: opportunistic infections, bacteremia

Source: humans, sewage

POI: skin

Treatment: vancomycin, rifampin, ciprofloxacin. No prophylaxis or
vaccines.

Streptococcus pneumoniae: Communicable Bacteria—Airborne Class 2

The leading cause of death in the world. This microorganism is commonly
known as pneumococcus and is the prime agent of lobar pneumonia, which
predominantly affects children. It is commonly carried asymptomatically in
healthy individuals. Carriage rates among children are high—about 30% for
children and 10% for adolescents. *S. pneumoniae* is a well-recognized cause
of nosocomial infection, and nursing home outbreaks are well documented
(Crossley 1999). Standard Precautions. Droplets Precautions for drug-resistant strains.

Disease: lobar pneumonia, sinusitis, meningitis, otitis media, toxic
reactions.

Source: humans

POI: upper respiratory tract

Treatment: penicillin, erythromycin. Prophylaxis with antibiotics.
Vaccine available.

Survival outside host: 1–25 days

Streptococcus pyogenes: Communicable Bacteria—Airborne Class 1

Streptococcus pyogenes is part of the normal flora of the human body and only
results in disease when host immunity is compromised. Also called Group
A *Streptococcus* (GAS). Often occurs as nosocomial infections in wounds,
and lung infections can also result. GAS is a common cause of community-
acquired pharyngitis and skin infection. Epidemics once swept through
Europe and the United States periodically. Infections are most common in
the 5- to 15-year age group, and from December to May. *Streptococcus agalac-
ticae* (GBS) is also encountered in infections, and infants acquire this bacteria
from the hands of HCWs, not from their mothers (Crossley 1999). GAS is frequently carried in the respiratory tract of HCWs but little nosocomial transmission from this source has been documented. Rectal or vaginal carriage of
S. pyogenes is the most commonly reported source of outbreaks in surgical site
infections, with skin friction and breaking wind two possible modes of aerosolization (Schaffner et al. 1969; Sula 2002). Settle plates were used to demonstrate that GAS was aerosolized after a carrier exercised in a room (McKee
et al. 1966). Viglionese et al. (1991) describe an outbreak of postpartum infections traced to anal carriage in an obstetrician. Aerosolization of GAS with
motion or activity followed by contamination of surgical sites is the usual

mode of transmission, and cases have occurred in operating rooms adjacent to the one in which the source was working (Berkelman et al. 1982; Mastro et al. 1990). Whitby et al. (1984) describe an outbreak that began in a burn center and spread to an ICU in an associated hospital. Standard Precautions. Droplet Precautions for UTIs. Contact Precautions for skin wounds.

Disease: scarlet fever, pharyngitis, toxic reactions

Source: humans

POI: upper respiratory tract, burns, wounds

Treatment: penicillin, clindamycin, cephalosporin

Prophylaxis: penicillin; no vaccine

Survival outside host: up to 195 days in dust, 9 days on metal

Trichosporon: Noncontagious Fungal Yeast—Airborne Class 2

Reportedly a cause of summertime hypersensitivity pneumonitis as a result of growth on damp wood and matting material. Although it is a yeast and does not occur in the spore form, hyphae and fragments may be released and become airborne. A cause of white piedra and onychomycosis in humans. Localized infections with *Trichosporon beiglii* may include endocarditis, meningitis, pneumonia, and ocular infections. An agent of bronchial and pulmonary infections in immunocompromised hosts. Standard Precautions.

Disease: hypersensitivity pneumonitis, white piedra, onychomycosis, opportunistic infections

Source: environmental, soil, water, vegetation

POI: upper respiratory tract, skin, hair shafts

Treatment: NA. No prophylaxis, no vaccine.

Survival outside host: survives outdoors

Varicella-Zoster Virus: Contagious Virus—Airborne Class 1

Varicella-Zoster virus (VZV) causes varicella (chickenpox) in almost everyone by the age of 10 and is highly contagious. Infections can recur for those who are immunodeficient, especially bone marrow transplant patients. Occurs worldwide chiefly as a disease of children (75% of population by age 15; 90% of young adults have had disease). More frequent in winter and early spring in temperate zones. It also causes herpes zoster, which is a reactivation of latent VZV in a dorsal nerve ganglion (Zaia 1999). It is more common in adults. Major transmission mode is respiratory, but direct contact with pustules can also produce infection. VZV may involve bacterial coinfections of the lower respiratory tract, producing pneumonia. Coinfections can involve *S. pneumoniae, H. influenza,* and *S. aureus* (Bullowa and Wishik 1935). Various studies have shown that VZV/chickenpox is an airborne disease (Habel 1945; Nelson and St. Geme 1966). VZV is spread by airborne droplets from nasopharyngeal secretions, which can be carried by air currents, or it can be exchanged by face-to-face exposure (Leclair et al. 1982; Gustafson et al. 1982; Asano, Iwayama, and Miyata 1980; Anderson et al. 1985; Greene, Barenberg, and Greenburg 1941; Tsujino, Sako, and Takahashi 1984). Herpes zoster is spread by direct contact or by exposure to airborne infectious virus particles (Josephson and Gombert 1988). VZV has been detected in hospital air samples (Sawyer et al. 1994). Airborne Precautions are often recommended for exposed susceptible patients as well as Contact Precautions.

Disease: chickenpox

Source: humans

POI: upper respiratory tract

Treatment: vidarabine and acyclovir. Prophylaxis possible. Vaccine available.

Survival outside host: for short periods

References

Adams, D. J., Spendlove, J. C., Spendlove, R. S., and Barnett, B. B. (1982). Aerosol stability of infectious and potentially infectious Reovirus particles. *Appl Environ Microbiol* 44(4), 903–908.

Agranovski, I. E., Safarov, A. S., Agafonov, A. P., Pyankov, O. V., and Sergeev, A. N. (2008). Monitoring of airborne mumps and measles viruses in a hospital. *Clean Soil, Air, Water* 36(10-11), 845–849.

Alford, R. H., Kasel, J. A., Gerone, P. J., and Knight, V. (1966). Human influenza resulting from aerosol inhalation. *Proc Soc Exp Biol Med* 122(3), 800–804.

Allen, K., and Green, H. (1987). Hospital outbreak of multi-resistant *Acinetobacter anitratus*: An airborne mode of spread? *J Hosp Infect* 9, 110–119.

Anderson, J. D., Bonner, M., Scheifele, D. W., and Schneider, B. C. (1985). Lack of nosocomial spread of varicella in a pediatric hospital with negative pressure ventilated patient rooms. *Infect Control* 6, 120–121.

Arnow, P. M., and Flaherty, J. P. (1999). Nonfermentative Gram-negative bacilli; in *Hospital Epidemiology and Infection Control*, C. G. Mayhall, ed., Lippincott Williams & Wilkins, Philadelphia, 431–452.

Asano, Y., Iwayama, S., and Miyata, T. (1980). Spread of varicella in hospitalized children having no direct contact with an indicator zoster case and its prevention by a live vaccine. *Biken J* 23, 157–161.

Bahlke, A. M., Silverman, H. F., and Ingraham, H. S. (1949). Effect of ultra-violet irradiation of classrooms on spread of mumps and chickenpox in large rural central schools. *Am J Pub Health* 41, 1321–1330.

Baron, S. (1996). *Medical Microbiology*. University of Texas Medical Branch, Galveston, TX.

Bartlett, M. S., and Lee, C.-H. (2010). Airborne spread of *Pneumocystis jirovecii*. *Oxford J Med Clin Inf Dis* 51(3), 266.

Bartlett, M. S., Vermund, S. H., Jacobs, R., Durant, P. J., Shaw, M. M., Smith, J. W., Tang, X., Lu, J.-J., Li, B., Jin, S., and Lee, C.-H. (1997). Detection of *Pneumocystis carinii* DNA in air samples: Likely environmental risk to susceptible persons. *J Clin Microbiol* 35(10), 2511–2513.

Berkelman, R. L., Martin, D., Graham, D. R., Mowry, J., Freisem, R., Weber, J. A., Ho, J. L., and Allen, J. R. (1982). Streptococcal wound infections caused by a vaginal carrier. *JAMA* 247, 2680–2682.

Bernards, A. T., Frenay, H. M. E., Lim, B. T., Hendriks, W. D. H., Dijkshoorn, L., and vanBoven, C. P. A. (1998). Methicillin-resistant *Staphylococcus aureus* and *Acinetobacter baumanii*: An unexpected difference in epidemiologic behavior. *Am J Infect Contr* 26, 544–551.

Best, E. L., Fawley, W. N., Parnell, P., and Wilcox, M. H. (2010). The potential for airborne dispersal of *Clostridium difficile* from symptomatic patients. *Clin Inf Dis* 50, 1450–1457.

Bischoff, W. E. (2010). Transmission route of rhinovirus type 39 in a monodispersed airborne aerosol. *Inf Contr Hosp Epidemiol* 31(8), 857–859.

Bloch, A. B., Orenstein, W. A., and Ewing, W. M. (1985). Measles outbreak in a pediatric practice: Airborne transmission in an office setting. *Pediatrics* 75, 676–683.

Bonten, M. J. M., Hariharan, R., and Weinstein, R. A. (1999). Enterobacteriaceae; in *Hospital Epidemiology and Infection Control*, C. G. Mayhall, ed., Lippincott Williams & Wilkins, Philadelphia, 407–430.

Boyce, J. M. (1999). Coagulase-negative staphylococci; in *Hospital Epidemiology and Infection Control*, C. G. Mayhall, ed., Lippincott Williams & Wilkins, Philadelphia, 365–384.

Brooks, S. E., Walczak, M. A., and Hameed, R. (2000). Are we doing enough to contain *Acinetobacter* infections? *Inf Contr Hosp Epidemiol* May, 304.

Buckland, F. E., and Tyrrell, D. A. J. (1962). Loss of infectivity on drying various viruses. *Nature* 195, 1063–1064.

Bullowa, J. G. M., and Wishik, S. M. (1935). Complications of varicella: I. Their occurrence among 2534 patients. *Am J Dis Child* 49, 923–926.

Caul, E. O. (1994). Small, round structured viruses: Airborne transmission and hospital control. *Lancet* 343, 1240–1242.

CDC (1983). Imported measles with subsequent airborne transmission in a pediatrician's office—Michigan. Centers for Disease Control. *MMWR* 32, 401–402.

_____. (2005). *Guidelines for Preventing the Transmission of* Mycobacterium tuberculosis *in Health-Care Facilities*. Centers for Disease Control, Atlanta, GA.

Chadwick, P. (1994). Airborne transmission of a small round structured virus [Letter-reply]. *Lancet* 343, 609.

Chenoweth, C. E., and Schaberg, D. R. (1999). *Enterococcus* species; in *Hospital Epidemiology and Infection Control*, C. G. Mayhall, ed., Lippincott Williams & Wilkins, Philadelphia, 395–406.

Choukri, F. (2010). Quantification and spread of *Pneumocystis jirovecii* in the surrounding air of patients with Pneumocystis pneumonia. *Clin Inf Dis* 51(3), 259–265.

Couch, R. B., Cate, T. R., Douglas, R. G., Gerone, P. J., and Knight, V. (1966). Effect of route of inoculation on experimental respiratory viral disease in volunteers and evidence for airborne transmission. *Bact Rev* 30, 517–529.

Crossley, K. B. (1999). Streptococci; in *Hospital Epidemiology and Infection Control*, C. G. Mayhall, ed., Lippincott Williams & Wilkins, Philadelphia, 385–394.

Crowe, M., Towner, K. J., and Humphreys, H. (1995). Clinical and epidemiological features of an outbreak of *Acinetobacter* infection in an intensive therapy unit. *J Med Microbiol* 43, 55–62.

Decker, M. D., and Schaffner, W. (1999). Nosocomial diseases of healthcare workers spread by the airborne or contact routes (other than TB); in *Hospital Epidemiology and Infection Control*, C. G. Mayhall, ed., Lippincott Williams & Wilkins, Philadelphia, 1101–1126.

Dennehy, P. H., Nelson, S. M., Crowley, B. A., and Saracen, C. L. (1998). Detection of rotavirus RNA in hospital air samples by polymerase chain reaction (PCR). *Pediatr Res* 43(4), 143A.

Dick, E. C., Blumer, C. R., and Evans, A. S. (1967). Epidemiology of infections with rhinovirus types 43 and 55 in a group of University of Wisconsin student families. *Am J Epid* 86, 386–400.

Dick, E. C., Jennings, L. C., Mink, K. A., Wartgow, C. D., and Inhorn, S. L. (1987). Aerosol transmission of rhinovirus colds. *J Infect Dis* 156, 442–448.

Douglas, R. G. J. (1975). Influenza in man; in *The Influenza Viruses and Influenza*, E. D. Kilbourne, ed., Academic Press, New York, 395–447.

Ehrenkranz, N. J., and Kicklighter, J. L. (1972). Tuberculosis outbreak in a general hospital: Evidence for airborne spread of infection. *Ann Intern Med* 77, 377–382.

Ehresmann, K. R., Hedberg, C. W., Grimm, M. B., Norton, C. A., MacDonald, K. L., and Osterholm, M. T. (1995). An outbreak of measles at an international sporting event with airborne transmission in a domed stadium. *J Infect Dis* 171(3), 679–683.

Eichenwald, H., Kotsevalov, O., and Fasso, L. A. (1960). The cloud baby: An example of bacterial-viral interaction. *Am J Dis Child* 100, 161–173.

Fabian, P., McDevitt, J. J., Dehaan, W. H., and Fung, R. (2008). Influenza virus in human exhaled breath: An observational study. *PLoS ONE* 3(7), e2691.

Farrington, M., Ling, T., and French, G. (1990). Outbreaks of infection with methicillin-resistant *Staphylococcus aureus* on neonatal and burns units of a new hospital. *Epidem Infect* 105, 215–228.

Fekety, R., Kim, K.-H., Brown, D., Batts, D. H., Cudmore, M., and Silvia, J. Jr. (1981). Epidemiology of antibiotic-associated colitis. Isolation of *Clostridium difficile* from the hospital environment. *Am J Med* 70, 906–908.

Flanningan, B., McCabe, E. M., and McGarry, F. (1991). Allergenic and toxigenic microorganisms in houses; in *Pathogens in the Environment*, B. Austin, ed., Blackwell Scientific, Oxford.

Garrett, D. O., Dooley, S. W., Snider, D. E., and Jarvis, W. R. (1999). *Mycobacterium tuberculosis*; in *Hospital Epidemiology and Infection Control*, C. G. Mayhall, ed., Lippincott Williams & Wilkins, Philadelphia, 477–503.

Gellert, G. A., Waterman, S. H., and Ewert, D. (1990). An outbreak of acute gastroenteritis caused by a small round structured virus in a geriatric convalescent facility. *Inf Contr Hosp Epidemiol* 11, 459–464.

Graman, P. S., and Hall, C. B. (1989). Epidemiology and control of nosocomial virus infections in nosocomial infections. *Infect Dis Clin North Am* 3, 4.

Greene, D., Barenberg, L. H., and Greenburg, B. (1941). Effect of irradiation of the air in a ward on the incidence of infections of the respiratory tract. *Am J Dis Child* 61, 273–275.

Grieble, H., Bird, T., Nidea, H., and Miller, C. (1974). Chute-hydropulping waste disposal system: A reservoir of enteric bacilli and *Pseudomonas* in a modern hospital. *J Infect Dis* 130, 602.

Gustafson, T. L., Lavely, G. B., Brawner, E. R., Hutcheson, R. H., Wright, P. F., and Schaffner, W. (1982). An outbreak of airborne nosocomial varicella. *Lancet* 70, 550–556.

Gwaltney, J. M., Moskalski, P. B., and Hendley, J. O. (1978). Hand-to-hand transmission of rhinovirus colds. *Ann Int Med* 88, 463–467.

Habel, K. (1945). Mumps and chickenpox as air-borne diseases. *Am J Med Sci* 209, 75–78.

Hall, C. B., and Douglas, R. G. (1981). Modes of transmission of respiratory syncytial virus. *J Pediatr* 99, 100–103.

Hall, C. B., Douglas, R. G., and Geiman, J. M. (1980). Possible transmission by fomites of respiratory syncytial virus. *J Infect Dis* 141(1), 98–102.

Hare, R., and Thomas, C. G. A. (1999). The transmission of *Staphylococcus aureus*. *Br Med J* 2, 840–844.

Hartstein, A. I., and Mulligan, M. E. (1999). Methicillin-resistant *Staphylococcus aureus*; in *Hospital Epidemiology and Infection Control*, C. G. Mayhall, ed., Lippincott Williams & Wilkins, Philadelphia, 347–364.

Hierholzer, W. J., and Archibald, L. K. (1999). Principles of infectious disease epidemiology; in *Hospital Epidemiology and Infection Control*, C. G. Mayhall, ed., Lippincott Williams & Wilkins, Philadelphia, 3–14.

Ho, M. S., Glass, R. I., Monroe, S. S., Madore, H. P., Stine, S., Pinsky, P. F., and Cubitt, D. (1989). Viral gastroenteritis aboard a cruise ship. *Lancet* 2, 961–965.

HWFB (2003). SARS Bulletin (24 April 2003). Health, Welfare, and Food Bureau, Government of the Hong Kong Special Administrative Region. Hong Kong.

Jawad, A., Heritage, J., Snelling, A. M., and Gascoyne-Binzi, D. M. (1996). Influence of relative humidity and suspending menstrua on survival of *Acinetobacter* spp. on dry surfaces. *J Clin Microbiol* 34, 2881–2887.

John, J. F., and Barg, N. L. (1999). *Staphylococcus aureus*; in *Hospital Epidemiology and Infection Control*, C. G. Mayhall, ed., Lippincott Williams & Wilkins, Philadelphia, 325–346.

Johnson, S., and Gerding, D. N. (1999). *Clostridium difficile*; in *Hospital Epidemiology and Infection Control*, C. G. Mayhall, ed., Lippincott Williams & Wilkins, Philadelphia, 467–476.

Josephson, A., and Gombert, M. E. (1988). Airborne transmission of nosocomial varicella from localized zoster. *J Infect Dis* 158, 238–241.

Kaatz, G. W., Gitlin, S. D., Schaberg, D. R., Wilson, K. H., Kaufman, C. A., Seo, S. M., and Fekety, R. (1988). Acquisition of *Clostridium difficile* from the hospital environment. *Am J Epidemiol* 127, 1289–1294.

Kelsen, S. G., and McGuckin, M. (1980). The role of airborne bacteria in the contamination of fine particle neutralizers and the development of nosocomial pneumonia. *Ann NY Acad Sci* 353, 218.

Keswick, B. H., Pickering, L. K., Dupont, H. L., and Woodward, W. E. (1983). Survival and detection of rotaviruses on environmental surfaces in day care centers. *Appl Environ Microbiol* 46, 813–816.

Kowalski, W. J. (2006). *Aerobiological Engineering Handbook: A Guide to Airborne Disease Control Technologies*. McGraw-Hill, New York.

Leclair, J. M., Zaia, J. A., Levin, M. J., Congdon, R. G., and Goldman, D. A. (1982). Airborne transmission of chickenpox in a hospital. *N Engl J Med* 302, 450–453.

Lindsley, W. G., Blachere, F. M., Davis, K. A., Pearce, T. A., Fisher, M. A., Khakoo, R., Davis, S. M., Rogers, M. E., Thewlis, R. E., Posada, J. A., Redrow, J. B., Celik, I. B., Chen, B. T., and Beezhold, D. H. (2010). Distribution of airborne influenza virus and respiratory syncytial virus in an urgent medical care clinic. *CID* 50, 693–698.

Lowbury, E. J. L., and Fox, J. (1954). The epidemiology of infection with *Pseudomonas pyocyanea* in a burn unit. *J Hyg* 52, 403–416.

Marks, P. J., Vipond, I. B., Carlisle, D., Deakin, D., Fey, R. E., and Caul, E. O. (2000). Evidence for airborne transmission of Norwalk-like virus (NLV) in a hotel restaurant. *Epidemiol Infect* 124(3), 481–487.

Mastro, T. D., Farley, T. A., Elliott, J. A., Facklam, R. R., Perks, J. R., Hadler, J. L., Good, R. C., and Spika, J. S. (1990). An outbreak of surgical wound infections due to group A *Streptococcus* carried on the scalp. *N Engl J Med* 323, 968–972.

Mayhall, C. G. (1999a). *Hospital Epidemiology and Infection Control*. Lippincott Williams & Wilkins. Philadelphia.

_____. (1999b). Nosocomial burn wound infections; in *Hospital Epidemiology and Infection Control*, C. G. Mayhall, ed., Lippincott Williams & Wilkins, Philadelphia, 275–286.

McKee, W. M., DiCaprio, J. M., Roberts, C. E., and Sherris, J. C. (1966). Anal carriage as the probable source of a streptococcal epidemic. *Lancet* 2, 1007–1009.

McLean, R. (1961). The effect of ultraviolet radiation upon the transmission of epidemic influenza in long-term hospital patients. *Am Rev Resp Dis* 83, 36–38.

Mortimer, E. A., Wolinsky, E., Gonzaga, A. J., and Rammelkamp, C. H. (1966). Role of airborne transmission in staphylococcal infections. *Br Med J* 5483, 319–322.

Moser, M. R., Bender, T. R., Margolis, H. S., Noble, G. R., Kendal, A. P., and Ritter, D. G. (1979). An outbreak of influenza aboard a commercial airliner. *Am J Epidemiol* 110(1), 1–6.

Mulligan, M. E., Rolfe, R. D., Finegold, S. M., and George, W. L. (1979). Contamination of a hospital environment by *Clostridium difficile*. *Curr Microbiol* 3, 173–175.

NCIRD (2011). Mumps; in *Epidemiology and Prevention of Vaccine-Preventable Diseases*, Centers for Disease Control and Prevention (CDC), Atlanta, 189–198.

Nelson, A. M., and St. Geme, J. W. (1966). On the respiratory spread of varicella-zoster virus. *Pediatrics* 37, 1007–1009.

Nielsen, P. (2008). *Clostridium difficile* aerobiology and nosocomial transmission. Northwick Park Hospital, Harrow, Middlesex, UK.

Pennington, J. E. (1980). *Aspergillus* lung disease. *Med Clin North Am* 64, 475.

Ransjo, U. (1979). Attempts to control clothes-borne infection in a burn unit. *J Hyg* 82, 369–384.

Remington, P. L., Hall, W. N., Davis, I. H., Herald, A., and Gunn, R. A. (1985). Airborne transmission of measles in a physician's office. *JAMA* 253(11), 1574–1577.

Riley, E. C., Murphy, G., and Riley, R. L. (1978). Airborne spread of measles in a suburban elementary school. *Am J Epidemiol* 107, 421–432.

Riley, R., Wells, W., Mills, C., Nyka, W., and McLean, R. (1957). Air hygiene in tuberculosis: Quantitative studies of infectivity and control in a pilot ward. *Am Rev Tuberc Pulmon* 75, 420–431.

Rutala, W. A., Katz, E. B. S., Sherertz, R. J., and F. A. Sarubbi (1983). Environmental study of a methicillin-resistant *Staphylococcus aureus* epidemic in a burn unit. *J Clin Microbiol* 18, 683–688.

Sawyer, L. A., Murphy, J. J., Kaplan, J. E., Pinsky, P. F., Chacon, D., Walmsley, S., Schonberger, L. B., Phillips, A., Forward, K., Goldman, C., Brunton, J., Fralick, R. A., Carter, A. O., Gary, W. G., Glass, R. I., and Low, D. E. (1988). 25- to 30-nm virus particle associated with a hospital outbreak of acute gastroenteritis with evidence for airborne transmission. *Am J Epidemiol* 127, 1261–1271.

Sawyer, M. H., Chamberlin, C. J., Wu, Y. N., Aintablian, N., and Wallace, M. R. (1994). Detection of varicella-zoster virus DNA in air samples from hospital rooms. *J Infect Dis* 169, 91–94.

Schaffner, W., Lefkowicz, L. B., Goodman, J. S., and Koenig, M. G. (1969). Hospital outbreak of infections with Group A Streptococci traced to an asymptomatic anal carrier. *N Engl J Med* 280, 1224–1225.

Schlapbach, L. J., Agyeman, P., Hutter, D., Aebi, C., Wagner, B. P., and Reidel, T. (2011). Human metapneumovirus infection as an emerging pathogen causing acute respiratory distress syndrome. *J Infect Dis* 203(2), 294–295.

Sherertz, R. J., Bassetti, S., and Bassetti-Wyss, B. (2001). "Cloud" Health Care Workers. *Emerg Inf Dis* 7(2), 241–244.

Sherertz, R. J., Regan, D. R., Hampton, K. D., Robertson, K. L., Streed, S. A., Hoen, H. M., Thomas, R., and J. R. Gwaltney (1996). A cloud adult: The *Staphylococcus aureus*-virus interaction. *Ann Intern Med* 124, 539–547.

Sienko, D. G., Friedman, C., McGee, H. B., Allen, M. J., Simonsen, W. F., Wentworth, B. B., Shope, T. C., and Orenstein, W. A. (1987). A measles outbreak at university medical settings involving health care workers. *Am J Pub Health* 77, 1222–1224.

Simmons, B. P., and Gelfand, M. S. (1999). Uncommon causes of nosocomial infection; in *Hospital Epidemiology and Infection Control*, C. G. Mayhall, ed., Lippincott Williams & Wilkins, Philadelphia, 593–604.

Snelling, A. M., Beggs, C. B., Kerr, K. G., and Sheperd, S. J. (2011). Spores of *Clostridium difficile* in hospital air. *Clin Infect Dis* 51, 1104–1105.

Stout, J. E., and Yu, V. L. (1999). Nosocomial *Legionella* infection; in *Hospital Epidemiology and Infection Control*, C. G. Mayhall, ed., Lippincott Williams & Wilkins, Philadelphia, 453–466.

Sula, M. (2002). Killer on the loose: A gruesome outbreak of flesh-eating bacteria at Evanston Hospital sends doctors scrambling to find the source before other patients get infected. *Reader* 31(33), 1.

Thompson, R. L., Cabezudo, I., and Wenzel, R. P. (1982). Epidemiology of nosocomial infections caused by methicillin-resistant *Staphylococcus aureus*. *Ann Intern Med* 97, 309–317.

Tsujino, G., Sako, M., and Takahashi, M. (1984). Varicella infection in a children's hospital: Prevention by vaccine and an episode of airborne transmission. *Biken J* 27, 129–132.

Turner, R. B. (1999). Nosocomial viral respiratory infections in pediatric patients; in *Hospital Epidemiology and Infection Control*, C. G. Mayhall, ed., Lippincott Williams & Wilkins, Philadelphia, 607–614.

Valenti, W. M. (1999). Influenza viruses; in *Hospital Epidemiology and Infection Control*, C. G. Mayhall, ed., Lippincott Williams & Wilkins, Philadelphia, 535–542.

Viglionese, A., Nottebart, V. F., Bodman, H. A., and Platt, R. (1991). Recurrent group A streptococcal carriage in a health care worker associated with widely separated nosocomial outbreaks. *Am J Med* 91(suppl), 329–333.

Wainwright, S. H., Singleton, J. A., Torok, T. J., and Williams, W. W. (1999). Nosocomial measles, mumps, rubella, varicella, and human parvovirus B19; in *Hospital Epidemiology and Infection Control*, C. G. Mayhall, ed., Lippincott Williams & Wilkins, Philadelphia, 649–664.

Walter, C. W., Kundsin, R. B., and Brubaker, M. M. (1963). The incidence of airborne wound infection during operation. *JAMA* 186, 908–913.

Wells, W. F., and Holla, W. A. (1950). Ventilation in flow of measles and chickenpox through community. Progress report, Jan. 1, 1946 to June 15, 1949, airborne infection study. *J Am Med Assoc* 142, 1337–1344.

Whitby, M., Sleigh, J. D., Reid, W., MacGregor, I., and Colman, G. (1984). Streptococcal infection in regional burns centre and a plastic surgery unit. *J Hosp Infect* 5, 63–69.

Williams, R. E. O. (1966). Epidemiology of airborne staphylococcal infection. *Bact Rev* 30(3), 660–672.

Williams, R., Blowers, R., Garod, L., and Shooter, R. (1966). Staphylococcal infections: Introduction; in *Hospital Infection Causes and Prevention*, R. E. O. Williams, ed., Lloyd-Luke, London, 22–41.

Zaia, J. A. (1999). Varicella-zoster virus; in *Hospital Epidemiology and Infection Control*, C. G. Mayhall, ed., Lippincott Williams & Wilkins, Philadelphia, 543–553.

5

Airborne Nosocomial Etiology

Introduction

Etiology is the study of the detailed factors of infectious disease. The etiology of airborne pathogenic disease is multifaceted, involving airborne transport of pathogens, surface contamination and resuspension of fomites, hand contact, direct contact with droplets, and inhalation in the case of respiratory pathogens. In the case of nonrespiratory pathogens, shedding of endogenous bacteria and pathogens is the initial route of aerosolization. A number of these elements of etiology have been introduced in previous chapters, where a distinction was made between airborne transmission and airborne transport, the former term referring to the causative route of infection while the latter refers simply to the physical transport of a microbe via air currents, such as when spores enter a hospital environment. Another concept of use in this chapter is the classification of airborne pathogens into seven categories, these being respiratory infections, nonrespiratory infections, SSIs, burn wound infections, immunocompromised infections, pediatric infections, and nursing home infections. The etiologies of these categories are addressed in the following sections and the modes of transport are summarized in Table 5.1. These are the primary or most common modes because virtually every mode of transport may contribute to each type of infection. Each of the airborne modes of transport will be discussed in the following sections after a review of aerosolization. The modes of droplet spray, hand contact, and ingestion were discussed in previous chapters or are self-explanatory and need no further discussion.

In each of the modes of transport listed in Table 5.1 airborne transmission or airborne transport plays a role in contributing to the infection. The exception is when the droplet spray route results in a pathogen achieving direct access to the mucosa of a new host. This is considered to be a form of direct contact. The odds of a droplet produced by a cough or sneeze traveling a few feet through the air and hitting a bull's-eye on the back of another person's throat or nose with an infectious dose seems less probable than inhalation of the aerosol cloud produced by the same action, but this seems to be the prevailing view of the droplet mode of spread. In any event, when droplets

TABLE 5.1

Etiology of Airborne Nosocomial Infections

Category	Modes of Transport
Respiratory Infections	Airborne inhalation, droplet spray, hand contact, fomites
Nonrespiratory Infections	Airborne settling, hand contact, fomites, ingestion
Surgical Site Infections (SSIs)	Airborne settling, hand contact, fomites
Burn Wound Infections	Airborne settling, hand contact, fomites
Immunocompromised Infections	Airborne inhalation and settling, hand contact, fomites, ingestion
NICU and Nursery Infections	Airborne inhalation, droplet spray, hand contact, fomites, ingestion
Nursing Home Infections	Airborne inhalation, droplet spray, hand contact, fomites, ingestion

are sprayed an aerosol is also produced and fomites are created on surfaces, and all these modes of transport contribute to the infection process.

Pathogen Aerosolization

Pathogens come from two main sources, humans and the environment. If the building itself is generating pathogens (i.e., a sick building) then it can be considered part of the environment, although no hospital building should be creating microbial contamination. Environmental sources of airborne pathogens will typically mean the outside air or the generation of aerosols from water.

Five classic mechanisms have been described for the transmission of infections: (1) direct transmission, (2) vertical transmission, (3) indirect transmission, (4) vector transmission, and (5) airborne transmission. Vertical transmission and vector transmission must be excluded because they are definitively not airborne. Direct hand contact is not airborne transmission, but if the hands should pick up fomites that have settled then this is fomite contact and may have an airborne component. Therefore, for the purposes of studying indoor airborne pathogens in hospitals, the following three transmission modes are relevant:

1. Aerosol clouds
2. Droplet spray transmission
3. Fomites on hands, surfaces, or equipment

Droplet spray transmission, in which droplets are projected several feet through the air from coughing and sneezing, is considered a form of direct contact when the droplets land directly on mucosal surfaces. Droplets may also land on surfaces to produce fomites, which is considered an indirect route of transmission. These fomites may be picked up by the hands of an HCW and cause him or her an infection, or else they can be carried to other patients where further direct contact leads to an infection. Although droplet spray could arguably be a form of nonairborne transmission, this distinction is more semantic than practical because the act of generating droplets also generates aerosols and fomites, and aerosol clouds may also settle out on surfaces and equipment to produce fomites. As originally noted, it is virtually impossible to separate the problem of airborne infections from surface-borne infections and so they must be treated together.

It has been shown that violent coughing is efficient at creating infectious droplet nuclei (Menzies et al. 2003). The main difference between droplet spray and aerosol clouds created when a patient coughs or sneezes is that the droplets may travel only a few feet while the aerosol cloud may travel much greater distances beyond the vicinity of the patient.

For infection to occur, pathogens must be transferred from a reservoir to a susceptible host in sufficient numbers to induce an infection (Hierholzer and Archibald 1999). The total infectious dose of a pathogen received by a new host, or at a surgical site, will be the sum of those that arrive from aerosols, those that arrive directly from droplet spray or shedding, and those that are brought into direct contact via hands, equipment, or surface contact. The total infectious dose received, T_{ID}, can be written in equation form as follows:

$$T_{ID} = D_a + D_d + D_c \tag{5.1}$$

where
D_a = dose transmitted by aerosol (inhalation or settling)
D_d = dose transmitted by direct contact (mucosa or wound)
D_c = dose transmitted by indirect contact (hands, equipment, etc.)

In some cases, like TB, influenza, or clean surgical site infections, it is possible that the majority of microbes arrive by the aerosol route. In most cases, however, the largest dose is probably transmitted by direct or indirect contact, in which the latter component will involve fomites that may have been transported via air.

Shedding of microbes can result in both aerosolization and the production of fomites and therefore is analogous to the process of droplet spray and aerosol clouds. Particles that are shed from a patient may fall with shorter trajectories than droplets that are forcefully expelled. Nonrespiratory infections and endogenous microbes may be shed from patients, such as *Staphylococcus*. Skin squames or particles may be shed locally within a few feet around a patient and smaller particles (less than 5 microns) may become suspended in air. Shedding can also produce fomites and indirect transmission may result. Shedding is a critical factor for SSIs, while droplets are a critical factor for respiratory infections.

Aerosolization of pathogens can result from other sources, such as humidifiers, spray devices, or laboratory processes. Aerosolized blood pathogens can present hazards to surgeons against which facemasks may not protect them. Aerosols in the respirable size range (less than 5 µm) containing blood can be generated in an operating room during surgery through the use of common surgical power tools (Jewett et al. 1992).

Measures for the prevention and control of airborne nosocomial infections are directed at various links in the chain of causation. These include measures to (a) eliminate or isolate the reservoirs of pathogens, (b) protect or immunize the host against disease, or (c) interrupt the transmission of infection. Isolation of patients who are shedding or aerosolizing infectious agents is an effective strategy provided they are diagnosed in time to make a difference. Immunization is not available for every disease. Interrupting the transmission by disinfection of the surroundings or the air is the final option and often the most important recourse once an infection is spreading in the hospital.

There is a potential health risk to hospital workers who handle medical waste that may aerosolize microbes in processes like compaction, grinding, or shredding (Gordon et al. 1999). Such processes may require the use of facemasks or be performed in appropriately ventilated areas.

Aerobiological Pathways

Once an agent is aerosolized it may take various pathways to a host before it causes an infection. The aerobiological pathways of infectious agents in health care settings can be subdivided into airborne transmission (usually meaning inhalation) and airborne transport. A generalized flowchart of the major aerobiological pathways by which patients become infected with airborne nosocomial pathogens is shown in Figure 5.1.

Resuspension results when surface or equipment contamination becomes reaerosolized from use or activity. All five categories of infections are included if we assume ingestion accounts for nonrespiratory infections. It is assumed that settling occurs only on surfaces, although it is possible for microbes to settle directly on hands. Figure 5.1 is a simplistic view of what probably happens, but it illustrates the fact that air and surface contamination are virtually inseparable and a complete etiology of airborne nosocomial infections must address both issues. It is possible for a microbe to become reaerosolized several times or pass from person to person before causing an infection (routes that can't be shown here), and of course health care workers as well as patients may be subject to inhalation of pathogens.

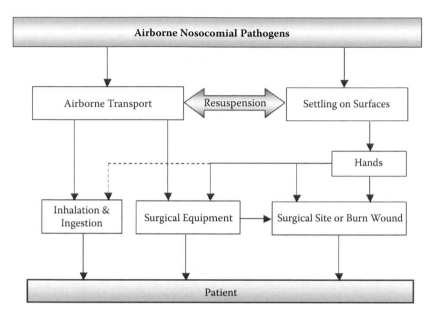

FIGURE 5.1
The major aerobiological pathways of airborne nosocomial pathogens, including respiratory, SSI, burn wound, and immunocompromised infections.

Having generalized all aerobiological pathways, we can now focus on some of the seven categories in more detail with representative examples from the literature. Four examples are selected, an airborne respiratory infection (SARS virus), airborne *Aspergillus* spores, shedding of *Staphylococcus aureus*, and shedding of *Clostridium difficile* spores. Most, if not all, other pathogens will be represented by these four examples.

Airborne Transmission of Respiratory Infections

SARS coronavirus, one of the newest nosocomial agents, is also one of the most hazardous for hospital personnel and serves as a representative example of a contagious respiratory virus. In a study by He et al. (2003) it was found that index patients were the first-generation source of transmission and they infected inpatients and medical staff, creating second-generation patients. The major transmission routes were close-proximity airborne droplet infection and close contact infection. There was also evidence for the likelihood of aerosol transmission of infections through the ventilation system, which spread the infection to other hospital floors. A similar report comes from Ho, Tang, and Seto (2003), who found that hospital outbreaks of SARS typically occurred within the first week after admission of the first SARS cases before recognition and before isolation measures were implemented. In the majority of hospital infections, there was close contact with a SARS patient, and transmission occurred via large droplets, direct contact with infectious fluids, or by contact with fomites from infectious fluids. In some instances, potential airborne transmission was reported in association with endotracheal intubation, nebulized medications, and noninvasive positive pressure ventilation of SARS patients. Nosocomial transmission was effectively halted by enforcement of standard routines, contact and droplet precautions in all clinical areas, and additional airborne precautions in high-risk areas.

Airborne Transport of Fungal Spores

Activity and traffic can aerosolize spores that may be resident on surfaces, especially floors or shadowed areas (ceiling cavities and walls) that are uncovered by construction. Fungal spores of *Aspergillus* released during construction activities can cause nosocomial infections. In a special care unit (SCU) adjacent to an area under renovation, nosocomial fungal pulmonary infections developed in two premature infants (Krasinski et al. 1985). Inspection showed that inadequate barriers permitted the passage of airborne particles between the two areas, and a significantly higher density of mold spores were found in the SCU compared to a construction-free area. The major source of mold was dust above the hospital's suspended ceiling. Physical barriers, pressurization, and air filtration are essential for controlling airborne fungal spore concentrations in high-risk areas during renovation projects. Such projects can produce bursts of spores and a sudden spike in local airborne concentrations that may overwhelm barriers and filters. Such spikes in airborne levels have also been noted to occur outdoors, where spores may travel in concentrated clouds or waves. Such spikes in outdoor ambient levels may pass entryways when doors are briefly opened, resulting in large numbers of spores gaining access to indoor areas of hospitals.

Mahieu et al. (2000) studied the relationship between air contamination with fungal spores, especially *Aspergillus*, in three renovation areas of a neonatal intensive care unit (NICU) and colonization and infection rates in a high care area equipped with HEPA filtration. Renovation work increased airborne concentrations of *Aspergillus* spores significantly in the medium care area and resulted in a significant increase in NICU concentrations. On the other hand, the use of a mobile HEPA air filtration system caused a significant decrease in the spore levels. There was no evidence of invasive aspergillosis during the renovation.

Aspergillus can cause fatal surgical site infections, as in bone marrow transplants (Petersen et al. 1983). ORs are typically filtered with HEPA filters or 95% filters that should not permit the passage of spores if the filters are properly maintained and leaktight. Contaminated filters can also disperse *Aspergillus* spores (Lentino et al 1982). The application of portable HEPA filtration units in one hospital building reduced the airborne concentrations of *Aspergillus* from a maximum of 0.4 cfu/m^3 to 0.009 cfu/m^3 (Sherertz et al. 1987). HEPA filters will completely remove all spores in the *Aspergillus* size range, but it must be ensured that when portable filtration units are used they are positioned and sized correctly to make for a sufficient air exchange rate. Incomplete air mixing in a room will diminish the effectiveness of air filtration.

Shedding of Bacteria

Bischoff et al. (2007) examined the shedding of *S. aureus* in volunteer subjects and what effect that gowns, scrubs, and facemasks had on the release rate. Results did not show a significant decrease in the airborne concentrations of *S. aureus* when facemasks were used, suggesting that contamination of the skin and hands played a more important role. The mean rate at which *S. aureus* were spread into the environment was 0.12 cfu/m^3 per minute of occupation. They found that the use of scrubs reduced the quantity of *S. aureus* dispersed into the air, but that the addition of gowns did not improve the efficiency of scrubs.

Dust particles, skin squames, and respiratory secretions from operating room personnel can increase airborne bacterial concentrations in the OR. These particles can fall or settle quickly and contaminate surgical sites or burn wounds located close to the contaminant source (Wong 1999). Some 10% of skin squames carry bacteria (Noble 1975). It has been found that the dispersal of *S. aureus* is increased in persons with common colds or upper respiratory infections, forming clouds around adult and infant dispersers (Sherertz et al. 1996). The airborne dispersal of coagulase-negative staphylococci (CONS) was studied by Bischoff et al. (2004) to see if other staphylococci, besides *S. aureus*, were also subject to increased airborne dispersal during infections with common colds. These CONS are usually found on the skin and mucous membranes, and results indicate that their airborne dispersal rate more than doubled during a cold. The primary isolates of CONS include *S. epidermis, S. capitus, S. hominis,* and *S. haemolyticus.* Other respiratory bacteria such as *Haemophilus influenzae, Neisseria meningitidis,* and *Streptococcus pyogenes* (GAS) may also be dispersed into the air at increased rates during an upper respiratory tract viral infection.

Shedding of *Clostridium difficile* Spores

Clostridium difficile is a bacterial spore acquired from human sources, which may include patients as well as asymptomatic HCWs (Johnson and Gerding 1999). Shedding of spores is common but sporadic in patients with symptomatic infection (Best et al. 2010). Spores can survive for weeks, which allows them time to spread far and wide throughout a hospital environment via airborne transport. Floors and bathrooms tend to be the most contaminated sites, and reaerosolization from floors may allow spores to travel beyond the ward of the original patient source. Traffic through wards and hallways, as well as local air currents, can cause spores to spread throughout a hospital and cause infections far from the source. The environment of infected patients is more frequently contaminated than other environments (Mulligan et al. 1979). Contaminated equipment also serves as a reservoir for dissemination, and environmental disinfection can interrupt an outbreak (Kaatz et al. 1988). The mean settling time of *Clostridium* spores, based on their size, is about 8 minutes, which is sufficient for them to be captured by strong air currents. In the absence of any activity, spores will ultimately settle close to the infectious source patient. *Clostridium* spores must be ingested to cause intestinal disease and so the spores must be brought into contact with the mouth (i.e., by the hands or equipment), or settle on food, or be aspirated and then ingested after mucociliary lung clearance. Both isolation of the patient and disinfection of the locale should be effective measures to control further dissemination in a hospital.

Viruses such as Norwalk virus and rotavirus are shed in a similar manner and will contaminate the local environment around an infected patient. The degree to which these pathogens spread in a hospital depends on the microbe's ability to survive outside a host and also on local air currents and activity.

References

Best, E. L., Fawley, W. N., Parnell, P., and Wilcox, M. H. (2010). The potential for airborne dispersal of *Clostridium difficile* from symptomatic patients. *Clin Inf Dis* 50, 1450–1457.

Bischoff, W. E., Bassetti, S., Bassetti-Wyss, B. A., Wallis, M. L., Tucker, B. K., Reboussin, B. A., D'Agostino, R. B., Pfaller, M. A. Jr., J. M. G., and Sherertz, R. J. (2004). Airborne dispersal as a novel transmission mechanism route of coagulase-negative staphylococci: Interaction between coagulase-negative staphylococci and rhinovirus infection. *Inf Contr Hosp Epidemiol* 25(6), 504–511.

Bischoff, W. E., Tucker, B. K., Wallis, M. L., Reboussin, B. A., Pfaller, M. A., Hayden, F. G., and Sherertz, R. J. (2007). Preventing the airborne spread of *Staphylococcus aureus* by persons with the common cold: Effect of surgical scrubs, gowns, and facemasks. *Inf Contr Hosp Epidemiol* 28(10), 1148–1154.

Gordon, J. G., Reinhardt, P. A., Denys, G. A., and Alvarado, C. J. (1999). Medical waste management; in *Hospital Epidemiology and Infection Control*, C. G. Mayhall, ed., Lippincott Williams & Wilkins, Philadelphia, 1387–1398.

He, Y., Jiang, Y., Xing, Y. B., Zhong, G. L., Wang, L., Sun, Z. J., Jia, H., Chang, Q., Wang, Y., Ni, B., and Chen, S. P. (2003). Preliminary result on the nosocomial infection of severe acute respiratory syndrome in one hospital of Beijing. *Zhonghua Liu Xing Bing Xue Za Zhi* 24(7), 554–556.

Hierholzer, W. J., and Archibald, L. K. (1999). Principles of infectious disease epidemiology; in *Hospital Epidemiology and Infection Control*, C. G. Mayhall, ed., Lippincott Williams & Wilkins, Philadelphia, 3–14.

Ho, P. L., Tang, X. P., and Seto, W. H. (2003). SARS: Hospital infection control and admission strategies. *Respirology* 8(Suppl), S41–S45.

Jewett, D. L., Heinsohn, P., Bennett, C., Rosen, A., and Neuilly, C. (1992). Blood-containing aerosols generated by surgical techniques: A possible infectious hazard. *Am Ind Hyg Assoc J* 53(4), 228–231.

Johnson, S., and Gerding, D. N. (1999). *Clostridium difficile*; in *Hospital Epidemiology and Infection Control*, C. G. Mayhall, ed., Lippincott Williams & Wilkins, Philadelphia, 467–476.

Kaatz, G. W., Gitlin, S. D., Schaberg, D. R., Wilson, K. H., Kaufman, C. A., Seo, S. M., and Fekety, R. (1988). Acquisition of *Clostridium difficile* from the hospital environment. *Am J Epidemiol* 127, 1289–1294.

Krasinski, K., Holman, R. S., Hanna, B., Greco, M. A., Graff, M., and Bhogal, M. (1985). Nosocomial fungal infection during hospital renovation. *Infect Control* 6(7), 278–282.

Lentino, J. R., Rosenkrantz, M. A., Michels, J. A., Kurup, V. P., Rose, H. D., and Rytel, M. W. (1982). Nosocomial aspergillosis: A retrospective review of airborne disease secondary to road construction and contaminated air conditioners. *Am J Epidemiol* 116(8), 430–437.

Mahieu, L. M., Dooy, J. J. D., Laer, F. A. V., Jansens, H., and Ieven, M. M. (2000). A prospective study on factors influencing aspergillus spore load in the air during renovation works in a neonatal intensive care unit. *J Hosp Infect* 45(3), 191–197.

Menzies, D., Adhikari, N., Arietta, M., and Loo, V. (2003). Efficacy of environmental measures in reducing potentially infectious bioaerosols during sputum induction. *Inf Contr Hosp Epidemiol* 24, 483–489.

Mulligan, M. E., Rolfe, R. D., Finegold, S. M., and George, W. L. (1979). Contamination of a hospital environment by *Clostridium difficile*. *Curr Microbiol* 3, 173–175.

Noble, W. C. (1975). Dispersal of skin microorganisms. *Br J Dermatol* 93, 477–485.

Petersen, P. K., McGlave, P., Ramsay, N. K., Rhame, F. S., Cohen, E., G. S. Perry, Goldman, A. I., and Kersey, J. (1983). A prospective study of infectious diseases following bone marrow transplantation: Emergence of *Aspergillus* and *Cytomegalovirus* as the major causes of mortality. *Infect Control* 42(2), 81–89.

Sherertz, R. J., Belani, A., Kramer, B. S., Elfenbein, G. J., Weiner, R. S., Sullivan, M. L., Thomas, R. G., and Samsa, G. P. (1987). Impact of air filtration on nosocomial *Aspergillus* infections. *Amer J Medicine* 83, 709–718.

Sherertz, R. J., Regan, D. R., Hampton, K. D., Robertson, K. L., Streed, S. A., Hoen, H. M., Thomas, R., and Gwaltney, J. R. (1996). A cloud adult: The *Staphylococcus aureus*-virus interaction. *Ann Intern Med* 124, 539–547.

Wong, S. (1999). Surgical site infections; in *Hospital Epidemiology and Infection Control*, C. G. Mayhall, ed., Lippincott Williams & Wilkins, Philadelphia, 189–210.

6

Hospital Facilities

Introduction

Hospital facilities consist of the buildings, ventilation systems, and equipment necessary to maintain the hospital indoor environment, including air quality. Hospital buildings are normally subdivided into zones served by separate ventilation systems. The outer envelope of a hospital building provides the first line of defense against the intrusion of outdoor environmental microbes and, in conjunction with filtration and pressurization control, should be capable of maintaining a high degree of air quality if the systems meet applicable codes and guidelines.

Hospital facilities include a variety of wards, zones, and rooms for specific purposes including areas for surgery and critical care, nursing, diagnostics, laboratories, radiology, administration, sterilization, and services. Only those areas in which airborne pathogens have been responsible for infections and outbreaks are addressed in detail here, and these areas include general wards for patients, operating rooms, burn wards, isolation rooms for TB or AIDS patients, and laboratories. This chapter describes these physical facilities and their associated ventilation systems in terms of their function in controlling the transmission of airborne nosocomial infections. The actual performance and analysis of ventilation systems is addressed in Chapter 7 while filtration is addressed separately in Chapter 8.

Hospital Ventilation and Filtration Systems

There are five basic types of ventilation systems in use in hospitals today: natural ventilation, constant volume, variable air volume, 100% outside air, and displacement ventilation. Some additional types of specialized ventilation systems are used in operating rooms and these are discussed further in Chapter 12 on surgical site infections.

Natural ventilation is uncommon in developed countries. Natural ventilation uses no forced ventilation or ventilation equipment but relies on outdoor air flowing through windows and grilles to purge the internal environment of pathogens. Natural ventilation relies on leakage into and out of the building envelope to provide the air exchange necessary for human occupancy and to remove internally generated heat. In cold climates buildings ventilated naturally must be provided with separate heating systems. The actual air exchange rate achievable by natural ventilation is heavily dependent on how many windows are open, the outdoor air temperature, building height, and local wind conditions. The air exchange rate in naturally ventilated buildings can vary from 1 to 6 ACH and is often lower, on the average, than any building with forced ventilation. The main drawback with natural ventilation is that there is no control over outdoor environmental microbes entering the building. Pollen, fungal spores, and environmental bacteria will invariably enter the building and pose pathogenic threats to the immunocompromised. Indoor levels of environmental spores in naturally ventilated buildings tend to be about 30% of the outdoor levels (Fisk 1994).

Constant volume (CV) systems use forced air from a constant volume fan to ventilate the hospital building or to ventilate specific zones. CV systems may or may not include integral cooling and heating. CV systems provide a constant rate of both outside and return airflow to the building. The design and operation of CV systems is straightforward and, if well designed, can be both comfortable and cost-effective. CV systems normally employ dust filters and final filters in accordance with guidelines (AIA 2006; ASHRAE 2003). The use of cooling coils in a mechanically ventilated building will tend to greatly lower indoor spore levels due to the filtration effect of the cooling coils when moisture condenses on them (Seigel and Walker 2001). CV systems will typically draw 15–25% of outside air for use in maintaining air quality and in pressurizing the hospital building. Multizone systems are a type of constant volume system in which zone dampers are used to mix air to the required zone conditions.

100% outside air systems are simple and reliable, and are often found in hospitals, health care facilities, clinics, and laboratories. They are typically used in situations where the risk of internal contamination outweighs the costs associated with using large volumes of outside air in winter or summer. Because of the energy costs associated with this type of system, they may be used in conjunction with air-to-air heat exchangers or enthalpy wheels to provide more energy-efficient air quality control (Tamblyn 1995; Nardell et al. 1991). Filtration requirements are the same as those for other types of ventilation systems, and building pressurization may be controlled using exhaust dampers.

Variable air volume (VAV) systems adjust the amount of outside air in response to either outside air temperature or enthalpy to minimize energy consumption. In favorable climates the amount of outside air drawn in is maximized. These systems save energy by taking advantage of mild air conditions, and some fixed amount of minimum outside air (15–25%) will be maintained. Under minimum outside air conditions these systems operate much like constant volume systems, while under maximum outside air they will operate like 100% outside air systems. In some systems the maximum outside air is as low as 50% and these systems may not provide the same energy benefits as 100% outside air systems even under the best conditions. Both zone pressurization and whole building pressurization may vary with the amount of outside air, although balancing dampers should keep it under control.

Displacement ventilation systems are typically only used for certain zones like operating rooms. They are more commonly known as laminar flow systems, which is a misnomer because the air movement is not truly laminar. The pattern of supply and exhaust airflow is controlled such that piston-like flow is developed through a room to thereby remove contaminants more efficiently than normal supply air systems in which considerable air mixing occurs (Chen et al. 2003; Skistad 1994). In displacement ventilation systems air mixing is limited and the efficiency with which the room air is removed is high. Two basic types of displacement system have been in use in ORs, vertical laminar flow systems and horizontal laminar flow systems, both of which employ a large matrix of HEPA filters for the supply air. These systems are often designed to be ultraclean systems in which airborne particle counts are kept extremely low (Persson and van der Linden 2004). In a vertical system the air is supplied from the ceiling and exhausted from the floor. Solberg et al. (1971) describes a laminar airflow (LAF) in an OR in which the airborne concentrations of bacteria were about 270 times lower than in a conventional OR. Patients requiring bone marrow transplants are often housed in laminar airflow rooms, which may feature an entire wall of HEPA filters. Figure 6.1 illustrates examples of horizontal and vertical laminar flow rooms. In the vertical laminar flow room the exhaust air may either be removed through grilles on the floor or the lower part of the walls.

All mechanical ventilation systems drawing outside air typically use prefilters of approximately 30% dust spot efficiency (approximately MERV 6–8)

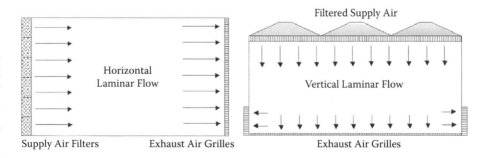

FIGURE 6.1
Schematic drawings of horizontal laminar flow room (left) and a vertical laminar flow room (right).

Hospital Airborne Infection Control

TABLE 6.1

Filter Ratings for Ventilation Systems in Hospitals

Area Designation	Filter Bed No. 1	Filter Bed No. 2
All areas for inpatient care, treatment, and diagnosis, and areas providing direct service such as sterile and clean processing	MERV 8	MERV 15
Protective environment room	MERV 8	HEPA
Laboratories	MERV 12	None
Administrative, bulk storage, soiled holding areas, food preparation areas, and laundries	MERV 8	None

immediately downstream of the mixing box or upstream of the heating and cooling coils (ASHRAE 2003). Requirements for filtration are summarized from the guidelines and shown in terms of minimum efficiency reporting value (MERV) in Table 6.1 (AIA 2006; ASHRAE 2003). MERV filters typically operate at a face velocity of 400–500 fpm and their performance is specified across a broad range of particle sizes (see Chapter 8). HEPA filters remove particles of 0.3 microns in size at a rate of at least 99.97% and operate at a face velocity of 250 fpm. HEPA filters are sometimes also referred to as MERV 17 filters.

Hospital Zones and Rooms

The hospital zones of primary interest in the control of airborne infections are the general wards, patient rooms, ICUs, NICUs, ORs, protective isolation rooms, airborne infection isolation rooms, laboratories, and the ancillary rooms and hallways associated with these areas. Often, hospitals are divided into zones that are supplied by separate ventilation systems and these zones may be positively or negatively pressurized relative to each other. This can be acceptable if contaminants move from cleaner areas to less clean areas, but if the reverse happens then doors between zones should be kept closed.

General wards and *patient rooms* are typically ventilated at a rate of 6 ACH total and 2 ACH of outside air, as are many of the support and administrative areas. Patient rooms should have a minimum of 100 ft² of floor area per each bed and no more than two beds per room. Each patient room should have a window and handwashing facilities in the toilet room.

Protective environment rooms maintain a positive pressure relative to other patient rooms and adjoining spaces with all supply air passing through HEPA filters. They are sometimes called positive pressure isolation rooms. These rooms protect immunocompromised and other kinds of patients from ambient microbes in the hospital air (Bagshawe, Blowers, and Lidwell 1978;

HICPAC 2007). They may include anterooms, which are separately ventilated and pressurized and designed to provide extra protection during opening of doors. The room is provided with more supply volume than exhaust and flow is typically regulated by modulating the main supply and exhaust dampers based on a signal from a pressure transducer located inside the isolation room (Gill 1994; van Enk 2006). The control point is set to about 0.1 in.w.g. (inches water gauge). The minimum airflow is 15 ACH. The most common application of positive pressure isolation rooms is for patients with various types of immunodeficiency (Linscomb 1994). Because ambient environmental fungi and human commensal bacteria may pose a serious threat to HIV patients, it is critical to prevent the ingress of any such microorganisms. Design criteria for HIV Rooms are similar to those for TB Rooms.

Airborne infection isolation (AII) rooms are for patients infected with a contagious airborne disease such as TB. The AII room contains patient-generated infectious agents within the room to protect other patients and HCWs. It is kept under negative pressure relative to adjoining spaces and has an internal air distribution pattern to assist removal of pathogens and reduce the exposure risk of uninfected HCWs. The room is provided with more exhaust volume than supply and flow is typically regulated with a sensor that monitors relative pressure and modulates supply and exhaust control dampers. Isolating TB patients is perhaps the most common application of negative pressure isolation rooms in the health industry (CDC 1984). The minimum airflow is 12 ACH. The pressure control is set to about 0.1 in.w.g. The exact air pressure differential that must be maintained is nominal only, as it merely indicates the airflow direction (Galson 1995). The relative pressure is not always practical for measuring purposes, and therefore other criteria such as maintaining an inward velocity of 100 fpm or exhausting 10% of the airflow or exhausting 50 cfm more than the supply is often specified (ASHRAE 1999). Figure 6.2 shows the pressure relationships for isolation rooms with anterooms, and indicates the equivalence of exhaust and supply airflows. In a typical isolation ward, a floor with individual isolation rooms

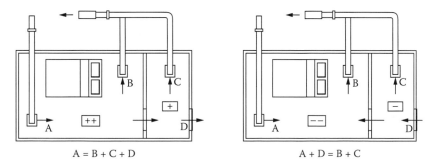

$$A = B + C + D \qquad\qquad A + D = B + C$$

FIGURE 6.2
Positive (left) and negative (right) pressurized isolation rooms with anterooms. Letters indicate airflows.

has a corridor separating it from the other adjacent areas and transfer of air occurs between the corridor and the other rooms. Facilities often differ significantly in their layouts and the pressurization scheme must be adapted individually for each facility (Ruys 1990). Negative pressure isolation rooms can lose integrity due to a number of factors, including faulty monitoring systems, turbulent airflow patterns, improper ventilation system balancing, airflow patterns that direct supply air out the door, shared anterooms, poorly sealed rooms, and disruption of air pressure by external exhaust systems (Pavelchak et al. 2000).

Approximately 15% of AIDS patients also suffer from TB, and this presents a unique design problem (ASHRAE 1999). One possible solution is to surround a positive pressure (HIV) room within a negative pressure (TB) room or vice versa (Gill 1994). An alternative is to use an entire house as a positive pressure zone with filtered outdoor air and internal recirculation of filtered and disinfected air.

Operating rooms (ORs) or surgical suites, and C-section rooms must be maintained with a high degree of room and air cleanliness to protect patients on the operating table. Microorganisms shed by HCWs and patients in the OR are the most common airborne pathogens occurring in a properly designed operating room with appropriate filtration (Hambraeus, Bengtsson, and Laurell 1977). A high volume of filtered air is provided from supply registers in the ceiling to create a downwash of air over the operating table (Streifel 1999). This displacement air should be delivered in such a manner that infectious particles shed by the operating team are swept away toward the exhaust ducts. Downward or vertical airflow is preferred over horizontal airflow for infection control and space management. Positive pressure airflow in the OR will prevent ambient contamination from entering the operating theater. Some operating rooms use a sterile core from which equipment and supplies are staged into the ORs (see Figure 6.3), but these areas may not be as sterile as the ORs themselves and may bring in unwanted contamination if the sterile core is at a higher relative pressure.

Opportunistic environmental spores such as *Aspergillus* or *Clostridium perfringens* should be minimized in ORs, if not eliminated completely. These soil microbes are readily filtered from outside air if the filters are installed and maintained properly (Luciano 1977; Streifel 1999). Providing spore-free air through filtration and ventilation, and local activity control, is the best method for preventing infections transmitted by fungal spores (Petersen et al. 1983).

The operating parameters of OR ventilation systems should be periodically checked because system airflow and pressure differentials may change over time. Positive pressure is maintained using 10–15% of air volume. Parameters that should be checked as part of commissioning or maintenance include total airflow, total outside airflow, pressurization and direction of airflow, and filter tightness. Filter performance can be checked with particle detectors, and air sampling can be used in ORs and isolation rooms to verify

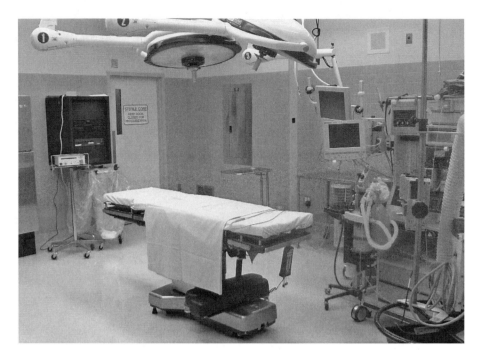

FIGURE 6.3
Typical operating room adjacent to a sterile core with supply air from ceiling diffusers and exhaust air grilles on walls near the floor.

that airborne contamination levels are acceptable or below any applicable guidelines or recommendations (see Chapter 18).

Intensive care units (ICUs) have requirements similar to those of patient rooms. Wound intensive care units, also called burn units, require the humidity to be controlled to 40–60% RH. Air is delivered unidirectionally throughout the room using nonaspirating ceiling diffusers with HEPA filters inside the diffusers operating at 250 fpm.

Neonatal intensive care units (NICUs) should have 150 square feet per patient with at least 6 feet between incubators because overcrowding can facilitate infectious outbreaks (Moore 1999). Ventilation in the NICU should provide positive pressure with air flowing from the ceiling to returns near the floor, and 12 air changes per hour (ASHRAE 2003). Central air handling units should use filters with at least 90% efficiency (i.e., MERV 13).

Laboratories in hospitals are provided for a variety of functions but in general must have a minimum of six total air changes per hour and must employ MERV 12 filters. Pressurization requirements are similar to those for isolation rooms (Ruys 1990). Laboratories are graded according to their Biosafety Level (BSL), which defines their capabilities in terms of handling dangerous pathogens (DHHS 1993; Richmond and McKinney 1999). Table 6.2 defines the

TABLE 6.2

Basic Requirements for BSL Containment Laboratories

BSL	Requirements	Recommendations	Application
1	No specific HVAC requirements	3–4 ACH, slight negative pressure	Agents of no known or minimal hazard
2	No specific HVAC requirements	100% OA, 6–15 ACH, slight negative pressure, use of safety cabinets	Agents of moderate potential hazard
3	Physical barrier, double doors, no recirculation, maintain negative pressure	Exhaust may require HEPA filtration	Agents that pose a serious hazard via inhalation
4	Physical barrier, double doors, no recirculation, maintain negative pressure, etc.	Requirements determined by biological safety officer	Agents that pose a high risk of lethality via inhalation

basic requirements for the four types of BSL rated laboratories. Laboratories typically have biosafety cabinets to control hazards and these cabinets are also subject to guidelines.

Pressurization Control

Controlling the pressurization of rooms, wards, or entire buildings can be used as one strategy to keep microbes out or to keep them isolated in certain areas, and the principle of pressurization can be applied on a larger scale to protect the entire building. The building envelope is essential to keeping the hospital free of environmental microbes, these being mainly fungal spores and bacteria from the outside air. All fungal spores (see Table 4.1) come ultimately from outdoors, and also bacteria such as *Acinetobacter, Clostridium perfringens* spores, *Legionella, M. avium, Nocardia asteroides, Pseudomonas aeruginosa,* and *Serratia marcescens.*

Positive pressurization with a separate air-handling unit can be used to isolate the first floor lobby and make it an anteroom for the whole hospital, as shown in Figure 6.4. The buffer zone serves as an anteroom for the entire building, the upper floors in this example, and is separately ventilated and pressurized. The remaining floors will be pressurized relative to the outdoors and may be neutral or pressurized relative to the buffer zone. The building envelope must be tight in order for the building to be pressurized against the outdoor environment because winds can upset the pressurization locally. The filters must also perform according to specifications.

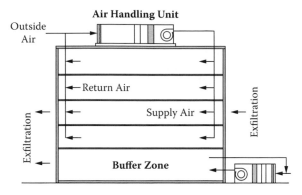

FIGURE 6.4
Separate pressurization of the lobby or first floor creates a buffer zone that can help keep airborne environmental pathogens out of the upper floors of the hospital.

References

AIA (2006). *Guidelines for Construction and Equipment of Hospital and Medical Facilities.* Mechanical Standards American Inst. of Architects, ed., Washington.

ASHRAE (1999). Chapter 7: Health care facilities; in *ASHRAE Handbook of Applications,* ASHRAE, ed., American Society of Heating, Refrigerating and Air Conditioning Engineers, Atlanta, GA, 7.1–7.13.

———. (2003). *HVAC Design Manual for Hospitals and Clinics.* American Society of Heating, Ventilating, and Air Conditioning Engineers, Atlanta, GA.

Bagshawe, K., Blowers, R., and Lidwell, O. (1978). Isolating patients in hospital to control infection. Part III—Design and construction of isolation accommodation. *Brit Med J* 2, 744–748.

CDC (1984). *Isolation Techniques for Use in Hospitals.* Centers for Disease Control, Atlanta, GA.

Chen, K., Kromin, A., Ulmer, M., Wessels, B., and Backman, V. (2003). Nanoparticle sizing with a resolution beyond the diffraction limit using UV light scattering spectroscopy. *Optics Commun* 228, 1–7.

DHHS (1993). *Biosafety in Microbiological and Biomedical Laboratories.* U.S. Department of Health and Human Services, Cincinnati, OH.

Fisk, W. (1994). The California healthy buildings study. *Center for Building Science News* Spring, 7, 13.

Galson, E., and Guisbond, J. (1995). Hospital sepsis control and TB transmission. *ASHRAE* May.

Gill, K. E. (1994). HVAC design for isolation rooms. *HPAC* July, 45–52.

Hambraeus, A., Bengtsson, S., and Laurell, G. (1977). Bacterial contamination in a modern operating suite. *J Hyg* 79, 121–132.

HICPAC (2007). *Guideline for Isolation Precautions: Preventing Transmission of Infectious Agents in Healthcare Settings.* Centers for Disease Control, Atlanta, GA.

Linscomb, M. (1994). AIDS clinic HVAC system limits spread of TB. *HPAC* February.

Luciano, J. R. (1977). *Air Contamination Control in Hospitals.* Plenum Press, New York.

Moore, D. L. (1999). Nosocomial infections in newborn nurseries and neonatal intensive care clinics; in *Hospital Epidemiology and Infection Control*, C. G. Mayhall, ed., Lippincott Williams & Wilkins, Philadelphia, 665–694.

Nardell, E. A., Keegan, J., Cheney, S. A., and Etkind, S. C. (1991). Airborne infection: Theoretical limits of protection achievable by building ventilation. *Am Rev Resp Dis* 144, 302–306.

Pavelchak, N., DePersis, R. P., London, M., Stricof, R., Oxtoby, M., G. DiFerdinando, and Marshall, E. (2000). Identification of factors that disrupt negative air pressurization of respiratory isolation rooms. *Inf Contr Hosp Epidemiol* 21(3), 191–195.

Persson, M., and van der Linden, J. (2004). Wound ventilation with ultraclean air for prevention of direct contamination during surgery. *Inf Contr Hosp Epidemiol* 25(4), 297–301.

Petersen, P. K., McGlave, P., Ramsay, N. K., Rhame, F. S., Cohen, E., III, G. S. P., Goldman, A. I., and Kersey, J. (1983). A prospective study of infectious diseases following bone marrow transplantation: Emergence of *Aspergillus* and *Cytomegalovirus* as the major causes of mortality. *Infect Control* 42(2), 81–89.

Richmond, J. Y., and McKinney, R. W. (1999). *Biosafety in Microbiological and Biomedical Laboratories, 4th Edition*. US Government Printing Office, Washington, DC.

Ruys, T. (1990). *Handbook of Facilities Planning Volume I: Laboratory Facilities*. Van Nostrand Reinhold, New York.

Seigel, J. A., and Walker, I. S. (2001). Deposition of biological aerosols on HVAC heat exchangers. Report No. LBNL-47669. Lawrence Berkeley National Laboratory, Berkeley, CA.

Skistad, H. (1994). *Displacement Ventilation*. John Wiley & Sons, New York.

Solberg et al. (1971). Laminar airflow protection in bone marrow transplantation. *Am Soc Microbiol* 21, 209–216.

Streifel, A. J. (1999). Design and maintenance of hospital ventilation systems and the prevention of airborne nosocomial infections; in *Hospital Epidemiology and Infection Control*, C. G. Mayhall, ed., Lippincott Williams & Wilkins, Philadelphia, 1211–1221.

Tamblyn, R. T. (1995). Toward zero complaints for office air conditioning. *Heating, Piping & Air Conditioning* March, 67–72.

van Enk, R. (2006). Modern hospital design for infection control. *Healthcare Design* September.

7

Ventilation and Dilution

Introduction

Ventilation and filtration are critical for controlling airborne microbial flora, and various guidelines provide detailed recommendations for both. The use of ventilation air to provide pressurization control for isolation hospital rooms and wards has been addressed in the previous chapter. Ventilation air also serves to dilute airborne microbial concentrations with outdoor air and to remove microbes generated internally in the hospital environment via outside air exhaust. High efficiency filtration protects against the intrusion and spread of airborne pathogens and allergens in the hospital environment. This chapter addresses the effectiveness of these technologies for controlling air quality based on both empirical evidence and analytical evaluation.

Ventilation Dilution

Hospital ventilation systems provide breathing air as well as heating, cooling, and humidity control for the comfort of patients and staff. They also provide for removal of airborne contaminants by purging with fresh outdoor air or by recirculating air through filters. The various types of hospital ventilation systems were described in the previous chapter.

The air exchange rate is often considered to be the primary determinant of how well biological contaminants are removed from indoor air (Morey, Feeley, and Otten 1990; Walter 1969). Another factor of comparable importance is the actual efficiency with which the interior air of any room is displaced and removed (Nardell et al. 1991). In some cases ventilation systems may also contribute to the spread of disease by disseminating pathogens and allergens, especially if they are poorly or incorrectly designed or have moisture and condensation problems. Examples of ventilation systems that have disseminated pathogens have been discussed in the literature (Ager and Tickner 1983; Ahearn et al. 1991; Brief and Bernath 1988; Neumeister et al. 1997; Price et al. 1994; Zeterberg 1973; He et al. 2003; Thornsberry et al.

1984; Parat et al. 1997). It is assumed that hospital ventilation systems are designed in accordance with appropriate guidelines and codes, and are not subject to internal mold growth and excessive condensation, and that they do not have outside air intakes too close to roof exhausts or cooling towers. Given these basic requirements, code conformance and absence of unusual microbial problems, any ventilation system can be modeled to determine its effectiveness at removing airborne pathogens from the building.

Any ventilation system that provides outside air purges the indoor air in a process called dilution ventilation (McDermott 1985). Dilution occurs at a rate that depends on the building or zone volume, the outside air flow rate, and the degree of mixing. In general, indoor conditions approach complete mixing and complete mixing models are appropriate to use for analysis. Dilution ventilation removes all airborne pathogens at approximately equal rates when complete mixing occurs, but some larger microbes and spores may settle out and remain on surfaces. Ventilation air may also add microbes from the outdoor air if there is no filtration or if filtration is inadequate. Modeling of dilution ventilation is fairly straightforward and may be accomplished using simple calculus methods or numerical integration on a spreadsheet. The latter method is discussed here in application to single zones and multiple zones. Computational fluid dynamics (CFD) software may also be used to study the details of airflow within zones, but the reader is referred to other texts for information on CFD software (FDI 1998; Kundu 1990).

The Steady State Model

Any area of a building can be modeled as a single zone in which indoor air is exchanged with fresh outdoor air. A steady state (SS) model represents the stabilized condition inside the zone after the system has been operated for some time. Under SS conditions, the number of pathogens brought into the zone equals the number purged from the zone by ventilation air. Assuming that the infiltration or exfiltration through the zone envelope is not significant, we can write Equation (7.1) in terms of the airflow concentrations in the outside air (OA) provided at the inlet, and the outlet or exhaust air:

$$Q_i C_a = Q_o C_{SS} \tag{7.1}$$

where
  Q_i = inlet airflow (OA), m³/min
  Q_o = outlet airflow (exhaust), m³/min
  C_a = ambient OA concentration of pathogens, cfu/m³
  C_{SS} = concentration of airborne pathogens indoors, cfu/m³

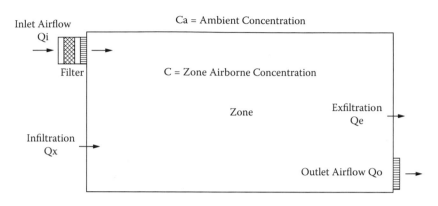

FIGURE 7.1
Single-zone model showing entering airflows and exhaust airflows.

The concentrations of pathogens in air are given in units of cfu/m^3, which are colony-forming units (cfu) of bacteria or fungi. The term plaque-forming units (pfu) is used for viruses, but this distinction is not made here for purposes of simplicity, and the units cfu/m^3 will implicitly refer to all pathogens except where stated otherwise. If it is assumed that infiltration and exfiltration occur simultaneously, which is a realistic assumption for modern buildings, the equation can be written as follows:

$$(Q_i + Q_x)C_a = (Q_o + Q_e)C_{SS} \qquad (7.2)$$

where
Q_e = exfiltration through envelope, m^3/min
Q_x = infiltration through envelope, m^3/min

Figure 7.1 depicts the balance of airflows in the single-zone model. Note that in the steady state model the zone volume will not appear in the calculations.

Some of the aerosolized pathogens will tend to settle out or adsorb to internal surfaces in a process known as plate-out. The removal rate of microbes due to plate-out is given in terms of a rate constant, K_d, and the unit area, A_d. In effect, this makes the zone act like a large low-efficiency filter. Equation (7.2) can be written to account for plate-out as follows:

$$(Q_i + Q_x)C_a = (Q_o + Q_e)C_{SS} + K_d A_d \qquad (7.3)$$

where
K_d = deposition rate constant, $cfu/min\text{-}m^2$
A_d = deposition area, m^2

Finally, internal generation of microbes must be included. Typically the source of most indoor pathogens like viruses and bacteria will be patients

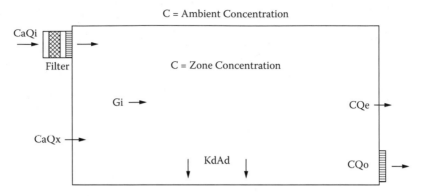

FIGURE 7.2
Flow diagram for microbes entering and exiting the single-zone model with internal generation (G_i) and plate-out (K_dA_d).

and workers. In some buildings that act as amplifiers, the building itself may be generating fungal spores or bacteria. The generation rate can be included in Equation (7.3) as follows:

$$(Q_i + Q_x)C_a + G_i = (Q_o + Q_e)C_{SS} + K_dA_d \tag{7.4}$$

where
$\quad G_i =$ microbes released internally to indoor air, cfu

Figure 7.2 depicts the total balance in terms of the entering and exiting microbes. The generation rate, G_i, can be multifaceted as there can be multiple source paths: occupant exhalation, patient shedding, resuspension from floors, bedding material, duct system surfaces, etc. The more accurately the sources in any single zone are quantified, the more accurate the model will be in predicting zone concentrations.

If the inlet filter has a penetration P_i, which is the complement of the efficiency, the SS concentration can be written as:

$$C_x = \frac{(Q_iP_i + Q_x)C_a + G_i - K_dA_d}{Q_i + Q_x} \tag{7.5}$$

Guidelines for the design of health care ventilation systems have evolved over the years and have proven adequate in providing the highest levels of air cleanliness in hospital wards, operating rooms, and associated facilities. Much of the focus of these design guidelines is on providing high rates of air exchange using 100% outside air supplied through high efficiency filters. Figure 7.3 illustrates the effectiveness of purging contaminants with various rates of filtered outside air with complete air mixing in a room. It

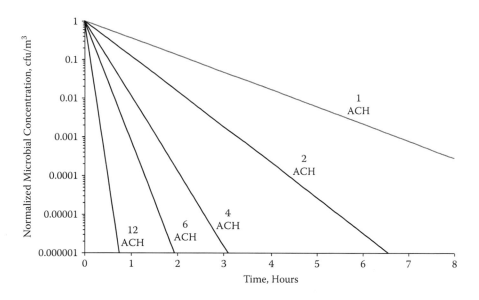

FIGURE 7.3
Effect of filtered outside air change per hour (ACH) on an initial level of airborne microbial contamination over an 8-hour period.

can be observed that virtual sterility can be achieved in a room assuming complete air mixing, sterile supply air, and no source of airborne microbes in the room. Of course, there will always be some leakage and generation of microbes in an occupied hospital building, and neither air mixing nor filtration is ever quite perfect, but the example illustrates the potential effectiveness of this approach. Treatment and operating rooms typically have an ACH of 12–15, whereas patient and intensive care rooms typically have an ACH of 2. It can be observed that, at an ACH of 12, contaminants will be purged rapidly and increases in the air exchange rate will tend to produce diminishing returns.

Transient Modeling

The concentrations of airborne contaminants inside hospital buildings vary continuously due to internal generation and outdoor seasonal variations. The previously described steady state model can only predict the final airborne concentrations after conditions have stabilized. The time it takes to stabilize depends primarily on the building volume, the airflow rates, and the degree of mixing. A transient model can predict how the indoor concentrations vary over time and will provide a more realistic view of indoor aerobiology.

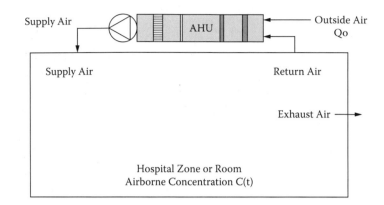

FIGURE 7.4
Single-zone model with recirculation, exfiltration, and mixed outside air.

The single-zone transient model presented here uses finite time steps (one minute) to estimate indoor airborne concentrations. This computational approach can easily be performed on a spreadsheet, as in the examples shown here, or via a programming language such as C++ or Basic. The single-zone transient model incorporates several assumptions, including complete mixing. The air in the single-zone model is assumed to mix completely on a minute-by-minute basis. Complete air mixing will slow the removal of airborne pathogens exponentially, as opposed to plug flow in which the removal rate is a linear function of time. Complete mixing represents the limiting case for normal buildings, and is a reasonable model to use for evaluating the removal rate of airborne pathogens.

Assuming complete mixing, the primary factor that determines the removal rate of airborne pathogens is the air change rate (air changes per hour or ACH). Figure 7.4 shows a schematic of a typical constant volume ventilation system with recirculation and outside air. The total air exhausted and exfiltrated will be equal to Q_o, the volume of outside air displaced.

A time step of one minute provides a close approximation of continuous flow without requiring excessive computations. Therefore, for each minute a finite volume of fresh supply air replaces an equal volume of mixed room air. The removal rate of microorganisms is given by the airborne concentration, $C(t)$ at any given time t, multiplied by the outside air flow rate, Q_o. For each minute the number of microorganisms removed is computed as

$$N_{out} = C(t) \cdot Q_o \tag{7.6}$$

Likewise, the number of microbes added will be the outside air flow rate, Q_o, multiplied by the concentration of microbes (bacteria and spores) in the outdoor air. For each minute of analysis, the number of microbes added is given by

$$N_{in} = C_a(t) \cdot Q_o \qquad (7.7)$$

The number of microbes generated internally may include bacteria, viruses, and fungi. The total population of microbes, $N(t)$, that will exist in the building, for any given minute t, will then be the previous minute's population plus the current minute's additions, minus the number exhausted to the outside air:

$$N(t) = N(t-1) + N_{in} - N_{out} \qquad (7.8)$$

Because complete mixing is assumed, the building microbial concentration will be defined as the building microbial concentration divided by the building volume, V, in cubic meters, at any given minute, or

$$C(t) = \frac{N(t)}{V} \qquad (7.9)$$

A good check on the results of any spreadsheet or program is to verify the steady state condition. The steady state concentration can be computed by dividing the rate of input of contaminants by the outside airflow. If there were any internal generation, at a rate of G_i cfu/min, the steady state would be computed as follows:

$$C_{ss} = \frac{G_i + Q_o C_a}{Q_o} \qquad (7.10)$$

where
C_{ss} = steady state concentration, cfu/m^3

Results for four representative cases spanning the range from 15% to 100% OA are shown graphically in Figure 7.5 for an initial airborne concentration of 10,000 cfu/m^3. It can be seen that zone concentrations are reduced rapidly at first and then more slowly in classic exponential decay. In these examples the final airborne concentration in all four cases reaches a steady state due to the fact that the outside air has a constant concentration and the indoor air cannot be brought down any lower without filters or air cleaning devices. These examples included no filtration or air disinfection.

Whether to use outdoor air for dilution and purging of indoor airborne contaminants hinges on economic and energy factors. If the climate is favorable it can be economical to use large quantities of outdoor air for purging. In cold climates the economics of heat exchangers may need to be studied to determine what percentage of outdoor air to use. Because filtration of out-

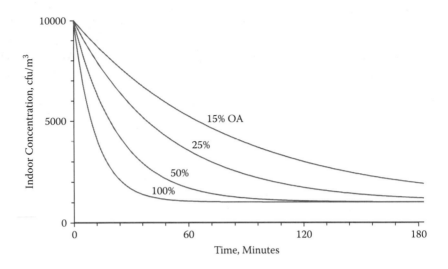

FIGURE 7.5
Transient concentrations for single-zone model.

door air will almost always be appropriate, the cost of filtration should be accounted for also.

Multizone Transient Modeling

When a hospital facility is composed of distinct and separate zones with separate airflows and even separate ventilation systems, a more accurate account of indoor concentrations can only be obtained by modeling the building with multiple zones. Multizone modeling generally requires the use of software packages. Some sophisticated software tools are available that perform multizone modeling and incorporate a variety of factors such as interior leakage, exterior leakage, wind pressure effects, stack effects, and plate-out effects. Some of the more common packages include CONTAMW, DOE-2, ESP-r, Risk V1.0, and others (Axley 1987; Sparks 1995; Dols, Walton, and Denton 2000; Demokritou 2001).

The CONTAMW program is public domain and has been used in a number of published studies on contaminant dispersion inside buildings (NIST 2006). The CONTAMW program has also been used to successfully model airborne microbial concentration in indoor air (Kowalski 2003). The CONTAMW program simulates the release and distribution of contaminants in a ventilated building in much the same way as the previously described methods. It has a variety of advantages over the previous methods

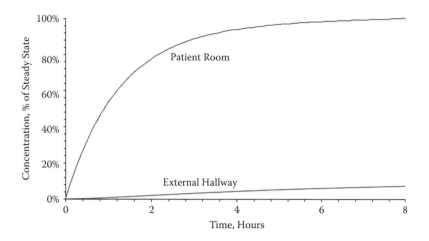

FIGURE 7.6
Airborne concentrations in a patient room and external hallway, shown in terms of the percentage of steady state conditions.

in that it can model multiple zones, vary infiltration and exfiltration rates, model wind pressure or buoyancy effects, and the contaminant source can be moved around and manipulated.

Figure 7.6 shows the results of a CONTAMW simulation of a general ward consisting of several patient rooms and a hallway. The rooms have standard dimensions and airflows with 2 ACH of filtered outside air. All rooms exhaust to the hallway. In this simulation an index patient is placed in a room at time zero and continuously releases infectious microorganisms into the air for 8 hours. The airborne concentration in the patient room climbs for several hours before approaching steady state conditions. The other patient rooms, which have no direct connection to the index patient room, remain at zero and do not show on the graph. The hallway, however, experiences a low level of airborne contamination that climbs continuously for 8 hours and does not achieve steady state. This graph is generic for all airborne pathogens and similar results could be expected for rooms with similar air exchange rates and dimensions. The CONTAMW results, which produce airborne levels in cfu/m³, can be used to predict the inhalation of airborne microbes and, using the epidemiological methods detailed in Chapter 2, can predict the infection rate in the patient room and other areas depending on breathing rate and time of occupation.

For information on modeling more sophisticated scenarios with transient releases in multistory buildings see Kowalski (2003) and Kowalski, Bahnfleth, and Musser (2003). In tall buildings, stack effects and wind effects may cause excessive infiltration, which can bring in environmental microbes, and exfiltration, which will cause energy losses. For information on modeling stack effects see Musser, Kowalski, and Bahnfleth (2002).

References

Ager, B. P., and Tickner, J. A. (1983). The control of microbiological hazards associated with air-conditioning and ventilation systems. *Ann Occup Hyg* 27(4), 341–358.

Ahearn, D. G., Simmons, R. B., Switzer, K. F., Ajello, L., and Pierson, D. L. (1991). Colonization by *Cladosporium* spp. of painted metal surfaces associated with heating and air conditioning systems. *J Ind Microb* 8, 277–280.

Axley, J. W. (1987). Indoor Air Quality Modeling. *NBSIR 87-3661*, NBS, Gaithersburg, MD.

Brief, R. S., and Bernath, T. (1988). Indoor pollution: Guidelines for prevention and control of microbiological respiratory hazards associated with air conditioning and ventilation systems. *Appl Indust Hyg* 3, 5–10.

Demokritou, P. (2001). Modeling IAQ and building dynamics; in *Indoor Air Quality Handbook*, J. D. Spengler, J. M. Samet, and J. F. McCarthy, eds., McGraw-Hill, New York.

Dols, W. S., Walton, G. N., and Denton, K. R. (2000). CONTAMW 1.0 Users Manual. NTIS Springfield, VA. http://fire.nist.gov/bfrlpubs/build00/art041.html

FDI (1998). *FIDAP 8*. Fluid Dynamics International, Inc. Lebanon, NH.

He, Y., Jiang, Y., Xing, Y. B., Zhong, G. L., Wang, L., Sun, Z. J., Jia, H., Chang, Q., Wang, Y., Ni, B., and Chen, S. P. (2003). Preliminary result on the nosocomial infection of severe acute respiratory syndrome in one hospital of Beijing. *Zhonghua Liu Xing Bing Xue Za Zhi* 24(7), 554–556.

Kowalski, W. J. (2003). *Immune Building Systems Technology*. McGraw-Hill, New York.

Kowalski, W. J., Bahnfleth, W. P., and Musser, A. (2003). Modeling immune building systems for bioterrorism defense. *J Arch Eng* June, 86–96.

Kundu, P. K. (1990). *Fluid Mechanics*. Academic Press, San Diego.

McDermott, H. J. (1985). *Handbook of Ventilation for Contaminant Control*. Butterworth, Boston.

Morey, P. R., Feeley, J. C., and Otten, J. A. (1990). *Biological Contaminants in Indoor Environments*. ASTM, Philadelphia.

Musser, A., Kowalski, W., and Bahnfleth, W. (2002). Stack and mechanical system effects on dispersion of biological agents in a tall building. *9th Symposium on Measurement and Modeling of Environmental Flows, International Mechanical Engineering Congress and Exposition*, New Orleans.

Nardell, E. A., Keegan, J., Cheney, S. A., and Etkind, S. C. (1991). Airborne infection: Theoretical limits of protection acheivable by building ventilation. *Am Rev Resp Dis* 144, 302–306.

Neumeister, H. G., Kemp, P. C., Kircheis, U., Schleibinger, H. W., and Ruden, H. (1997). Fungal growth on air filtration media in heating ventilation and air conditioning systems. *Healthy Buildings/IAQ'97*, Bethesda, MD, 569–574.

NIST (2006). Strategies to Reduce the Spread of Airborne Infections in Hospitals. *NIST GCR 06-887*, National Institute of Standards and Technology, Gaithersburg, MD.

Parat, S., Perdrix, A., Fricker-Hidalgo, H., Saude, I., Grillot, R., and Baconnier, P. (1997). Multivariate analysis comparing microbial air content of an air-conditioned building and a naturally ventilated building over one year. *Atmos Environ* 31(3), 441–449.

Price, D. L., Simmons, R. B., Ezeonu, I. M., Crow, S. A., and Ahearn, D. G. (1994). Colonization of fiberglass insulation used in heating, ventilation and air conditioning systems. *J Ind Microbiol* 13, 154–158.

Sparks, L. E. (1995). IAQ Model for Windows: Risk Version 1.0 User Manual. *EPA-600/R-96-037*, US Environmental Protection Agency.

Thornsberry, C., Balows, A., Feeley, J., and Jakubowski, W. (1984). Legionella: Proceedings of the 2nd International Symposium. Atlanta, GA.

Walter, C. W. (1969). Ventilation and air conditioning as bacteriologic engineering. *Anesthesiology* 31, 186–192.

Zeterberg, J. M. (1973). A review of respiratory virology and the spread of virulent and possibly antigenic viruses via air conditioning systems. *Annals of Allergy* 31, 228–299.

8

Air Filtration

Introduction

High efficiency filtration protects against the intrusion and spread of airborne pathogens and allergens in the hospital environment. Given that the ventilation system performs according to guidelines and design parameters, filtration is a critical component for controlling airborne pathogens. Filters remove ambient environmental microbes from the outdoor air and microbes produced indoors from human reservoirs and other sources. Filters are required for a variety of hospital facilities including general wards, isolation rooms, and operating rooms. This chapter addresses the effectiveness of filtration for controlling hospital air quality based on both empirical evidence and analytical evaluation.

Filters and Performance Curves

The effectiveness of a filter against airborne microbes depends primarily on the filter characteristics, the air velocity, and the size and type of microbe being intercepted and removed from the airstream. Filters are not sieves for micron sized particles because the gaps between fibers may average at least 20 microns and nosocomial microbes are smaller than this. Filtration efficiency is a probabilistic function—the massive number of microscopic fibers in any filter ensures that the probability of a particle being intercepted by a fiber increases exponentially with the thickness of the filter. The thickness of the filter pleats and the number of pleats, as well as the face velocity, therefore determines the performance of the filter. Filter characteristics vary from one manufacturer or model to another, but their performance can be generalized, as with the models presented here.

The varieties of filters available today meet many specialized needs. The basic types range from the lowest-efficiency dust filters, such as roll-type filters used in commercial buildings, to high efficiency particulate air (HEPA)

TABLE 8.1

Filter Types and Approximate Ratings

Filter Type	Applicable Size Range (µm)	Dust Spot Efficiency (%)	Total Arrestance (%)	MERV Rating (Estimated)
Dust Filters	>10	<20	<65	1
		<20	65–70	2
		<20	70–75	3
		<20	75–80	4
High Efficiency	3–10	<20	80–85	5
		<20	85–90	6
		25–35	**>90**	**7–8**
	1–3	40–45	>90	9
		50–55	>95	10
		60–65	**>95**	**11**
		70–75	>98	12
	0.3–1	**80–90**	**>98**	**13**
		90–95	NA	14–15
		>95	NA	**16**
HEPA	<0.3	NA	NA	**17–18**
ULPA	<0.3	NA	NA	**19–20**

and ultra-low penetration air (ULPA) filters used in clean rooms and operating rooms. Filters used in health care facilities include those with nominal dust spot efficiencies of 25%, 60%, 80%, 90%, and HEPA filters, otherwise collectively known as Group III filters (ASHRAE 1992). ASHRAE Standard 52.2-1999 provided new designations for these types of high efficiency filters in terms of their "minimum efficiency reporting value" (MERV), and these are the preferred ratings for current use (ASHRAE 1999, 2003). Dust spot efficiencies do not necessarily correlate exactly with MERV ratings, but Table 8.1 shows the approximate correlations, with typical hospital filters highlighted in bold.

A performance curve defines the removal rate versus particle size for any particular filter model. Performance curves from filter manufacturer catalogs typically cut off below 0.1–0.3 microns, as does the MERV rating system, and the curve must be extended to predict the filtration of viruses and small bacteria. Mathematical models of filters are used to extend performance curves into the virus size range based on Kowalski, Bahnfleth, and Whittam (1999) and Kowalski and Bahnfleth (2002, 2002a). Figure 8.1 shows some representative examples of MERV-rated filters with their performance curves extended below 0.1 microns. HEPA filters cannot be adequately displayed on this figure because their minimum efficiency, 99.97%, would form a straight horizontal line, and these are treated separately in a later section.

When filters are operated above their design velocity, filter efficiency decreases below that specified by their performance curves and penetration of small bacteria and viruses can greatly increase. The models presented

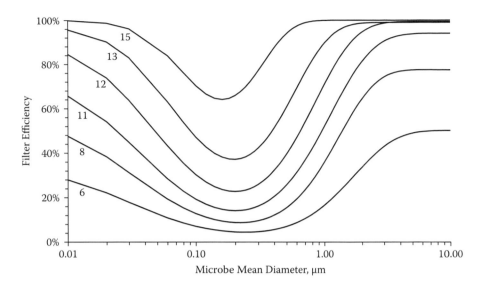

FIGURE 8.1
Filter performance curves for MERV-rated filters used in health care, extended into the virus size range.

assume that airflow is within the standard operating velocity of filters, about 500 fpm, or is within the range of 400–600 fpm. Figure 8.2 shows the impact of various face velocities on filter performance for a MERV 12 filter compared to normal operating air velocity (500 fpm). It can be observed that the main impact is in the virus size range below about 0.2 microns, and that operating a filter above its design velocity may permit more viruses to penetrate.

When air is recirculated through a filter as in room recirculation units the effective filtration rate is increased because of the multiple passes. The net filtration efficiency increases with the number of passes through the filter. With a large number of passes, an 80% or 90% filter can approach 100% efficiency.

Outside air is filtered before being supplied to hospital areas with the required levels of filtration based on ASHRAE (1999a, 1999b). The filters are located in sequence as indicated and often have a dust filter preceding them. Although these filters are typically about MERV 6–8, they should be capable of removing most environmental bacteria and spores, which tend to be relatively large and easy to filter.

Filtration of Airborne Nosocomial Pathogens

The rate at which airborne microbes are removed by filters depends on the filter performance curve and the mean diameters of the microbes. Each

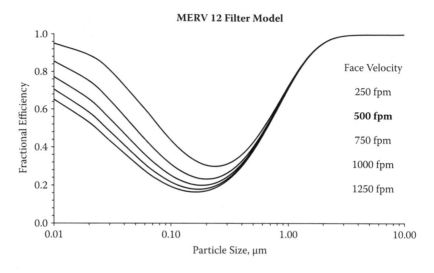

FIGURE 8.2
Impact of face velocity on filter performance for a MERV 12 filter. Dark line shows normal operating air velocity (500 fpm).

microbial species has a characteristic range of sizes that forms a lognormal distribution between the minimum and the maximum. Almost all nosocomial airborne microbes are spherical or ovoid and can be approximated as spheres. Some bacteria and spores are rod shaped and can be conservatively approximated by spheres representing their minimum dimensions. A few exceptions to this rule include rod-shaped bacteria that are smaller than the most penetrating particle size range of filters, and bacteria that have aspect ratios greater than about 3.5. Above this aspect ratio empirical correction factors can be used to adjust the maximum dimension (Kowalski, Bahnfleth, and Whittam 1999). Table 8.2 identifies the equivalent logmean diameter of all airborne nosocomial pathogens. Sources for logmean diameters are available from the references (Kowalski 2006; Kowalski, Bahnfleth, and Whittam 1999). These logmean diameters are used along with MERV filter models to estimate removal rates under standard operating velocity. However, the particular models apply only to the specific manufacturer's filters on which they were based and will not necessarily be accurate predictors for different manufacturer's filter models even at the same MERV rating. To be sure, the manufacturer's performance curve should be obtained and used with the specified logmean diameters.

Airborne microbes may be present in clumps, skin squames, dust particles, or droplet nuclei containing from a few to a few thousand microbes. The natural forces that hold these particles together are weak and upon impact with filter fibers they will break apart (Kowalski, Bahnfleth, and Whittam 1999). After a sufficient number of such impacts within filter media, particles will be reduced to bacterial cells, spores, or individual virions, at which point they

TABLE 8.2

Airborne Nosocomial Pathogen Removal Rates

Microbe	Size (µm)	MERV Filter Model Filtration Efficiency (%)								
		6	8	9	10	11	12	13	14	15
Parvovirus B19	0.022	21	32	35	40	52	72	89	97	98
Rhinovirus	0.023	21	31	34	39	51	70	88	97	98
Coxsackievirus	0.027	19	29	31	36	47	66	85	96	97
Norwalk virus	0.029	18	27	30	35	45	64	84	95	97
Rubella virus	0.061	11	16	18	21	28	43	62	82	84
Rotavirus	0.073	9	14	15	18	24	38	57	77	79
Reovirus	0.075	9	14	15	17	24	37	56	77	79
Adenovirus	0.079	9	13	14	17	23	36	54	75	77
Influenza A virus	0.098	7	11	12	14	19	31	48	69	71
Coronavirus (SARS)	0.11	6	10	11	13	18	28	45	66	68
Measles virus	0.158	5	8	9	10	15	24	38	59	63
Mumps virus	0.164	5	8	9	10	14	23	38	58	63
VZV	0.173	5	8	8	10	14	23	37	58	63
Mycoplasma pneumoniae	0.177	5	7	8	10	14	23	37	58	63
RSV	0.19	5	7	8	9	14	23	37	58	64
Parainfluenza virus	0.194	4	7	8	9	14	23	37	58	64
Bordetella pertussis	0.245	4	7	8	9	14	23	38	61	68
Haemophilus influenzae	0.285	4	8	9	10	16	25	41	64	73
Proteus mirabilis	0.494	7	13	15	16	25	39	60	84	92
Pseudomonas aeruginosa	0.494	7	13	15	16	25	39	60	84	92
Legionella pneumophila	0.52	7	14	16	17	27	41	62	86	93
Serratia marcescens	0.632	9	17	21	22	33	49	71	92	97
Mycobacterium tuberculosis	0.637	9	18	21	22	33	49	72	92	97
Klebsiella pneumoniae	0.671	10	19	22	24	35	52	74	93	98
Corynebacterium diphtheriae	0.698	10	20	24	25	37	54	76	94	98
Streptococcus pneumoniae	0.707	11	20	24	26	37	54	77	94	98
Alcaligenes	0.775	12	23	27	29	41	59	81	96	99
Neisseria meningitidis	0.775	12	23	27	29	41	59	81	96	99
Staphylococcus aureus	0.866	14	26	31	33	45	64	85	97	99
Staphylococcus epidermis	0.866	14	26	31	33	45	64	85	97	99
Staphylococcus pyogenes	0.894	14	27	32	34	47	66	86	97	99.5
Mycobacterium avium	1.118	19	35	41	44	57	76	93	99	99.8
Nocardia asteroides	1.118	19	35	41	44	57	76	93	99	99.8
Acinetobacter	1.225	21	39	45	48	61	80	94	99	99.9
Enterobacter cloacae	1.414	24	45	52	55	68	85	97	99	99.9
Enterococcus	1.414	24	45	52	55	68	85	97	99	99.9
Haemophilus parainfluenzae	1.732	30	53	61	65	76	92	98	99	99.9
Clostridium difficile	2	34	59	66	71	81	95	99	99	99.9

Continued

TABLE 8.2 (*Continued*)

Airborne Nosocomial Pathogen Removal Rates

Microbe	Size (µm)	MERV Filter Model Filtration Efficiency (%)								
		6	8	9	10	11	12	13	14	15
Pneumocystis carinii	2	34	59	66	71	81	95	99	99	99.9
Fugomyces cyanescens	2.12	35	61	69	73	83	96	99	99	99.9
Histoplasma capsulatum	2.236	36	63	70	76	85	96	99	99	99.9
Pseudallescheria boydii	3.162	44	71	78	86	91	99	99	99	99.9
Scedosporium	3.162	44	71	78	86	91	99	99	99	99.9
Penicillium	3.262	44	72	79	87	91	99	99	99	99.9
Aspergillus	3.354	45	72	79	87	92	99	99	99	99.9
Coccidioides immitis	3.464	45	73	80	88	92	99	99	99	99.9
Cryptococcus neoformans	4.899	49	75	82	91	94	99	99	99	99.9
Clostridium perfringens	5	49	75	82	91	94	99	99	99	99.9
Rhizopus	6.928	50	75	82	92	94	99	99	99	99.9
Mucor	7.071	50	75	82	92	94	99	99	99	99.9
Trichosporon	8.775	50	75	82	92	94	99	99	99	99.9
Altemaria alternata	11.225	50	75	82	92	94	99	99	99	99.9
Fusarium	11.225	50	75	82	92	94	99	99	99	99.9
Blastomyces dermatitidis	12.649	50	75	82	92	94	99	99	99	99.9

Note: Size = logmean diameter. Filter performance may vary with manufacturer's model.

are subject to filtration at the efficiencies specified in Table 8.2. If clumps of bacteria or spores do not break apart on impact, they will be subject to higher filtration efficiencies because they are above the most penetrating particle size range of about 0.2 microns. Clumps of viruses below this size range will have even weaker forces holding them together and they will break up on impact with fibers. There are limited data on virus filtration but the data presented later (see HEPA and ULPA Filters) indicates that viruses below 0.2 microns are filtered out within 1% of the efficiencies predicted by these filter models.

Typical outdoor air filters (MERV 6–8) are capable of removing the larger fungal spores at rates of about 50–75%. Figure 8.3 shows how microbes line up in terms of size on a MERV 13 filter performance curve. It can be seen that the fungal and bacterial spores are removed at approximately 100% efficiency by a MERV 13 filter. Smaller bacteria and viruses occupy the most penetrating particle size range at about 0.2 microns. Note also that the smallest viruses are removed less efficiently than larger viruses—this is due to the fact that particles in this size range are subject to diffusional capture (rather than impaction), and the smaller the particle the more it will diffuse.

Some tests have found that the filtration of microorganisms is closely approximated by particles of identical size (Ginestet et al. 1996). Other tests have shown that microbes are removed at rates slightly lower than those of particles of equivalent diameters (Sinclair 1976). In some of the earliest tests on the filtration

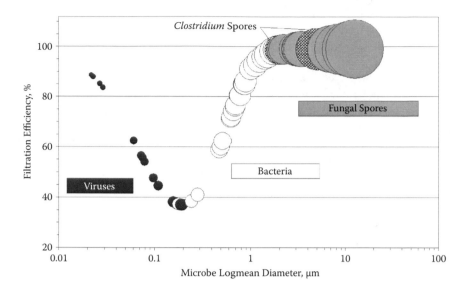

FIGURE 8.3
Airborne nosocomial pathogens shown in relative size on a MERV 13 filter performance curve. Bacterial spores are highlighted.

of microorganisms by glass fiber filters, Decker et al. (1954) demonstrated a 99% removal rate for *Serratia indica, E. coli,* and the virus T-3 bacteriophage.

Studies have generally shown that bacteria captured by filters die rapidly from dehydration, even under high relative humidity. A study by Ruden and Botzenhart (1974) found that HEPA filters do not act as a growth medium for microorganisms even at 70% and 90% RH. Microbes tested included *Staphylococcus, Streptococcus, Pseudomonas, Corynebacterium, Bacillus, Clostridium, Penicillium,* and *Aspergillus,* which showed no growth, and *Candida,* which showed slight growth. Maus, Goppelsroder, and Umhauer (1997) found that *Mycobacterium luteus* and *E. coli* collected in air filter media and exposed to low air humidity in the range of 30–60% lost their viability in less than one hour. Maus, Goppelsroder, and Umhauer (2001) found that spores of *Bacillus subtilis* and *Aspergillus niger* experienced no growth regardless of relative humidity in most filter media but that some filter media produced mold growth when exposed to 98% RH. In general, the lack of available nutrients on clean filters precludes most mold growth even under high relative humidity, but this may not be true for filters that have been in operation and accumulated dust and organic debris. Studies have reported or shown that microbes can survive or grow on filters under the right conditions, including Samson (1994); Pasanen et al. (1992); Fuoad et al. (1997); Chang, Foarde, and Van Osdell (1996); and Neumeister et al. (1997). Kemp et al. (1995) reports that microbes can grow on untreated air filters and even grow through the filters and release spores downstream. In general, water damage, condensation, or very high humidity is required for microbes to grow on filters.

Re-entrainment of microbes from filters should not occur in a properly installed filter. Studies have shown that the shedding of microorganisms from filters is generally low and independent of the loading of the filters, air velocity, or humidity (Ginestet et al. 1996). Antimicrobial filters are available that inhibit the growth of microorganisms, but they are not always effective (Foarde, Hanley, and Veeck 2000). Filters contaminated with microorganisms may pose a risk to maintenance workers, and many facilities now have procedures for handling used filters and filter installations designed to facilitate filter removal safely.

Moritz et al. (2001) studied the ability of medium efficiency air filters to retain airborne outdoor microorganisms in two HVAC systems. The filters reduced bacteria levels by approximately 70% and mold spores by over 80%. However, when humidity exceeded 80%, a proliferation of bacteria on air filters resulted in a subsequent release into the filtered air.

In an early hospital field trial in which filtered air was used in a dressing station for burns, *Staphylococcus* isolates were reduced by approximately 58% (Lowbury 1954). Isolates of other bacteria experienced similar reductions from the use of air filtration.

HEPA and ULPA Filters

High efficiency particulate air (HEPA) filters are used to filter supply air in isolation rooms and exhaust air from isolation rooms, laboratories, and other facilities. HEPA filters are often used when air is recirculated in isolation rooms, TB rooms, and other areas (ASHRAE 2003a). HEPA filters must be installed with airtight seals and operated at design velocity, 250 fpm, to perform according to specifications. Damage to HEPA filters, such as bending, breaking, or punctures, will have a major impact on filtration efficiency. Figure 8.4 shows a HEPA filter performance curve modeled with the same methods used for the previous MERV filters. This model is compared against test data for viruses from several studies (Harstad and Filler 1969; Jensen 1967; Roelants, Boon, and Lhoest 1968; Thorne and Burrows 1960; Washam 1966). Note the excellent agreement between the HEPA filter model and the test results in the lower size range. This corroborates the expectation that viruses below the most penetrating particle size range will be filtered out at the efficiency predicted based on particle size, and that clumping has no effect on predicted microbial filter efficiencies. No test data exist for the larger viruses or for velocities outside the design range, but this model is likely to be a reasonable predictor of filtration efficiency in these size ranges also.

HEPA filters do not necessarily remove all microbes, and it is possible that some microbes in the most penetrating particle size range of a HEPA could penetrate the filter in small numbers. This is unlikely to be a problem

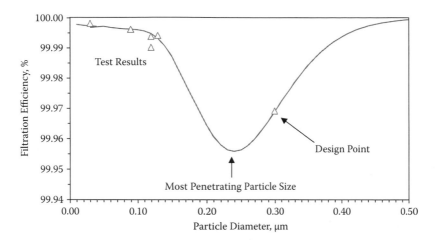

FIGURE 8.4
HEPA filter model shown with test data for viruses and virus-sized particles.

unless very high concentrations appear at the filter inlet. Figure 8.5 shows the results of modeling to identify the most penetrating airborne nosocomial agents for a HEPA filter. This example illustrates a characteristic of all filters—that certain microbes penetrate more effectively than others. The greatest penetrations will occur primarily in the most penetrating particle size range. In a study on the potential for microbes to penetrate HEPA filters, 888 bacteria and fungi samples were collected from installed filters over a 26-month period and no microbial penetration was detected in any samples (Rechzeh and Dontenwill 1974).

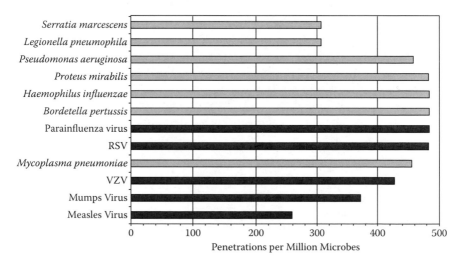

FIGURE 8.5
Most penetrating airborne nosocomial microorganisms for a HEPA filter under design velocity.

HEPA filters were originally designed for use in filtering radioactive particles, hence the specification of 99.97% efficiency at 0.3 microns. This level of performance may be adequate for radioactive isotopes, but for the purpose of filtering airborne microbes it would appear to be overkill (Luciano 1977). It can be seen from Table 8.2 that MERV-rated filters can have very high removal rates for airborne nosocomial pathogens. The added benefit of using HEPA filters in health care is minimal while the energy cost and first costs are high. It will also be seen in later chapters that filtration augmented with UVGI can have near equivalent performance. Furthermore, the basis for requiring removal rates as high as those of HEPA filters is debatable—a simple reduction of 70–90% of airborne pathogens may be sufficient to reduce risks to an acceptable minimum. Because HEPA filters require one-half the normal face velocity of MERV filters, they require twice the face area, and hence twice the number of filters. Attempting to retrofit HEPA filters may require extensive ductwork changes or may even be infeasible. HEPA filters are sometimes mistakenly operated at 500 fpm. A HEPA filter operated at a face velocity of 500 fpm will allow over 10 times as many microbial penetrations as one operated at the design velocity of 250 fpm, and will more closely resemble a MERV 15–16 filter than a HEPA in terms of performance. Some health care facilities mistakenly believe they are using HEPA filters when they are designated "HEPA-like" filters, an unofficial term, which usually means a 95% filter (MERV 15–16).

Ultra low penetration air (ULPA) filters are designed for use in clean rooms where extreme air cleanliness is necessary. ULPA filter performance is typically stated as 99.9995% efficiency at 0.12 microns. ULPA filters are sometimes used in ultraclean operating room ventilation systems, but the added benefit of using these filters in hospitals is debatable because even HEPA filters are over-designed for microbial removal applications (Luciano 1977). The penetration rate of HEPA filters is negligible and the same would be true of ULPA filters. The energy and maintenance costs of both these types of filters is also quite high and possibly unjustifiable when it is considered that a MERV 15 filter is likely to reduce the infection risk just as much as a HEPA filter. It will also be shown in Chapter 19 that coupling a MERV filter with UVGI can produce performance comparable to a HEPA filter at a fraction of the cost.

Face Masks and Respirators

Surgical face masks and respirators are used by HCWs in accordance with requisite precautions. Masks are required wherever body fluids may be generated that could contact mucous membranes. Masks have limited ability to filter airborne microbes. A study by Fairchild and Stampfer (1987) found that particles exhaled by healthy subjects ranged in size from 0.09 to 3 microns, while exhaled breath contained about 0.1–0.35 particles/m³ during normal breathing and 0.15–2 particles/m³ during coughing and talking. Endogenous microbes from HCWs tend to collect on the inner surface of masks while exogenous or environmental microbes tend to collect on the exterior. Medical face masks have greater leakage through the filter and around the face seal than respirators, but respirators have not shown any added benefit over face masks (Grinshpun et al. 2009). Surgical masks have no certification tests and have the highest rate of leakage, about 25–83%. A tightly fitted face mask should be capable of filtering most spores, but obtaining a tight fit is difficult. A study by Oberg and Brosseau (2008) on surgical mask filter performance showed that although manufacturers reported filtration efficiencies in the 95–99% range, none of the nine masks tested provided sufficient performance to protect HCWs from airborne particles generated by infectious patients. The performance of face masks is governed by ASTM standard F2100-07 (ASTM 2007). The FDA recommends two types of filter efficiency testing: (1) particulate filtration efficiency (PFE) using a non-neutralized aerosol of 0.1 micron latex spheres at a challenge velocity between 0.5 and 25 cm/s, and (2) bacterial filtration efficiency (BFE) using a non-neutralized 3 micron *Staphylococcus aureus* aerosol and a flow rate of 28.3 L/min. The FDA requires no minimum level of performance. See CDC (2005) for guidelines on the use of surgical masks in health care settings.

The Occupational Safety and Health Administration (OSHA) regulates the selection and use of respirators in workplaces. OSHA requires respirators to be provided when necessary to protect the health of the employee. There are several categories of National Institute for Occupational Safety and Health (NIOSH)-approved respirators including, in order of effectiveness, dust-mist, dust-mist-fume, N95, HEPA, and powered air purifying respirators (Decker and Schaffner 1999). NIOSH specifies three types of respirators, types P, R, and N, which are certified to perform at efficiencies of 99.97%, 99%, and 95%, respectively, at 0.3 microns.

References

ASHRAE (1992). *Chapter 25: Air Cleaners for Particulate Contaminants*, R. American Society of Heating, and Air Conditioning Engineers, ed., ASHRAE, Atlanta, GA.

_____. (1999). ASHRAE Standard 52.2-1999. Atlanta, GA.

_____. (1999a). Standard 62-1999: Ventilation for acceptable indoor air quality. ASHRAE, Atlanta, GA.

_____. (1999b). *Handbook of Systems and Equipment*. ASHRAE, Atlanta, GA.

_____. (2003). *Handbook of Applications*. American Society of Heating, Refrigerating, and Air-Conditioning Engineers, Atlanta, GA.

_____. (2003a). *HVAC Design Manual for Hospitals and Clinics*. American Society of Heating, Ventilating, and Air Conditioning Engineers, Atlanta, GA.

ASTM (2007). Standard Specification for Performance of Materials Used in Medical Face Masks. *ASTM F2100-07*, American Society for Testing and Materials, West Conshohocken, PA.

CDC (2005). *Guidelines for Preventing the Transmission of* Mycobacterium tuberculosis *in Health-Care Facilities*. Centers for Disease Control, Atlanta, GA.

Chang, J. C. S., Foarde, K. K., and Van Osdell, D. W. (1996). Assessment of fungal (*Penicillium chrysogenum*) growth on three HVAC duct materials. *Environment International* 22(4), 425.

Decker, H. M., Harstad, J. B., Piper, F. J., and Wilson, M. E. (1954). Filtration of microorganisms from air by glass fiber media. *HPAC* May, 155–158.

Decker, M. D., and Schaffner, W. (1999). Tuberculosis control in the hospital: Compliance with OSHA requirements; in *Hospital Epidemiology and Infection Control*, C. G. Mayhall, ed., Lippincott Williams & Wilkins, Philadelphia, 1091–1100.

Fairchild, C. I., and Stampfer, J. F. (1987). Particle concentration in exhaled breath. *Am Ind Hyg Assoc J* 48, 948–949.

FDA (2004). Guidance for industry and FDA staff: Surgical masks—premarket notification [510(k)] submissions. U.S. Food and Drug Administration (FDA), Center for Devices and Radiological Health, Washington, DC.

Foarde, K. K., Hanley, J. T., and Veeck, A. C. (2000). Efficacy of antimicrobial filter treatments. *ASHRAE J* Dec, 52–58.

Fuoad, H., Baird, G., Donn, M., and Isaacs, N. (1997). Indoor airborne bacteria and fungi in New Zealand office buildings. *Healthy Buildings/IAQ '97*, Bethesda, MD, 233.

Ginestet, A., Mann, S., Parat, S., Laplanche, S., Salazar, J. H., Pugnet, D., Ehrler, S., and Perdrix, A. (1996). Bioaerosol filtration efficiency of clean HVAC filters and shedding of micro-organisms from filters loaded with outdoor air. *J Aerosol Sci* 27(Suppl 1), S619–S620.

Grinshpun, S. A., Haruta, H., Eninger, R. M., Reponen, T., McKay, R. T., and Lee, S.-A. (2009). Performance of an N95 filtering facepiece particulate respirator and a surgical mask during human breathing: Two pathways for particle generation. *J Occup Environ Hyg* 6(10), 593–603.

Harstad, J. B., and Filler, M. E. (1969). Evaluation of air filters with submicron viral aerosols and bacterial aerosols. *AIHA J* 30, 280–290.

Jensen, M. (1967). Bacteriophage aerosol challenge of installed air contamination control systems. *Appl Microbiol* 15(6), 1447–1449.

Kemp, S. J., Kuehn, T. H., Pui, D. Y. H., Vesley, D., and Streifel, A. J. (1995). Filter collection efficiency and growth of microorganisms on filters loaded with outdoor air. *ASHRAE Transactions* 101(1), 228.

Kowalski, W. J. (2006). *Aerobiological Engineering Handbook: A Guide to Airborne Disease Control Technologies*. McGraw-Hill, New York.

Kowalski, W. J., and Bahnfleth, W. P. (2002). MERV filter models for aerobiological applications. *Air Media* Summer, 13–17.

_____. (2002a). Airborne-microbe filtration in indoor environments. *HPAC Engineering* 74(1), 57–69.

Kowalski, W. J., Bahnfleth, W. P., Whittam, T. S. (1999). Filtration of airborne microorganisms: Modeling and prediction. *ASHRAE Transactions* 105(2), 4–17.

Lowbury, E. J. L. (1954). Air-conditioning with filtered air for dressing burns. *Lancet* 1, 293–295.

Luciano, J. R. (1977). *Air Contamination Control in Hospitals.* Plenum Press, New York.

Maus, R., Goppelsroder, A., and Umhauer, H. (1997). Viability of bacteria in unused air filter media. *Atmos Environ* 31(15), 2305–2310.

_____. (2001). Survival of bacterial and mold spores in air filter media. *Atmos Environ* 35, 105–113.

Moritz, M., Peters, H., Nipko, B., and Ruden, H. (2001). Capability of air filters to retain airborne bacteria and molds in heating, ventilating and air-conditioning (HVAC) systems. *Int J Hyg Environ Health* 203(5-6), 401–409.

Neumeister, H. G., Kemp, P. C., Kircheis, U., Schleibinger, H. W., and Ruden, H. (1997). Fungal growth on air filtration media in heating ventilation and air conditioning systems. *Healthy Buildings/IAQ '97*, Bethesda, MD, 569–574.

Oberg, T., and Brosseau, L. M. (2008). Surgical mask filter and fit performance. *AJIC* 36(4), 276–282.

Pasanen, A.-L., Juutinen, T., Jantunen, M. J., and Kalliokoski, P. (1992). Occurrence and moisture requirements of microbial growth in building materials. *Intl Biodeterioration & Biodegradation* 30, 273–283.

Rechzeh, G., and Dontenwill, W. (1974). Contribution to the question of the contamination of suspended-substances filters by germs. *Zbl Bakt Hyg* 159, 272–283.

Roelants, P., Boon, B., and Lhoest, W. (1968). Evaluation of a commercial air filter for removal of viruses from the air. *Appl Microbiol* 16(10), 1465–1467.

Ruden, H., and Botzenhart, K. (1974). Experimental studies on the capacity of glass-fibre HEPA-filters to retain microorganisms. *Zbl Bakt Hyg* 159, 284–290.

Samson, R. A., ed. (1994). *Health Implications of Fungi in Indoor Environments.* Elsevier, Amsterdam.

Sinclair, D. (1976). Penetration of HEPA filters by submicron aerosols. *Journal of Aerosol Science* 7, 175–179.

Thorne, H. V., and Burrows, T. M. (1960). Aerosol sampling methods for the virus of foot-and-mouth disease and the measurement of virus penetration through aerosol filters. *Journal of Hygiene* 58, 409–417.

Washam, C. J. (1966). Evaluation of filters for removal of bacteriophages from air. *Appl Microbiol* 14(4), 497–505.

9

Hospital Disinfection

Introduction

The disinfection of hospital equipment and surfaces is essential for interrupting aerobiological pathways. The primary sources of airborne nosocomial pathogens are humans and the environment while the intermediate pathways of airborne infections can include surfaces, equipment, and hands. Bacteria shed from humans will tend to settle downward over time and accumulate on horizontal surfaces such as floors, tables, bed sheets, etc. Individual viruses (virions) or clumps of viruses may remain suspended in air indefinitely, but viruses that are attached to dust particles or droplet nuclei will tend to settle downward over time. Environmental microbes like *Aspergillus* spores may enter a hospital through doorways, leaky filters, or windows, or may be brought inside on shoes and clothing, and they will also tend to settle toward the floor. *Aspergillus* spores are often or occasionally detected in operating rooms, which have filters that are impermeable to fungal spores, and so it must be inferred that spores are traveling through hospital hallways or being carried by personnel from the outside to the ORs. Floors may well be a pathway to the OR for spores and other environmental bacteria, and microbes resting on horizontal surfaces such as floors and carpets can become re-aerosolized through activity and traffic (Ayliffe et al. 1967). Once re-aerosolized, these microbes may present an inhalation hazard or they may settle on wounds, catheters, and other equipment and result in infections. Interrupting the aerobiological pathway by disinfecting floors and other hospital surfaces can break the chain of causation of airborne infections and so the disinfection of hospital rooms, equipment, and surfaces can play a direct role in interdicting airborne pathogens.

Most human pathogens are mesophiles (growing at 35–37°C), having adapted to human body temperature and often having the ability to survive in environments designed for human comfort (or about 25°C). It would be impractical to raise or lower indoor temperatures to inhibit microbial survival indoors, but other factors may be manipulated. Humidity can be kept low to prevent microbial growth. Sources of moisture can be eliminated.

Abundant light can be provided everywhere, so as to deny shade to fungi. Entryways can be pressurized so that air leakage goes outward, and they can be designed such that conditions are unfavorable or intolerable to environmental microbes (i.e., no rugs, all glass exteriors, black floors that heat up in sunlight, antimicrobial surfaces, etc.).

Overall, bacteria account for more than 90% of hospital-acquired infections (Filetoth 2003). Although fungal spores from the outside environment may play a very minor role in hospital-acquired infections, certain bacterial spores pose major hazards, including *Clostridium difficile*. Nonsporulating bacteria from the environment such as *Pseudomonas* and *Acinetobacter* also play a role in hospital-acquired infections.

This chapter reviews the methods and technologies for disinfecting hospital surfaces that are in common use throughout the health care industry. Newer technologies, such as ozonation and whole room UV disinfection, are dealt with in more detail in later chapters. Because not every disinfectant will have the same effect on every microbe, exceptions to the disinfection methods are described herein and alternative methods targeting specific microbes are addressed. As with other chapters in this book, the focus here is on those airborne nosocomial pathogens (listed in Table 4.1) that have the potential to transmit or transport by the airborne route (Airborne Class 1 and 2), while other microbes that are strictly blood borne, waterborne, or food borne (Airborne Class 3) are specifically not addressed. Also not addressed in this chapter is human source disinfection or disinfection of patient skin, for which other protocols exist (see Chapter 10).

Cleaning, Disinfection, and Sterilization

The primary objective of inanimate hospital surface disinfection is to decontaminate the source of an infection or outbreak. Three levels of decontamination are possible. These are, in order of increasing decontamination level, cleaning, disinfection, and sterilization. The nature of the infectious agent or severity of the outbreak will dictate the level of decontamination required, and will also typically dictate the method of decontamination.

Cleaning removes microbes and organic matter from a surface by physical means such as scrubbing or by chemical means such as detergents. Cleaning represents the lowest level of disinfection, but for many common applications and microbes it is an adequate means of control. Cleaning does not guarantee complete removal of contaminant microbes and is used only for surfaces and items with a low risk of infection and not as a precursor to more efficient disinfection methods meant for intermediate or high-risk infections. Two types of cleaning are common in hospitals—dry cleaning and wet cleaning.

Dry cleaning uses a dry cloth or vacuum cleaner to remove only the superficial dust, but because clouds of dust can be raised by such methods they may create a hazard if pathogens become aerosolized. Dry cleaning methods are therefore not recommended for use in health care settings.

Wet cleaning with solvents like water or detergent can dissolve particulate matter and dust that may contain microorganisms without raising a cloud of dust. Wet cleaning methods are used for surfaces with a low risk of infection, such as floors, walls, and tabletops. It can greatly reduce concentrations of spores and bacteria on surfaces but cannot eliminate them. Wet cleaning can also serve as a preliminary to disinfecting items such as medical equipment. The solvent used, whether water or detergent, should be free of any microbial contaminants. Common tap water, however, is often found to contain various bacteria such as *Enterobacter, Pseudomonas, Acinetobacter, Serratia, Proteus, Klebsiella, Legionella,* and others, and it may be prudent to use disinfected potable water if available.

Disinfection is defined as the killing of microbes even if the destruction is incomplete. That is, disinfection is the reduction of a microbial population but not necessarily complete eradication. Ideally, disinfection reduces microbial contaminants to a level that is not hazardous to human health or that presents a greatly reduced risk of infection. A variety of methods are available for disinfecting hospital surfaces and items, but the most common method is to use chemical disinfectants. Equipment can be efficiently disinfected with heat but large room surfaces need to be disinfected with chemicals or else with gaseous decontamination or ultraviolet germicidal irradiation.

Chemical disinfectants will denature essential structural proteins and may damage the nucleic acids of the microbe, thereby hindering reproduction and growth. The degree of disinfection achievable by any chemical disinfectant depends on its toxicity, the amount of disinfectant used, and the time of exposure. Microbes can vary in their resistance to chemical disinfectants, with vegetative bacteria being the most susceptible and spores of fungi and bacteria being the most resistant.

If a disinfectant has a killing effect on microorganisms then it may carry the suffix "cide" as in fungicide. If a disinfectant merely inhibits or stops growth without actually killing microbes then it may have the suffix "static" as in bacteriostatic.

Three levels of disinfection have been proposed: high level, intermediate level, and low level disinfection, as follows (Filetoth 2003):

1. High Level Disinfection: killing all microbes including spores and *Mycobacteria*
2. Intermediate Level Disinfection: killing all microbes but with limited action against spores and some viruses (lipid and small viruses)
3. Low Level Disinfection: killing only vegetative bacteria, lipid and medium-sized viruses, and having limited action against fungi

Microorganisms rarely exist in isolation in the environment—they are often surrounded by organic matter or incorporated in droplet nuclei. Organic matter contains proteins, lipids, and other substances that offer protection against disinfectants and reduce the efficacy of the disinfection process. For this reason detergents are often added to disinfectants to dissolve the organic matter and thereby eliminate its protective effect. In the case of walls and floors, cleaning and disinfection are typically done in separate phases with detergents used to dissolve the organic matter and disinfectants used to kill or inactivate microorganisms.

Sterilization is the process of achieving complete disinfection or removal of all microorganisms, including bacterial and fungal spores. In other words, sterilization is complete disinfection with no survivors (Rutala 1999). In practice, the absence of survivors can only be established by sampling, and therefore the definition of sterility is limited by our ability to detect microorganisms and the accuracy of sampling equipment and methods. The inactivation of microbes by any means obeys the mathematics of exponential decay, which means that there is always a statistical probability of survivors. As a matter of practicality and convenience for air disinfection (only), the mathematical definition of sterility can be taken as a six log reduction in airborne concentrations. The rationale for this is that it would be exceedingly unusual to have airborne concentrations higher than about 10,000 cfu/m^3 in any indoor environment and therefore a six log reduction would leave no more than one in a million survivors (or 0.01 cfu/m^3). A six log reduction is therefore an adequate definition of sterilization for air disinfection applications and provides a mathematical cornerstone for the design of air disinfection systems. This definition does not necessarily apply to surfaces (or to water). There is not yet any accepted definition of surface sterility and a six log reduction, although adequate for air, may be meaningless for hospital surfaces (other than equipment surfaces) where the concentration of microbial contamination might be unlimited.

For equipment disinfection, the Sterility Assurance Level (SAL) is the probability of a nonsterile item remaining after a sterilization process. An SAL of 10^{-6} (one in a million survivors) is generally accepted as an indication of sterility for terminally sterilized items in Europe and the United States (Filetoth 2003). Sterilization is used for decontamination of high-risk items used for procedures on the human body, and for preventing procedure-associated infections if there is a high risk of developing diseases.

The decimal reduction value, or D value, is used as a mathematical expression of the resistance or susceptibility of a microbe to physical and chemical disinfection. It is either assumed to represent one log reduction (base 10 log) of a population (90% reduction) or it is coupled to the level of disinfection designated by a subscript (i.e., D_{90} = 90% reduction, D_{99} = 99% reduction, etc.). The D value is commonly used for steam sterilization applications where it is expressed in terms of the time of exposure. For chemical disinfection the D value is similarly expressed in units of time, but for ultraviolet disinfection the D value is specified in terms of the UV dose, which is the exposure time

TABLE 9.1

Decimal Equivalent of Log Reductions

Log Reduction	Reduction frac	Survival frac	Reduction %	Survival %
1	0.9	0.1	90	10
2	0.99	0.01	99	1
3	0.999	0.001	99.9	0.1
4	0.9999	1E-04	99.99	0.01
5	0.99999	1E-05	99.999	0.001
6	0.999999	1E-06	99.9999	0.0001

multiplied by the UV irradiance level (i.e., in W/m^2). Regardless of disinfection method, the percentage reduction for any given log reduction will be as shown in Table 9.1.

A room or environment may remain contaminated with microbes such as MRSA, *Clostridium difficile*, Norovirus, *Acinetobacter*, and VRE even after the infectious patient is removed (Huang, Datta, and Platt 2006; Weber et al. 2010; Drees et al. 2008). Improved cleaning and disinfection and hand hygiene can reduce the spread of these pathogens. Carling et al. (2008) studied the cleaning procedures and methods used in 36 acute care hospitals and suggest that improving cleaning methods can significantly reduce environmental contamination and may reduce infection rates. The authors recommend a structured approach using surface targeting methods and objective evaluation of the thoroughness of room disinfection cleaning as well as education and administrative intervention.

Decontamination Methods

Heat is the most common method of sterilization in health care facilities due to its ease of use and relatively low cost. Boiling is perhaps the oldest method of sterilization and typically employs 100°C (212°F) water for 30 minutes. Boiling has been all but forgotten since the advent of more convenient systems employing saturated or superheated steam. Burning or incineration is an ancient method of disinfection that is rarely used in hospitals today except for the disposal of waste.

Steam sterilization using saturated steam has become the most economical and efficient means of sterilizing instruments. It destroys microorganisms by denaturing their heat-sensitive cellular or soluble proteins. The condensation of saturated steam brings the water into direct contact with the surfaces, and microbes, to be sterilized. Dry heating with superheated steam is less effective in most cases and requires higher temperatures and longer exposure times to achieve the same degree of disinfection.

Gamma radiation uses high-energy gamma rays from radioactive sources like cobalt-60 to sterilize items by destroying the DNA or RNA of any contaminant microbes. Gamma rays penetrate all materials leaving no parts unaffected. Most disposable medical instruments are sterilized with gamma rays. Due to the expense and safety hazards, gamma radiation is not commonly used in hospitals.

Ultraviolet germicidal irradiation (UVGI) uses electromagnetic radiation in the germicidal range (200–400 nm) to inactivate microorganisms by damaging their DNA or RNA. It does not penetrate materials very well and so can only be used to disinfect air and surfaces, and to a certain depth, water. Its effectiveness depends on the level of irradiance produced and the length of the exposure time. It is commonly used in laboratories to decontaminate equipment but is also used increasingly today to decontaminate cooling coils in air handling units, to disinfect air in ducts, and to disinfect air and surfaces in rooms. UV disinfection of inanimate surfaces can generally only be performed in unoccupied rooms due to UV exposure hazards.

Filtration is a method of removing microbes from air or from water, which does not necessarily destroy them, although microbes trapped on filters tend to die from desiccation. For air cleaning applications, high efficiency filters are used to remove particulates and airborne microbes, thereby disinfecting the air. The size or filter rating (i.e., MERV or HEPA), airflow, and building and airflow characteristics determine the level of disinfection achieved (see Chapter 8).

A variety of gaseous disinfection methods are in current use and the selection of which to use depends on the intended target (microbe) and the nature of the room or facility to be disinfected. Gaseous disinfectants are typically hazardous to humans and can only be used in unoccupied rooms, or in sterilization chambers (for equipment). The ability of a gas to penetrate a surface depends on the type of gas and the surface material, and the presence of organic materials (soiling) may interfere with the disinfection process.

Disinfectants and Antiseptics

Chemical disinfectants can be used for scrubbing surfaces and equipment, and some can be vaporized and used as gaseous disinfectants. The more common disinfectants are reviewed here. Antiseptics are a class of disinfectants that can be used on skin.

Alcohol is widely used as a disinfectant and an antiseptic. Alcohol may be used for disinfecting surfaces and equipment depending on local disinfection protocols. It does not penetrate well into organic matter and should therefore be used only on surfaces that have already been cleaned physically.

Chlorine-based disinfectants are in wide use and include a variety of commercially available bleaches and cleaning compounds. Hypochlorites are the most widely used chlorine disinfectants and they are available in liquid or solid form. Hypochlorites are fast acting and have broad-spectrum antimicrobial activity but are readily inactivated by inorganic matter. They are effective against viruses and are used for environmental decontamination after blood spilling. They are also used for baths, sinks, kitchen cleaning, and for disinfecting instruments. They are not compatible with cationic detergents.

Phenolics or phenols are used for environmental disinfection and are the agent of choice for mycobacteria. They are not easily inactivated by organic matter, are absorbed by rubber and plastics, and are incompatible with cationic detergents. Phenolic disinfectants can be absorbed through the skin and appropriate protective clothing must be worn. Phenolics must not be used on any surfaces that come in contact with skin, incubators, or on any food preparation surfaces.

Aldehydes include the disinfectants glutaraldehyde and formaldehyde. *Glutaraldehyde* kills microbes by alkylating organic molecules, but its toxicity limits its use for room disinfection. It can only be used in ventilated rooms and may cause respiratory and eye irritation. *Formaldehyde* vapor is a disinfectant commonly used for disinfecting laboratory safety cabinets and for disinfecting rooms of patients with dangerous transmissible infections, but due to its toxicity it is not normally the agent of choice for sterilizing room surfaces.

Chlorine dioxide is a gaseous disinfectant that has been used to disinfect entire buildings contaminated with anthrax (Kowalski and Bahnfleth 2003). The disinfection ability of chlorine dioxide can be diminished by high relative humidity and can reportedly damage furnishings.

Ozone has also been used for hospital whole room disinfection, and it oxidizes proteins as well as other materials (Berrington and Pedler 1998; deBoer et al. 2006). The ability of ozone to destroy microorganisms has been well documented, but it also poses health hazards to humans (Kowalski, Bahnfleth, and Whittam 1998).

Hydrogen peroxide and related compounds have low toxicity. Hydrogen peroxide is sometimes used for cleaning spills and for disinfecting certain types of equipment. Vapor-phase hydrogen peroxide is highly sporicidal, but it is also corrosive and this limits its use for disinfecting room surfaces.

Ethylene oxide (EO) is highly sporicidal and is used for gaseous disinfection and sterilization of equipment. Due to its toxicity it is not used for disinfecting whole rooms and is used for equipment only for items that cannot be sterilized by other means.

Plasma gas is an alternative to EO and formaldehyde vapor. It employs free radicals to destroy nucleic acids. The end products are nontoxic, and therefore it is a viable method for decontaminating whole rooms or equipment (Burts et al. 2009).

Peracetic acid is an oxidizing agent that destroys proteins. Its use is limited to items that can be fully immersed and so it is not in wide use. It has the

TABLE 9.2

Antimicrobial Activity of Disinfectants

Compound Disinfectant	Antimicrobial Activity			
	Bacteria	Mycobacteria	Spores	Viruses
Alcohol 60–70%	Good	Good	None	Moderate
Chlorine 0.5–1%	Good	Good	Good	Good
Phenolics 1–2%	Good	Moderate	None	Poor
Glutaraldehyde 2%	Good	Good	Good	Good
Peracetic Acid 0.2–0.35%	Good	Good	Good	Good
Hydrogen peroxide 3–6%	Good	Varies	Varies	Varies

advantage of decomposing into harmless decomposition products. It is more effective than glutaraldehyde at penetrating biofilms and remains effective in the presence of organic matter.

SNL foam, from Sandia National Laboratories, is a recently developed alternative to ozone or chlorine dioxide for treating rooms and buildings contaminated with biological or chemical agents (Modec 2001). SNL foam consists of a combination of quaternary ammonium salts, cationic hydrotopes, and hydrogen peroxide. It is highly oxidative and toxic byproducts are limited. When the foam is pumped into a contaminated room it remains stable and in contact with the room's surfaces for several hours before evaporating.

Table 9.2 summarizes the antimicrobial activity of several of the most commonly used disinfectants and antiseptics. For some compounds, viruses may be further differentiated by whether they are enveloped or not—for enveloped viruses the activity of phenolics is moderate and for hydrogen peroxide compounds the activity against enveloped viruses is good.

Disinfection of the Inanimate Environment

The inanimate environment of the hospital can serve as the source or as a vector for the transmission of communicable pathogens or opportunistic infections. This includes hospital surfaces as well as the air, the water, and food supplies. Hospital surfaces may be contaminated by patients, personnel, and visitors, and ambient environmental microbes may find their way into the hospital via air currents, floors, shoes, clothing, or transport of supplies, to become opportunistic infectious agents. Nosocomial microbes that lie on horizontal surfaces or that adhere to other surfaces must be removed to minimize the infection risk to patients. These surfaces may include building surfaces such as floors and walls, medical equipment, clothing, bedcovers, furnishings, and food implements.

Surfaces may be contaminated with microbe-laden dust that may contain fungal spores, bacterial spores, mycobacteria, or other infectious agents. Any moisture or organic matter on these surfaces will protect pathogens. The surfaces in general wards and intensive care units (ICUs) may present a low risk of infection; the surfaces in operating rooms may contain dust or microbes that present a serious risk of a surgical site infection. Surfaces in operating rooms are classified as intermediate risk or high risk even though they do not contact the patient directly (Filetoth 2003). Operating room surfaces may often become contaminated with blood and other bodily fluids.

Surfaces that have a low risk of infection require regular cleaning and disinfection every day. Wet cleaning is sufficient because the objective is to remove dust and organic matter. Cleaning such surfaces may be done up to three times a day in areas where there are patients. Cleaning can itself cause aerosolization of microbes, and in a study by Braymen (1969) it was found that significant numbers of bacteria, spores, and viruses could be sampled from the air after using different disinfectants and cleaning methods. Airborne concentrations were always significantly lower after cleaning than before, except when water was used. High-pressure sprayers produced considerably more aerosolized microbes than hand scrubbing.

Cleaning should also be performed as necessary in the event of a spill or other sudden contamination. In such cases a high level of disinfection is normally required in order to kill a broad spectrum of microorganisms. Surfaces in high-risk areas should be disinfected with a high-level disinfectant because the risk of the infection is high if these surfaces are damp or dusty. In operating rooms the surfaces become soiled and contaminated in virtually every operation and disinfection should be performed immediately after each operation. Disinfectants spread on OR surfaces such as floors have a residual effect and will decontaminate ducts and organics, and also prevent the raising of clouds of dust. Bedcovers, clothes, and other textiles can be a dangerous source of infectious contamination. These are typically laundered with combinations of thermal disinfection and chemical disinfection.

Two types of equipment are used in hospitals—disposable equipment and reusable equipment. Disposable equipment is typically for single use and is often made of plastic. Disposability eliminates the possibility of cross-contamination between patients. Cleaning and disinfection of disposable equipment is not recommended because the equipment may be damaged, and furthermore it may not be possible to sterilize disposable equipment because it was not designed for such use. Once disposable equipment is used it must be assumed to be contaminated and must be disposed of as infectious waste. Infectious waste is collected separately in special bags or containers and then decontaminated by incineration or other methods.

Three categories for instrument disinfection have been adopted by agencies like the CDC: (1) Critical, (2) Semicritical, and (3) Noncritical (CDC 2008).

Critical items are those associated with a high risk of infection if the item is contaminated with any microorganism, including surgical instruments, cardiac and urinary catheters, implants, and ultrasound probes used in sterile body cavities. Semicritical items are those that come in contact with mucous membranes or nonintact skin, including respiratory therapy and anesthesia equipment, some endoscopes, laryngoscope blades, esophageal menometry probes, anorectal manometry catheters, and diaphragm-fitting rings. Noncritical items are those that come in close contact with intact skin but not mucous membranes.

Reusable equipment is intended to be decontaminated and reused, and is designed to be capable of being decontaminated to a high level using standard methods. Disinfection and cleaning of reusable equipment is typically performed in a disinfecting room that is isolated and equipped with disinfection equipment.

The process of disinfection may itself generate aerosols or may aerosolize dust, and it therefore requires special cautions and protocols. Personnel performing decontamination are typically required to wear a coat, goggles, gloves, and a face mask. Gloves and a coat alone may be sufficient if the disinfection equipment is completely enclosed.

The disinfection of water supplies in hospitals is required if the water is used for drinking or if clean water is required. If the hospital acquires its water from a remote location or central plant, the disinfection of the water is often performed at that location. Otherwise, hospitals may have to disinfect their own water. Water for swimming pools is commonly treated with chlorine for disinfection purposes. Water for humidifiers must be treated to ensure no *Legionella* contamination occurs. The topic of disinfecting water, however, has limited relation to air and surface disinfection because there are extremely few waterborne microbes that present airborne nosocomial threats via air and surfaces and so the subject of water disinfection is not addressed further in this text.

Disinfection protocols and procedures are available from a wide variety of sources and these should be consulted for more specific information on the proper use and application of disinfectants (Ayliffe, Collins, and Taylor 1990; Canada 1998; Castle and Ajemian 1987; CNO 2009; Cundy and Ball 1977; Damani 1977; ISID 2001; Kennamer 2007; Palmer, Giddens, and Palmer 1996; WHO 2004).

Table 9.3 provides a summary of the recommended disinfectants and moist heat inactivation temperatures and times for all the airborne nosocomial pathogens from Table 4.1. A variety of emerging pathogens cause concerns in health care facilities, but virtually all of these microbes have been studied and they have all been found to be susceptible to existing disinfectants (Rutala and Weber 2004).

TABLE 9.3

Disinfection of Airborne Nosocomial Pathogens

		Moist Heat	
Microbe	Disinfectants	°C	min
Acinetobacter	1% NaOCl, phenolics, HCHO, glutaraldehyde	121	30
Adenovirus	1% NaOCl	121	30
Alcaligenes	1% NaOCl, phenolics, HCHO, glutaraldehyde	121	30
Alternaria alternata	1% NaOCl, phenolics, HCHO, glutaraldehyde	121	30
Aspergillus	1% NaOCl, 2% glutaraldehyde	121	30
Blastomyces dermatitidis	1% NaOCl, phenolics, HCHO, 10% formalin	121	30
Bordetella pertussis	1% NaOCl, 70% ethanol, phenolics, glutaraldehyde, HCHO	121	15
Clostridium difficile	1% NaOCl, 70% ethanol	121	15
Clostridium perfringens	1% NaOCl, prolonged contact with glutaraldehyde	121	30
Coccidioides immitis	1% NaOCl, phenolics, glutaraldehyde, HCHO	121	30
Coronavirus (SARS)	1% NaOCl, 2% glutaraldehyde	121	30
Corynebacterium diphtheriae	1% NaOCl, phenolics, glutaraldehyde, HCHO	121	15
Coxsackievirus	70% ethanol, 5% lysol, 1% NaOCl	60	30
Cryptococcus neoformans	1% NaOCl, iodine, phenolics, glutaraldehyde	121	15
Enterobacter cloacae	1% NaOCl, phenolics, glutaraldehyde, HCHO	121	15
Enterococcus	1% NaOCl, 2% glutaraldehyde, HCHO, iodines	121	30
Fugomyces cyanescens	1% NaOCl, iodine, glutaraldehyde, HCHO	121	15
Fusarium	1% NaOCl, phenolics, HCHO, glutaraldehyde	121	30
Haemophilus influenzae	1% NaOCl, 70% ethanol, glutaraldehyde, HCHO	121	15
Haemophilus parainfluenzae	1% NaOCl, 70% ethanol, glutaraldehyde, HCHO	121	15
Histoplasma capsulatum	1% NaOCl, phenolics, HCHO, glutaraldehyde	121	15
Influenza A virus	1% NaOCl, 70% ethanol, glutaraldehyde, HCHO	56	30
Klebsiella pneumoniae	1% NaOCl, 70% ethanol, 2% glutaraldehyde, iodines	121	15
Legionella pneumophila	1% NaOCl, 70% ethanol, glutaraldehyde, HCHO	121	15
Measles virus	1% NaOCl, 70% ethanol, glutaraldehyde, HCHO	121	30
Mucor	1% NaOCl, phenolics, HCHO, glutaraldehyde	121	30
Mumps virus	1% NaOCl, 70% ethanol, glutaraldehyde, HCHO	121	30
Mycobacterium avium	5% phenol, 1% NaOCl, iodine, glutaraldehyde, HCHO	121	15
Mycobacterium tuberculosis	5% phenol, 1% NaOCl, iodine, glutaraldehyde, HCHO	121	15
Mycoplasma pneumoniae	1% NaOCl, 70% ethanol, glutaraldehyde, HCHO	121	15
Neisseria meningitidis	1% NaOCl, 70% ethanol, glutaraldehyde, HCHO	121	15
Nocarida asteroides	1% NaOCl, 2% glutaraldehyde, HCHO	121	15
Norwalk virus	Bleach, chlorine (i.e., 0.1% hypochlorite)	121	30

Continued

TABLE 9.3 (*Continued*)

Disinfection of Airborne Nosocomial Pathogens

Microbe	Disinfectants	Moist Heat	
		°C	min
Parainfluenza virus	1% NaOCl, 70% ethanol, glutaraldehyde, HCHO	121	30
Parvovirus B19	1% NaOCl, aldehydes	121	30
Penicillium	1% NaOCl, phenolics, HCHO, glutaraldehyde	121	30
Pneumocystis carinii	1% NaOCl, phenolics, HCHO, glutaraldehyde	121	30
Proteus mirabilis	1% NaOCl, 70% ethanol, glutaraldehyde, HCHO	121	30
Pseudallescheria boydii	1% sodium hypochlorite, 2% glutaraldehyde	121	30
Pseudomonas aeruginosa	1% NaOCl, 70% ethanol, 2% glutaraldehyde	121	30
Reovirus	1% NaOCl, 70% ethanol, glutaraldehyde, HCHO	121	30
Respiratory syncytial virus	1% NaOCl, 70% ethanol, 2% glutaraldehyde	55	30
Rhinovirus	1% NaOCl, iodine, phenol-alcohol, 2% glutaraldehyde	121	30
Rhizopus	1% NaOCl, phenolics, HCHO, glutaraldehyde	121	30
Rotavirus	1% NaOCl, 70% ethanol, 2% glutaraldehyde	121	30
Rubella virus	1% NaOCl, 70% ethanol, glutaraldehyde, HCHO	56	30
Scedosporium	1% NaOCl, phenolics, HCHO, glutaraldehyde	121	30
Serratia marcescens	1% NaOCl, 70% ethanol, glutaraldehyde, HCHO	121	15
Staphylococcus aureus	1% NaOCl, iodine/alcohol solutions, glutaraldehyde	121	15
Staphylococcus epidermis	1% NaOCl, iodine/alcohol solutions, glutaraldehyde	121	15
Streptococcus pneumoniae	1% NaOCl, 2% glutaraldehyde, HCHO, iodines, 70% ethanol	121	15
Streptococcus pyogenes	1% NaOCl, glutaraldehyde, HCHO, 70% ethanol	121	15
Trichosporon	Ethanol, NaOCl, chorhexidine gluconate	121	30
Varicella-zoster virus (VZV)	1% NaOCl, 2% glutaraldehyde, HCHO	121	30

Note: NaOCl = sodium hypochlorite; HCHO = formaldehyde.

References

Ayliffe, G. A., Collins, B. J., Lowbury, E. J., Babb, J. R., and Lilly, H. A. (1967). Ward floors and other surfaces as reservoirs of hospital infections. *J Hyg* 65(4), 515–536.

Ayliffe, G. A. J., Collins, B. J., and Taylor, L. J. (1990). *Hospital-Acquired Infection: Principles and Prevention.* Wright, London.

Berrington, A. W., and Pedler, A. J. (1998). Investigation of gaseous ozone for MRSA decontamination of hospital side-rooms. *J Hosp Inf* 40, 61–65.

Braymen, D. T. (1969). Survival of micro-organisms in aerosols produced in cleaning and disinfection. *Pub Health Rep* 84(6), 547–552.

Burts, M., Alexeff, I., Meek, E., and McCullers, J. (2009). Use of atmospheric non-thermal plasma as a disinfectant for objects contaminated with methicillin-resistant *Staphylococcus aureus. Am J Infect Contr* 37(9), 729–733.

Canada (1998). Hand washing, cleaning, disinfection and sterilization in health care. *Canada Communicable Disease Report Volume 24SB*, Health Canada, Laboratory Centre for Disease Control Ottawa, ON.

Carling, P. C., Parry, M. M., Rupp, M. E., Po, J. L., Dick, B., and Von Beheren, S. (2008). Improving cleaning of the environment surrounding patients in 36 acute care hospitals. *Inf Contr Hosp Epidemiol* 29(11):1–7.

Castle, M., and Ajemian, E. (1987). *Hospital Infection Control*. John Wiley & Sons, New York.

CDC (2008). *Guideline for Disinfection and Sterilization in Healthcare Facilities*. Centers for Disease Control, Atlanta, GA.

CNO (2009). *Infection Prevention and Control*. College of Nurses of Ontario, Toronto, ON.

Cundy, K. R., and Ball, W. (1977). *Infection Control in Health Care Facilities*. University Park Press, Baltimore.

Damani, N. N. (1997). *Manual of Infection Control Procedures*. Greenwich Medical Media, London.

deBoer, H. E. L., van Elzelingen-Dekker, C. M., van Rheenen-Verberg, C. M. F., and Spanjaard, L. (2006). Use of gaseous ozone for eradication of methicillin-resistant *Staphylococcus aureus* from the home environment of a colonized hospital employee. *Inf Contr and Hosp Epidemiol* 27, 1120–1122.

Drees, M., Snydman, D. R., Schmid, C. H., Barefoot, L., Hansjosten, K., Vue, P. M., Cronin, M., Nasraway, S. A., and Golan, Y. (2008). Prior environmental contamination increases the risk of acquisition of vancomycin-resistant enterococci. *Clin Infect Dis* 46, 678–685.

Filetoth, Z. (2003). *Hospital-Acquired Infection*. Whurr, London.

Huang, S. S., Datta, R., and Platt, R. (2006). Risk of acquiring antibiotic-resistant bacteria from prior room occupants. *Arch Int Med* 166, 1945–1951.

ISID (2001). *A Guide to Infection Control in the Hospital*. International Society for Infectious Diseases, Brookline, MA.

Kennamer, M. (2007). *Basic Infection Control for Health Care Providers*. Thomson Delmar Learning, Clifton Park, NY.

Kowalski, W. J., Bahnfleth, W. P., Whittam, T. S. (1998). Bactericidal effects of high airborne ozone concentrations on *Escherichia coli* and *Staphylococcus aureus*. *Ozone Sci Eng* 20(3):205–221.

Kowalski, W., and Bahnfleth, W. P. (1998). Airborne respiratory diseases and technologies for control of microbes. *HPAC* 70(6), 34–48.

_____. (2003). Immune-building technology and bioterrorism defense. *HPAC Engineering* 75 (Jan.)(1), 57–62.

Modec, I. (2001). Technical Report MDF2001-1002: Formulations for the Decontamination and Mitigation of CB Warfare Agents, Toxic Hazardous Materials, Viruses, Bacteria and Bacterial Spores. Denver, CO. http://www.reevesmfg.com/Literature/Foam_Technical_Report.pdf

Palmer, S., Giddens, J., and Palmer, D. (1996). *Infection Control*. Skidmore-Roth Publishing, Inc., El Paso.

Rutala, W. A. (1999). Selection and use of disinfectants in healthcare; in *Hospital Epidemiology and Infection Control*, C. G. Mayhall, ed., Lippincott Williams & Wilkins, Philadelphia, 1161–1188.

Rutala, W. A., and Weber, D. J. (2004). Disinfection and sterilization in health care facilities: What clinicians need to know. *Healthcare Epidemiol* 39, 702–709.

Weber, D. J., Rutala, W. A., Miller, M. B., Huslage, K., and Sickbert-Bennett, E. (2010). Role of hospital surfaces in the transmission of emerging health care-associated pathogens: Norovirus, *Clostridium difficile*, and *Acinetobacter* species. *AJIC* 38(5 Suppl 1), S25–S33.

WHO (2004). Practical Guidelines for Infection Control in Health Care Facilities. *SEARO Reg Pub 41*, World Health Organization, Geneva.

10

Hand Hygiene

Introduction

The hands seem to get much of the blame for nosocomial infections today but the exact degree to which they contribute remains uncertain. Hand hygiene has been cited as the single most important practice to reduce the transmission of infectious agents in health care settings and is an essential element of Standard Precautions (HICPAC 2007). The degree to which hands contaminated with fomites picked up from surfaces contributes to the transmission of airborne pathogens is probably significant, and is perhaps even more significant than direct airborne transmission or inhalation for many pathogens. An estimated 20–40% of nosocomial infections have been attributed to cross infection via the hands of HCWs (Weinstein 1991). In any event, the hands cannot be neglected as one of the aerobiological pathways of airborne nosocomial infections, and focus on hand hygiene, as well as skin hygiene, is an important component of any airborne infection control program. Hand antisepsis will reduce the incidence of health care–associated infections regardless of the ultimate source of contamination.

Decontamination of the inanimate hospital environment was covered in the previous chapter. This chapter addresses the skin flora that may become airborne, the etiology of hand contamination, strategies of hand hygiene, skin antiseptics, and handwashing protocols.

Skin Flora

Three groups of microorganisms may be distinguished on the skin: (1) microbes that reside on the skin (resident flora), (2) microbes that transiently colonize the skin (transient flora), and (3) infectious pathogens that transiently contaminate the skin (Rotter 1999). Microbes that reside on the skin include *Staphylococcus epidermis*, *S. aureus*, and *S. hominis*. Transient colonizers may include *Corynebacterium* species, *Acinetobacter*, *Enterobacter*, *Klebsiella*, and others. Infectious pathogens on the hands are typically viruses, including influenza, adenovirus, rhinovirus, RSV, and VZV.

The skin is normally contaminated with a variety of endogenous and transient flora and these can present infection-transmission problems if they are not controlled. Endogenous or resident microflora survive in the deep and superficial layers of the skin. Resident microflora consist mainly of gram-positive bacteria such as *Staphylococcus*, *Corynebacteria*, and *Propionibacterium* (Damani 1997). About one-fifth of humans carry staphylococci permanently on their skin. Resident microflora rarely cause any problems but under certain conditions, or when host resistance is weakened, they may multiply and cause various infections. Transient microflora are temporary contaminants of the skin and primarily occupy only the superficial layers (Canada 1998). The transient microflora includes most of the microbes responsible for cross-infections such as gram-negative bacilli (*E. coli*, *Klebsiella*, and *Pseudomonas* spp.), *Salmonella* spp., *Staphylococcus aureus*, and viruses (i.e., rotavirus).

Gram-negative bacteria such as *Acinetobacter* and species of *Enterobacter/Klebsiella* may be isolated from moist skin and from the hands of HCWs (Adams and Marrie 1982). Males are significantly more likely to be carriers than females. Persons who washed their hands less than eight times a day were consistently found to be more likely to carry gram-negative bacteria than those who washed more often.

Etiology of Hand Contamination

The normal microbial skin flora performs the function of resisting colonization by potentially pathogenic microorganisms. Transient floras are characterized by an inability to multiply on the skin and they do not usually survive for long. Transient floras are most commonly associated with hospital infections. Following contact with patients or with a contaminated environment, microorganisms can survive on hands for lengths of time between 2 and 60 minutes (WHO 2009). The hands of HCWs become progressively colonized with commensal flora as well as with potential pathogens during patient care (Pessoa-Silva et al. 2004).

Transient resident floras of the skin are mainly found on the hands or other skin surfaces that have been in recent contact with foreign or external sources of contamination. Some of these contaminated sources may be the result of settling of droplets and aerosols containing viruses and bacteria, the shedding of endogenous bacteria and skin squames, or contamination from environmental fungal spores and bacteria that were transported by air into the hospital environment or arrived from other sources. Endogenous microbes can survive on hospital surfaces for many hours. Figure 10.1 illustrates the

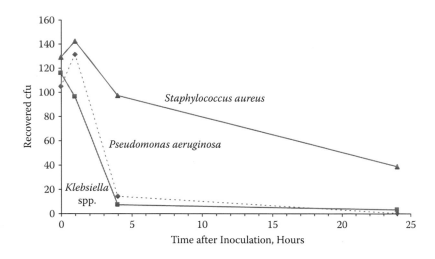

FIGURE 10.1
Survival of common skin bacteria on laminate surfaces. Based on data from Scott and Bloomfield (1990).

survival of some of these microbes on laminate surfaces, based on a surface sampling area of 25 cm².

HCWs with close patient contact are more likely to carry *Klebsiella* on their hands (Polk and Lopez 1972). This may be true of other nosocomial bacteria and so the patient and surfaces in the patient room may be the reservoir or ultimate source of hand contamination.

The population density of resident skin bacteria ranges from 100 to 1000 cfu/cm² (Rotter 1999). The pathogenic potential of normal skin flora is low, but in the cases of surgical sites and burn wounds, endogenous skin flora may prove pathogenic. The fingertips may hold from 0 to 300 cfu of bacteria based on a study by Pittett et al. (1999). This study also found that direct patient contact and respiratory-tract care were most likely to contaminate the fingers of care-givers. The hands of HCWs can become contaminated with gram-negative bacilli, *S. aureus*, *Enterococcus*, or *Clostridium difficile* by performing clean procedures or touching intact areas of the skin of hospitalized patients (CDC 2002). *S. aureus* may be recovered from the hands of about 21% of nurses and physicians, with median counts of greater than 1000 cfu. Similar percentages of HCWs carry gram-negative bacilli on their hands with median counts of up to 30,000 cfu or more. A study by Ehrenkranz and Alfonso (1991) showed that nurses who washed their hands after touching the moist skin of patients were still able to transfer bacteria to catheter materials.

Whole body disinfection is performed to disinfect the body of infectious agents and is usually done as part of general hygiene or before scheduled surgical procedures. Different areas of the body have varying counts of aerobic bacteria (CDC 2002). The scalp has about 1,000,000 cfu/cm², the armpit about 50,000 cfu/cm^2, the abdomen has about 40,000 cfu/cm^2, and the forearm has about 10,000 cfu/cm^2. Skin cannot be sterilized and therefore only a high degree of disinfection is possible. Therefore some risk always remains that endogenous or transient skin microflora may produce unwanted contamination during surgery.

Swenson et al. (2009) studied the effects of preoperative skin preparation on postoperative wound infection rates and found that the lowest infection rates were seen with iodine povacrylex in isopropyl alcohol. It is generally assumed that reduction of skin flora might translate into reduced SSI rates, but there appear to be no studies available that validate this assertion. The actual method of skin preparation and the disinfectants used may have more to do with SSI rates than reductions in skin flora.

Strategies of Hand Hygiene

The objective of *handwashing* is to remove dirt and organic debris, and loosely adhering microbial skin flora. The objective of *hygienic hand rubbing* is to reduce the release of transient pathogens with maximum efficiency to render the hands safe after suspected contamination. The objective of the *hygienic hand wash* is to reduce the release of transient flora using procedures that are significantly more effective than using ordinary soap.

When hands are washed for one minute with soap and water, the release of bacteria from the hands is reduced by two or three orders of magnitude. Cleaning may be largely ineffective for microflora in the deep layers of skin, but it can reduce the concentrations on the surface of the skin. Laboratory studies have demonstrated that handwashing with soap or water removes or kills about 90% of the transient microbes on the hands (Ayliffe et al. 1988). Cleaning agents and disinfectants can effectively reduce the concentrations of these bacteria to safe levels. Skin invariably contains organic matter that can inhibit the effects of disinfectants. Therefore the skin should be cleaned of organic matter first and then disinfected, but these processes are often done in a single phase.

Handwashing is considered the most important practice in preventing hospital-acquired infections today (Kennamer 2007). Hand disinfection prevents transmission of an infectious agent from a source, be it a patient or otherwise, to a health care worker or another patient. The techniques of hand disinfection include (1) handwashing, (2) complete immersion, and (3) scrubbing and drying the disinfectant on the skin. The handwashing method employs tap water and a detergent mixed with a disinfectant. After the solution is thoroughly applied the hands are rinsed with tap water, which poses a slight risk of recontamination. Liquid soaps are recommended because the soap cannot easily become contaminated after use. The effect of the disinfectant is limited to the period of time that the hands are washed.

Hand scrubbing or rubbing using chemical disinfectants does not require water because the disinfectant is left to dry on the skin. This method is more effective at killing microorganisms than the use of soaps containing only detergent. Scrubbing solutions are only available in liquid form and so avoid the problem of cross-contamination. The risk of using contaminated tap water is also eliminated. Immersion involves soaking the hands in a basin of chemical disinfectant for a specified time. Because the solution becomes contaminated it must be changed each time it is used. This method is not commonly used today.

The choice of which preventive measures are to be used to control hand-associated microbial transfer depends on which group of microbial flora is to be attacked (Rotter 1999). It is much easier to reduce the release of transient flora from the hands than to remove resident flora. No-touch techniques (using instruments rather than fingers) and protective gloves are suitable approaches, but instruments and gloves must then be changed after each patient is treated. Transient bacteria are more easily removed from gloves than from skin and can be cleaned before reuse. Hands must also be cleaned after the removal of gloves.

Meers and Yeo (1978) report that washing the hands with unmedicated soap makes it easier for skin squames carrying viable bacteria to escape into the air, and increases in the number of squames released after handwashing was significant. Besides gravitational settling onto clean or sterile surfaces, particles released in this manner are attracted by, and adhere to, objects with a static charge, such as plastic objects. The dissemination of viable bacteria is reduced to normal levels before washing if an antiseptic surgical scrub is used instead of soap.

Skin Antiseptics

This section summarizes those antiseptic agents used most often for hand disinfection. Additionally, there are some phenol compounds that were once in common use but are no longer popular due to ecologic concerns.

Alcohol is widely used as a disinfectant and an antiseptic. It can be used as a base for other disinfectants like chlorhexidine and iodine for preoperative skin disinfection.

Chlorhexidine is relatively nontoxic and is often combined with detergents or alcohol for skin disinfection. It is used exclusively as an antiseptic and is inactivated by soap, organic matter, and anionic detergents.

Iodine and *iodophors* are typically used as antiseptics. They are inactivated by organic matter and may corrode metals. Combinations of iodine or iodophors and alcohol are suitable for preoperative skin preparation. Povidone iodine detergent preparations are suitable for surgical hand disinfection.

Quaternary ammonium compounds (QAC) include a variety of compounds with detergent properties that inhibit the growth of bacteria but do not kill them. They are inactivated by soaps, anionic detergents, and organic matter. QACs may be used for cleaning wounds but are not recommended as environmental disinfectants. Contamination of QAC fluids is a potential problem and used solutions should be discarded.

Hexachlorophane is a chlorinated bisphenol useful for skin disinfection and for surgical hand disinfection. Although it has a limited killing effect, it is persistent, with soap and other organic materials having little effect on it.

Triclosan is a diphenyl ether. It has an intermediate killing effect but is very persistent on the skin. Organic matter does not significantly diminish its antiseptic activity.

Table 10.1 provides a summary of the antimicrobic activity of antiseptics against groups of pathogens, based on Rotter (1999).

Handwashing Protocols

There are two major types of handwashing that have been detailed as procedures or protocols—hygienic and surgical. The purpose of *hygienic hand disinfection* is to prevent transmission of infectious agents from one patient to another, and for the protection of health care workers. Either handwashing or scrubbing is considered effective and it should be performed before and after each patient is treated or examined by the health care personnel.

Surgical hand disinfection is performed to prevent transmission of both endogenous and transient microflora from the surgeon's hands to the surgical wound in the event the sterile gloves are breached. First the hands are cleaned with a liquid soap to remove moisture and organic debris. Then the hands are scrubbed with chemical disinfectants to kill both resident and transient microorganisms. The residual disinfectant that dries on the skin will provide further protection against recontamination from deeper skin layers.

Handwashing protocols and procedures have been widely published and are all rather similar, but the specific procedures may be uniquely prescribed for each institution. Compliance with handwashing procedures is essential and noncompliance is often cited as a risk factor for disease transmission. A typical handwashing protocol is provided following.

Generic Handwashing Protocol

1. Hands must be washed
    a. Between direct contact with patients
    b. Before performing invasive procedures

TABLE 10.1

Antimicrobial Activity of Common Antiseptics

| Compound Antiseptic | Gram-Positive Bacteria | Gram-Negative Bacteria | Mycobacteria | Antimicrobial Activity | | Fungi | Spores |
				Enveloped Viruses	Nonenveloped Viruses		
Alcohol	Good	Good	Good	Good	Moderate	Good	None
Chloroxylenol	Good	Poor	Poor	Poor	Variable	Poor	None
Chlorhexidine	Good	Moderate	Poor	Moderate	Poor	Poor	None
Hexachlorophane 3%	Good	Poor	Poor	—	—	Poor	None
Iodine and iodophors	Good	Good	Moderate	Moderate	Moderate	Moderate	Variable
Triclosan	Good	Moderate	Variable	—	—	Variable	None
QACs	Moderate	Poor	Variable	Poor	—	Variable	None

    c. Before caring for patients in ICUs or immunocompromised patients

    d. When hands are visibly soiled

    e. After procedures in which microbial or blood contamination of hands is likely

    f. After removing gloves

    g. After using the toilet or blowing the nose

2. Do not use fingernail polish or artificial nails.

3. Remove jewelry before washing hands.

4. Rinse hands under warm running water.

5. Lather with soap and use friction to cover all surfaces of the hands and fingers.

6. Dry hands thoroughly with a disposable towel or forced air dryer.

7. Turn off faucet without recontaminating hands (i.e., with a foot pedal).

8. Handwashing with an antiseptic agent should be performed in the following situations:

    a. When there is heavy microbial soiling

    b. When there is contact with infected tissue or organic matter

    c. When there is contact with feces or blood

    d. Prior to performing invasive procedures

    e. Prior to inserting intravascular or urinary catheters

    f. Before contact with immunocompromised patients

    g. Before contact with wounds or burn wounds

    h. Before and after direct contact with patients who have antimicrobial resistant microbes

9. Hands must be dry before applying waterless alcohol-based agents.

10. Hands must be rubbed or scrubbed with the antiseptic agent all around.

11. The disinfectant must be allowed to dry and remain on the surface of the skin.

For more specific and graphic step-by-step procedures for hand washing see, for example, Damani (1997); APIC (1995); WHO (2009); and UTMB (2009). Also see CDC (2002) and Canada (1998) for more information and for categorized recommendations designed to improve hand hygiene practices.

References

Adams, B. G., and Marrie, T. J. (1982). Hand carriage of aerobic gram-negative rods by health care personnel. *J Hyg* 89(1), 23–31.

APIC (1995). APIC guideline for handwashing and antisepsis in health care settings. *Am J Inf Control* 23, 251–269.

Ayliffe, G. A. J., Babb, J. R., Davies, J. G., and Lilly, H. A. (1988). Hand disinfection: A comparison of various agents in laboratory and ward studies. *J Hosp Inf* 11, 226–243.

Canada (1998). Hand Washing, Cleaning, Disinfection and Sterilization in Health Care. *Canada Communicable Disease Report Volume 24SB*, Health Canada, Laboratory Centre for Disease Control Ottawa, ON.

CDC (2002). *Guideline for Hand Hygiene in Health-Care Settings*. Centers for Disease Control, Atlanta, GA.

Damani, N. N. (1997). *Manual of Infection Control Procedures*. Greenwich Medical Media, Ltd., London.

Ehrenkranz, N. J., and Alfonso, B. C. (1991). Failure of bland soap handwash to prevent hand transfer of patient bacteria to urethral catheters. *Inf Contr Hosp Epidemiol* 12, 654–662.

HICPAC (2007). Guideline for Isolation Precautions: Preventing Transmission of Infectious Agents in Healthcare Settings. Centers for Disease Control, Atlanta, GA.

Kennamer, M. (2007). *Basic Infection Control for Health Care Providers*. Thomson Delmar Learning, Clifton Park, NY.

Meers, P. D., and Yeo, G. A. (1978). Shedding of bacteria and skin squames after handwashing. *J Hyg (Camb)* 81, 99-105.

Pessoa-Silva, C. L., Dharan, S., Hugonnet, S., Touveneau, S., Posfay-Barbe, K., Pfister, R., and Pittet, D. (2004). Dynamics of bacterial hand contamination during routine neonatal care. *Infect Contr Hosp Epidem* 25, 192–197.

Pittett, D., Dharan, S., Touveneau, S., Sauvan, V., and Perneger, T. V. (1999). Bacterial contamination of the hands of hospital staff during routine patient care. *Arch Intern Med* 159, 821–826.

Polk, H. C., and Lopez, J. F. (1972). Bacterial ecology of hands of intensive care unit nurses cleansed with povidone-iodine; in *Medical and Surgical Antispesis with Betadine Microbiocides*, H. C. Polk and N. J. Ehrenkranz, eds., Purdue Frederick, New York.

Rotter, M. L. (1999). Hand washing and hand disinfection; in *Hospital Epidemiology and Infection Control*, C. G. Mayhall, ed., Lippincott Williams & Wilkins, Philadelphia, 1339–1356.

Scott, E., and Bloomfield, S. F. (1990). The survival and transfer of microbial contamination via cloths, hands and utensils. *J Appl Bacteriol* 68(3), 271–278.

Swenson, B., Hedrick, T., Metzger, R., Bonatti, H., Pruett, T., and Sawyer, R. (2009). Effects of preoperative skin preparation on postoperative wound infection rates: A prospective study of 3 skin preparation protocols. *Infect Contr Hosp Epidem* 30(10), 964–971.

UTMB (2009). *Institutional Handbook of Operating Procedures*. University of Texas Medical Branch, Galveston.

Weinstein, R. A. (1991). Epidemiology and control of nosocomial infections in adult intensive care units. *Am J Med* 91(Suppl 3B), S179–S184.

WHO (2009). WHO Guidelines on Hand Hygiene in Health Care: A Summary. World Health Organization, Geneva.

11

Respiratory Infections

Introduction

Respiratory tract infections include all those infections that normally transmit in the community as well as those that predominate in the health care industry. They are due to a variety of pathogens including viruses, bacteria, and fungi. Because most respiratory tract infections can be caused by inhalation as well as by direct contact, an unknown but possibly significant fraction of these infections can be ascribed to airborne microorganisms. This chapter focuses on respiratory infections that can be attributed to airborne pathogens but because it is difficult, if not impossible, to separate airborne from direct contact infections, all respiratory tract infections are included regardless of transmission mode.

Nosocomial outbreaks of respiratory infections often follow the seasonal patterns of infections in the community. All community-acquired respiratory tract infections are addressed here first to provide a background for addressing hospital-acquired respiratory tract infections, which are caused by a different spectrum of microbes and have a different etiology. Nosocomial respiratory tract infections are also largely due to microbes that can transmit via the airborne route, although most likely transmit by direct contact in hospital environments. Some nosocomial respiratory infections transmit only by direct contact, such as ventilator-assisted pneumonia (VAP), which is largely if not completely due to contamination of equipment with endogenous or exogenous microbes. Special focus is also provided for tuberculosis in this chapter because of its importance as a nosocomial infection and because it provides a model for bacterial airborne transmission and general airborne infection control.

Respiratory Tract Infections (RTIs)

Respiratory tract infections, community acquired or otherwise, can be divided into three categories: (1) upper respiratory tract infections (URIs),

(2) middle respiratory tract infections, and (3) lower respiratory tract infections. The common etiologic agents of upper respiratory infections in the community are identified in Table 11.1, which also identifies the airborne class, Airborne Class 1 being demonstrably airborne in hospital infections. Upper (U), middle (M), and lower (L) RTI sites are indicated as well as the source, human (H) or environmental (E). Occurrence of ventilator-assisted pneumonia (VAP) is noted.

Upper respiratory tract infections (URIs) usually involve the nasal cavity and pharynx. In the community, over 80% of these infections are caused by pathogenic viruses (Ryan 1994). The diseases of the upper respiratory tract are named according to the site of infection. Rhinitis refers to inflammation of the nasal mucosa. Pharyngitis refers to pharyngeal infection. Tonsillitis refers to inflammation of the tonsils. Stomatitis refers to infections of the mucous membranes of the oral cavity. There are several additional types of infections that are less common or are combinations of the above, including rhinopharyngitis and tonsillopharyngitis.

Middle respiratory tract infections can include the epiglottis, surrounding aryepiglottic tissues, and the larynx, trachea, and bronchi. Infections can include laryngitis, bronchitis, laryngotracheitis, laryngotracheobronchitis, and infection of the epiglottis. Laryngitis can be considered to include its more severe form, croup.

Lower respiratory tract infections result from diseases of the lung, including the alveolar spaces, the interstitium, and the terminal bronchioles. Infection may occur as a result of a middle respiratory tract infection working its way into the lower respiratory tract, or as a result of inhalation or aspiration past the upper airway defenses, or from a more remote site of infection. Acute pneumonia is an infection of the lung parenchyma that usually develops rapidly. Chronic pneumonia has a slow and insidious onset that may take weeks or months to develop.

Nosocomial Respiratory Infections

This section addresses nosocomial or hospital-acquired respiratory infections, which differ from the previously addressed community-acquired respiratory infections in terms of the causative agents and the etiology or modes of transmission. The incidence of nosocomial respiratory infections is seasonal and the risk of infection is greatest during community outbreaks (Turner 1999). Patients, HCWs, and visitors are all at risk for infection with a nosocomial respiratory virus and may serve as sources for infections. Table 11.1 summarizes those pathogens that cause hospital-acquired respiratory tract infections. A number of these agents do not cause infections in the community, or are extremely rare, but cause pneumonia or other infections

TABLE 11.1

Nosocomial RTIs Associated with Airborne Pathogens

Pathogen	Type	Airborne Class	Source	RTI Site	Pneumonia	VAP
Acinetobacter	Bacteria	2	E	L	Y	Y
Adenovirus	Virus	2	H	UML	Y	
Alcaligenes	Bacteria	2	HE	U		
Aspergillus	Fungi	1	E	L	Y	
Blastomyces dermatitidis	Fungi	2	E	L	Y	
Bordetella pertussis	Bacteria	1	H	M		
Coccidioides immitis	Fungi	2	E	L	Y	
Coronavirus (SARS)	Virus	1	H	U		
Corynebacterium diphtheria	Bacteria	2	H	UM		Y
Coxsackievirus	Virus	2	H	U		
Cryptococcus neoformans	Fungi	2	E	L	Y	
Enterobacter	Bacteria	2	HE	L	Y	Y
Enterococcus	Bacteria	2	H	L	Y	Y
Haemophilus influenzae	Bacteria	2	H	UML	Y	Y
Haemophilus parainfluenzae	Bacteria	2	H	U		Y
Histoplasma capsulatum	Fungi	1	E	L	Y	
Influenza	Virus	1	H	UML	Y	
Klebsiella pneumoniae	Bacteria	2	HE	L	Y	Y
Legionella	Bacteria	1	E	L	Y	
Measles	Virus	1	H	M		
Mucor plumbeus	Fungi	2	E	U		
Mumps virus	Virus	1	H	U		
Mycobacterium tuberculosis	Bacteria	1	H	L	Y	
Mycoplasma pneumoniae	Bacteria	2	H	ML	Y	
Nocardia	Bacteria	2	E	L	Y	
Parainfluenza	Virus	2	H	UML	Y	
Pneumocystis jirovecii	Fungi	2	HE	L	Y	
Proteus mirabilis	Bacteria	2	H	L	Y	Y
Pseudomonas aeruginosa	Bacteria	1	E	L	Y	Y
Respiratory syncytial virus	Virus	1	H	UML	Y	
Rhinovirus	Virus	2	H	UM		
Rhizopus stolonifer	Fungi	2	E	U		
Rubella virus	Virus	2	H	U		
Serratia marcescens	Bacteria	2	E	L	Y	Y
Staphylococcus aureus	Bacteria	1	H	UML	Y	Y
Streptococcus pneumoniae	Bacteria	2	H	ML	Y	Y
Streptococcus pyogenes	Bacteria	1	H	UM		Y
Varicella-zoster virus	Virus	1	H	U		

in hospital environments. Agents of respiratory tract infections (RTI) are distinguished from those that cause pneumonia, and from those that cause ventilator-assisted pneumonia (VAP).

Nosocomial respiratory viruses are transmitted from person to person by aerosols or by hand contact followed by self-inoculation (Turner 1999). Aerosols are readily produced by coughing and sneezing, and speech can also produce aerosols (Gerone et al. 1966). Small-particle aerosols composed of droplet nuclei between 2 and 3 microns account for approximately 95% of particles and 25% of the total volume produced by coughing and sneezing (Buckland, Bynoe and Tyrrell 1965). Once an air space is contaminated by small-particle aerosols, the risk of infection is a function of the air change rate and the duration a subject remains in the area. Large-particle aerosols are considered to be composed of particles 10 microns or larger, which tend to be sprayed outward by distances of a few feet, and therefore close proximity to infected patients is required for infection.

Transmission of viral respiratory infections by direct contact usually requires that susceptible individuals get their hands contaminated either directly from the patient or by touching fomites left on surfaces and equipment. Uninfected intermediates (i.e., HCWs) may also contribute to the spread of an infection in hospitals. Spread of viruses by hand contact is limited by the ability of viruses to survive on skin, and spread of viruses by fomites is limited by the ability of viruses to remain viable on surfaces. The longer a virus can survive on hands and surfaces the greater the likelihood that an infection may be transmitted by this route. Some studies have shown that goggles can reduce the infection rate of certain respiratory viruses, implying the eyes as a route of inoculation, probably from rubbing the eyes with the hands.

Most nosocomial upper respiratory tract infections present themselves soon after admission with RSV, adenovirus, and influenza accounting for most of these infections (Graman and Hall 1989). Transmission of influenza virus occurs by small-particle aerosols, and isolation of infected patients in negative-pressure isolation rooms interrupts the spread of the virus. Ultraviolet radiation has been used to reduce the incidence of influenza infections in the hospital setting, but this approach is not widely used (McLean 1961).

Nosocomial Pneumonia

Pneumonia accounts for about 15% of all hospital-acquired infections and is the second most common nosocomial infection after UTIs (CDC 2003). It is also the most common infection in ICUs (Bonten and Bergmans 1999). The primary risk factor for acquiring nosocomial pneumonia is mechanical ventilation and endotracheal intubation. Patients who are mechanically ventilated are at least six times more likely to develop pneumonia than those who do not receive mechanical ventilation. Ventilator-assisted pneumonia (VAP) is considered to result primarily from oropharyngeal colonization and secondarily from gastric colonization. Bacteria can also gain access to the lower respiratory tract via inhalation of aerosols generated by nebulization devices, and by aspiration of bacteria found in dental plaques. In all of these cases the transmission is essentially direct and not due to airborne transmission, and therefore VAP must be considered a nonairborne infection. The source of the microorganisms responsible for VAP is either endogenous transmission or via the hands of health care workers that become transiently colonized or contaminated (Adams and Marrie 1982; Dascher 1985). The prevention of VAP can then be pursued via handwashing, gloving, and decontamination of equipment and water supplies. Two types of VAP can be distinguished—early onset VAP and late onset VAP. *Early onset VAP* is mainly caused by *Streptococcus pneumoniae, Staphylococcus aureus,* and *Haemophilus influenzae*, pathogens that may already colonize the respiratory tract at the time of intubation (Bonten and Bergmans 1999). *Late onset VAP* is mainly caused by *S. aureus, Pseudomonas aeruginosa,* and Enterobacteriaceae (*Klebsiella, Enterobacter, Serratia, Citrobacter, Proteus,* etc.). Other pathogens that have been associated with nosocomial pneumonia include *Mycoplasma pneumoniae* and *Chlamydia pneumoniae.*

Pneumonia resulting from sources other than mechanical ventilation equipment may be due to airborne transmission or via direct contact, but the mode of transmission may be dependent on the species involved in the infection. The species of primary concern in non-VAP pneumonia include *Legionella, Bordetella pertussis,* influenza, parainfluenza, RSV, adenovirus, SARS virus, and *Aspergillus*. Viruses can be responsible for up to 20% of pneumonia in patients and outbreaks in health care facilities usually follow seasonal patterns in the community (CDC 2003).

Microbes colonizing the respiratory tract and causing VAP come from either endogenous or exogenous sources. Colonization usually precedes the development of pneumonia and the predominant mode of inoculation is aspiration. Inhalation of *Aspergillus* spores or other fungi can also cause pneumonia (Bonten and Bergmans 1999). Pathogens colonizing the upper respiratory tract (oropharynx, sinus cavities, nares, and dental plaque) may be aspirated with oropharyngeal fluid and may bring pathogens to the lungs. Exogenous sources of pathogens may include other colonized patients and the hospital environment (sinks, faucets, sheets, mechanical ventilation devices, ventilator circuits, etc.). Routes of colonization that transfer pathogens from endogenous sources to the upper respiratory tract include the gastropulmonary route and the rectopulmonary route. In the gastropulmonary route, bacteria reach the upper respiratory tract via the stomach and colonize the oropharynx and the trachea, after which they may be aspirated into the lower respiratory tract. In the rectopulmonary route, intestinal microbes spread from the rectal area via the patient's skin and hands to the upper respiratory tract. Transfer of pathogens from exogenous sources in the hospital most likely occurs via the hands of HCWs, through direct inoculation of microbes into the tracheobrochial tree during manipulation of ventilator circuits.

Patients themselves are major reservoirs of nosocomial pathogens (Maki 1978; Maki, Alvarado, and Hassemer 1982). Patient self-inoculation with endogenous pathogens can hardly be called airborne, nor can direct inoculation from HCWs hands. Therefore only exogenous sources and cross-contamination or cross-infection of non-VAP pneumonia could be the result of airborne transmission. According to Chetchotisakd, Phelps, and Hartstein (1994), some 10% of bacterial isolates from patients were suspected to have resulted from cross-transmission. Such cross-infections from other patients must be due to either HCW hand contamination or inhalation. Evidence of cross-colonization of *Pseudomonas aeruginosa* among patients in different ICU wards was presented by Bergmans et al. (1998) and this was attributed to HCW hand contamination. Although patients may be a source of *Pseudomonas*, this microbe originally hails from reservoirs in the environment and so must be airborne at some point when it enters the hospital, even if the inoculation itself is not an airborne process. Each of the etiologic agents of potentially airborne, non-VAP pneumonia is examined in the following paragraphs.

Legionella is the causative agent of Legionnaires' disease and has been responsible for a number of outbreaks. *Legionella* species are found in natural and man-made aquatic environments. The primary mechanism by which *Legionella* gains entry to the respiratory tract is via inhalation of aerosols of contaminated water. It has frequently been isolated in the water systems of hospitals (Vickers et al. 1991). Several hospital outbreaks have been traced to aerosols generated by cooling towers, showers, faucets, respiratory therapy equipment, and humidifiers (CDC 2003). Because of the proximity required for inhalation of *Legionella* from contaminated water sources, air cleaning and air disinfection do not provide for effective protection, except in the rare cases where bacteria circulates in the ventilation system. The best solution appears to be water decontamination programs combined with diagnostic testing.

Bordetella pertussis causes highly infectious acute respiratory tract infections that primarily affect children and infants. Infected adolescents and adults serve as a reservoir for continued outbreaks. Adults with cough, including health care workers and hospital visitors, can be a major source of nosocomial infections and can shed the microorganism for extended periods before detection or diagnosis (CDC 2003). Pertussis is transmitted during close contact with an infected person who may spray droplets into the vicinity from coughing or sneezing. Autoinoculation may also occur when the hands become contaminated with fomites. In one study *Bordetella* was detected in air samples taken up to 4 meters away from an infected patient, but airborne transmission over long distances (i.e., 6 feet, or beyond droplet spray distance) has not been demonstrated (Aintablian, Walpita, and Sawyer 1998). If airborne transmission does not occur, or occurs rarely, air disinfection would not be a suitable means of control, but the use of droplet precautions and surface disinfection technologies in conjunction with hand hygiene may provide options.

Influenza can cause pneumonia either directly or as a result of secondary infections with other pathogens. Nosocomial influenza follows seasonal outbreak patterns in the community. Influenza is primarily transmitted by droplets sprayed onto mucosal surfaces from coughing or sneezing, and by fomites on hands or surfaces. Airborne transmission also occurs but presumably to a lesser degree than direct contact (Alford et al. 1966; Henle et al. 1946; Moser et al. 1979). Infected persons are a reservoir for the virus and they can shed viral particles for up to seven days or longer (CDC 2003). HCWs have been implicated in the transmission of influenza to patients and vaccination can reduce illnesses and secondary infections (Valenti 1999).

Human parainfluenza virus causes respiratory infections of the upper and lower respiratory tract, including pneumonia and croup in children. Transmission is thought to be by direct contact and large droplet transmission. Droplets may be sprayed by coughing and sneezing, depositing fomites on surfaces and spraying droplets directly on another person's eyes or mucosa (CDC 2003). Fomites on surfaces and objects result in hand contamination.

Respiratory syncytial virus (RSV) causes infections among children and infants and is the leading cause of admissions in infants under one year of age. Nosocomial RSV outbreaks usually parallel community outbreaks and infected children are usually the source of RSV in hospitals, but visitors and HCWs who pick up infections in the community can also disseminate the virus. RSV is transmitted by close contact and most probably by autoinfection when infectious secretions are picked up by the hands. RSV is also disseminated by droplet spread when droplets are sprayed into the air by coughing and sneezing. RSV was detected in air samples from hospital rooms housing infected patients (Aintablian, Walpita, and Sawyer 1998). RSV can remain viable on surfaces for up to six hours (Hall, Douglas, and Geiman 1980). Implementation of standard precautions, including hand hygiene, can be effective in controlling spread.

Adenovirus infections occur mainly in childhood, causing upper respiratory infections and sometimes pneumonia. Nosocomial outbreaks of adenovirus have occurred in ICUs, pediatric care facilities, military hospitals, and other health care facilities (CDC 2003). Infections can be introduced into hospitals by HCWs, patients, and visitors who acquire them in the community. Transmission is thought to occur by autoinoculation from hands to mouth or eyes. Transmission by other routes, including airborne inhalation, has been suggested. Adenovirus can remain viable in fomites for up to 49 days on surfaces, and up to 10 days on cloth (Gordon et al. 1993).

SARS virus is a coronavirus and an emerging respiratory infection that has caused outbreaks in heath care settings (Ho, Tang, and Seto 2003; HWFB 2003). Transmission is believed to be by droplet spray from coughing and sneezing and by direct contact, but airborne transmission can also occur

(CDC 2003; He et al. 2003). SARS has been transmitted to HCWs during high-risk exposure associated with aerosolization of respiratory secretions.

Aspergillus species are ubiquitous in the environment and are common indoor contaminants. Spores of *Aspergillus* have been found in ventilation systems, especially those with unfiltered air, indoor horizontal surfaces, and other locations (Walsh and Dixon 1989; Cali et al. 2001). They can be aerosolized during construction and renovation in hospitals. *Aspergillus* can multiply in indoor environments (Arnow et al. 1991). *Aspergillus fumigatus* and *Aspergillus flavus* are the most frequently isolated in infected patients (CDC 2003). Immunocompromised patients are increasingly at the greatest risk. Pulmonary aspergillosis is acquired primarily by inhalation of the spores. Hospital outbreaks are often associated with activities that result in increasing airborne concentrations of *Aspergillus* spores. Control of indoor environmental contamination is a main focus of remediation attempts, and patients can be protected with air cleaning technologies such as air filtration and increased air exchange rates. High air exchange laminar flow systems have been used to control airborne *Aspergillus* concentrations. Air recirculation units using high efficiency filtration have also been used to control airborne *Aspergillus* (Sherertz et al. 1987). When spores on surfaces become resuspended in the air from activity, a low-lying cloud of particles may drift along the floor with local air currents and foot traffic. *Aspergillus* spores can survive long enough to distribute widely in the hospital environment.

Nosocomial Tuberculosis

Mycobacterium tuberculosis is the causative agent of TB and is carried in airborne particles generated by infectious TB patients who cough, sneeze, shout or sing (Wells 1955; Nardell 1990; CDC 2003). Droplets are sprayed into the immediate vicinity of the patient within a radius of up to six feet. Smaller droplets may remain suspended in local air currents but evaporate rapidly down to droplet nuclei (1–5 µm) or even to single bacilli (Duguid 1945; Liu 2000). A 1-micron droplet could hold about 1–10 TB bacilli, which is the same as the infectious dose (ID_{50}), so a single inhalation or a single contact with a fomite could conceivably produce an infection. Under ideal conditions, cells can reportedly survive 40–100 days, but they likely die more rapidly on normal indoor surfaces due to desiccation and sunlight. Aerosols containing individual TB bacilli could remain suspended for extended periods of time and can be carried great distances. The risk of infection is correlated with the airborne concentration of bacilli and the duration of exposure (Garrett et al. 1999). *M. bovis* is also infectious and is also airborne, but is not common in hospital settings.

The spread of TB by short-range droplet spray was recognized early in the age of microbiology, before the concept of airborne disease transmission was accepted (Decker and Schaffner 1999). The idea of droplet nuclei, which remain as the droplets evaporate, was introduced in relation to TB by Wells (1934). Droplets smaller than 1–2 microns will evaporate completely before they fall to the ground, if they are entirely water. Any microbes contained in the droplets may remain as airborne clumps or even individual microorganisms floating in the air for hours, settling gradually but getting stirred up continuously by activity and air currents. Therefore, two routes can be distinguished, *droplet spray* in which the droplets fall rapidly within 3–6 feet of a cough or sneeze, and *aerosol clouds* in which submicron particles may float for hours.

In health care facilities, the likelihood of exposure to *M. tuberculosis* is affected by factors that include the prevalence of infectious tuberculosis in the population served by the facility, the degree of crowding in the facility, the effectiveness of the TB infection control program, the training of HCWs, and the engineering controls (Garrett et al. 1999). Engineering controls may include isolation rooms (TB rooms) under negative pressure to contain the contaminated air with filtered supply air. Air recirculating units equipped with filters and ultraviolet germicidal irradiation (UVGI) are also commonly used to disinfect the air at the source. Tuberculosis infections among health care workers are strongly associated with inadequate ventilation in general patient rooms and with the type and duration of work (Menzies et al. 2000). There have been numerous cases of HCWs acquiring TB from occupational transmission, and the magnitude of the risk is partly dependent on and can vary considerably with the type of health care facility (Charney and Fragala 1999). Based on a three-year prospective study performed in Washington State between 1982 and 1984, the annual risk is 1.5 times the background rate for hospital employees, approximately 11 times the background rate for nursing home employees, 6 times the rate for home health care workers, and double the rate for home care employees.

Since 1960, there have been some 30 outbreaks of tuberculosis in health care facilities, and departments affected include emergency departments, inpatient medical wards, intensive care units, an operation room, radiology suites, HIV wards and clinics, a nursery, a maternity ward, and other areas (Garrett et al. 1999). Inadequate ventilation increases the likelihood of exposure. Figure 11.1 illustrates the results of a summary of studies on health care facility TB outbreaks in which some airflow-related contributing factors were identified (Garrett et al. 1999). The lack of positive pressure (43% of cases) allows the bacilli to spread out into the corridors and other areas, while inadequate ventilation (27%) allows airborne concentrations to accumulate.

Recirculation of potentially contaminated air from sputum induction chambers or isolation rooms has been a contributing factor in some hospital outbreaks. Failure to keep isolation room doors closed and not keeping patients in their rooms, both of which would have spread the infection

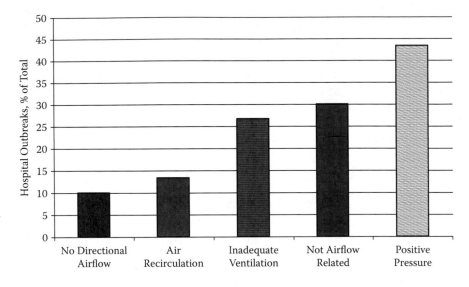

FIGURE 11.1
Airflow-related contributing factors in hospital TB outbreaks. Based on data from Garrett et al. (1999). Other non-airflow-related factors are not shown.

through other areas, have been identified as contributing factors. Performing procedures that generate aerosols like suctioning, surgical drainage, and irrigation of abscesses can also cause outbreaks. Increasingly, large outbreaks of TB involve multidrug-resistant (MDR) TB. Health care workers have been exposed and infected at facilities experiencing outbreaks of MDR TB. MDR TB outbreaks can happen in settings with immunocompromised patients or HIV-infected individuals, in which high attack rates occur.

CDC (2005) recommendations for control of TB in health care include the use of a single-pass ventilation system; airborne infection isolation (AII) rooms with at least 6 ACH, and preferably 12 ACH; HEPA filtration if air is recirculated; and the use of UVGI systems to increase the equivalent ACH. HEPA filters should be used to remove infectious droplet nuclei from air that is recirculated, and can also be used to filter air that is discharged to the environment. UVGI can be used in rooms or corridors in the form of upper-room systems and can be used to irradiate the airflow in ductwork. UVGI can be used in ducts that recirculate air back into the same room. UVGI should not be used as a replacement for filtration. AII rooms are used to (1) separate patients suspected of having TB from other patients, (2) control environmental factors to reduce the concentration of droplet nuclei, and (3) prevent the escape of droplet nuclei into adjacent areas. Portable room air recirculation units equipped with HEPA filters have been demonstrated to be effective in removing bioaerosols from room air and various commercial units are available (Cheng, Lu, and Chen 1998; Miller-Leiden, Lobascio, and Nazaroff 1996). HEPA filters remove particles in the size range of TB bacilli but no studies have been performed to demonstrate a decrease in risk of infection (Rutala et al. 1995). Portable units should be designed to achieve 12 ACH, an exchange rate that can depend on the room volume and existing room airflow. The use of anterooms can be effective in confining airborne contamination to an AII room because they will buffer the room from pressure fluctuations and door openings.

Research on UVGI has demonstrated that it is effective in inactivating *M. tuberculosis* in full-scale studies (Xu et al. 2003; Riley and Nardell 1989). UVGI is recommended by various sources as an adjunct to existing ventilation measures for the control of TB infections (ASHRAE 2003, 2009; CDC 2005; WHO 1999; USACE 2000). Three types of UVGI systems have been used for TB control: (1) in-duct air disinfection systems, (2) upper-room (or upper-air) UVGI systems, and (3) portable room air cleaners or recirculation units. The effectiveness of all these technologies can depend on factors like room air mixing and relative humidity, and care should be taken in the design and selection of such systems.

Personnel in areas where aerosol might be generated, including sputum booths, should wear respiratory protection. Respirators must be certified by CDC/NIOSH as a nonpowered particulate filter respirator (N-, R-, P-95, P-99, or P-100), or a PAPR with high efficiency filters (CDC 2005). Patients suspected of having tuberculosis should immediately be placed in an isolation room where air is either exhausted to the outside or filtered before being recirculated.

Ventilation is the most important environmental control measure for TB and it can involve high air exchange rates (12–15 ACH), in-room recirculation units employing HEPA filters, and UVGI systems (Garrett et al. 1999). Local exhaust ventilation systems that capture the contamination at the source and remove it before it is dispersed into the air are the most efficient type of control. Anterooms may serve as an extra level of protection. UVGI has been used to kill TB bacilli in experimental applications and has demonstrated reductions in airborne bacteria (Riley et al. 1957; Riley et al. 1962; Kundsin 1988; Riley and Nardell 1989). UVGI has been used with some reported effect in hospitals to disinfect the air of TB bacilli (Riley 1994; Iseman 1992; Stead et al. 1996). Recirculation units using HEPA filters have also proven useful for providing filtration of the air at the source (Marier and Nelson 1993).

Respiratory protection can be used to protect HCWs from airborne bacilli provided they meet certain standards. The original standard stated the performance requirements of respirator filters as the ability to filter particles one micron in size in the clean state with an efficiency of 95% at a flow rate of up to 50 liters a minute (Decker and Schaffner 1999). Three classes of particulate respirators are defined, types P, R, and N, which have minimum efficiencies of 99.97%, 99%, and 95%, respectively, at 0.3 mm. Facemasks act to filter the room air before it is inhaled, and to filter exhaled air. Facemasks will tend to become contaminated with endogenous microbes on the inside and exogenous or environmental microbes on the outside after use. A well-fitted facemask is capable of filtering out spores (Pippin et al. 1987).

The exposure limit for TB bacilli is 0 cfu/m³, because the minimum infective dose is not known, and it has been demonstrated experimentally that a single bacilli may initiate an infection (Riley and O'Grady 1961; Riley et al. 1962). In terms of removal efficiency, a MERV 15 filter should be able to filter out 97% of TB bacilli, and an ultraviolet light system with a UVGI rating value (URV) of 15 should be capable of removing virtually 100%, as would a HEPA filter. Combined, these technologies would provide a higher degree of air disinfection than either of them separately.

References

Adams, B. G., and Marrie, T. J. (1982). Hand carriage of aerobic Gram-negative rods by health care personnel. *J Hyg* 89(1), 23–31.

Aintablian, N., Walpita, P., and Sawyer, M. H. (1998). Detection of *Bordetella pertussis* and respiratory syncytial virus in air samples from hospital rooms. *Infect Control Hosp Epidemiol* 19(12), 918–923.

Alford, R. H., Kasel, J. A., Gerone, P. J., and Knight, V. (1966). Human influenza resulting from aerosol inhalation. *Proc Soc Exp Biol Med* 122(3), 800–804.

Arnow, P. M., Sadigh, M., Costas, C., Weil, D., and Chudy, R. (1991). Endemic and epidemic aspergillosis associated with in-hospital replication of *Aspergillus* organisms. *J Infect Dis* 164, 998–1002.

ASHRAE (2003). *HVAC Design Manual for Hospitals and Clinics.* American Society of Heating, Ventilating, and Air Conditioning Engineers, Atlanta GA.

_____. (2009). *ASHRAE Position Document on Airborne Infectious Diseases.* American Society of Heating, Refrigerating and Air-Conditioning Engineers, Atlanta, GA.

Bergmans, D. C. J. J., Bonten, M. J. M., van Tiel, F. H., Gaillard, C. A., van der Geest, S., Wilting, R. M., de Leeuw, P. W., and Stobberingh, E. E. (1998). Cross-colonization with *Pseudomonas aeruginosa* of patients in an intensive care unit. *Thorax* 53, 1053–1058.

Bonten, M. J. M., and Bergmans, D. C. J. J. (1999). Nosocomial pneumonia; in *Hospital Epidemiology and Infection Control*, C. G. Mayhall, ed., Lippincott Williams & Wilkins, Philadelphia, 211–238.

Buckland, F. E., Bynoe, M. L., and Tyrrell, D. A. J. (1965). Experiments on the spread of colds. *Journal of Hygiene* 63(3), 327–343.

Cali, S., Scheff, P., Conroy, L., Curtis, L., Baker, K., Ou, C.-H., and Norlock, F. (2001). *Aspergillus* surveillance project at an urban hospital; in *Environmental Health Risk*, D. Fajzieva and C. A. Brebbia, eds., WIT Press, Southampton.

CDC (2003). *Guidelines for Preventing Health-Care Associated Pneumonia.* Centers for Disease Control, Atlanta, GA.

_____. (2005). *Guidelines for Preventing the Transmission of* Mycobacterium tuberculosis *in Health-Care Facilities.* Centers for Disease Control, Atlanta, GA.

Charney, W., and Fragala, G. (1999). *The Epidemic of Health Care Worker Injury: An Epidemiology.* CRC Press, Boca Raton, FL.

Cheng, Y. S., Lu, J. C. and Chen, T. R. (1998). Efficiency of a portable air cleaner in removing pollen and fungal spores. *Aerosol Sci Technol* 29(2), 93–101.

Chetchotisakd, P., Phelps, C. L., and Hartstein, A. I. (1994). Assessment of bacterial cross-transmission as a cause of infections in patients in intensive care units. *Clin Infect Dis* 18, 929–937.

Dascher, F. D. (1985). The transmission of infections in hospitals by staff carriers, methods of prevention and control. *Infect Control* 6(3), 97–99.

Decker, M. D., and Schaffner, W. (1999). Tuberculosis control in the hospital: Compliance with OSHA requirements; in *Hospital Epidemiology and Infection Control*, C. G. Mayhall, ed., Lippincott Williams & Wilkins, Philadelphia, 1091–1100.

Duguid, J. P. (1945). The size and the duration of air-carriage of respiratory droplets and droplet-nuclei. *J Hygiene* 54, 471–479.

Garrett, D. O., Dooley, S. W., Snider, D. E., and Jarvis, W. R. (1999). *Mycobacterium tuberculosis*; in *Hospital Epidemiology and Infection Control*, C. G. Mayhall, ed., Lippincott Williams & Wilkins, Philadelphia, 477–503.

Gerone, P. J., Couch, R. B., Keefer, G. V., Douglas, R. G., Derrenbacher, E. B., and Knight, V. (1966). Assessment of experimental and natural viral aerosols. *Bacteriological Reviews* 30(3), 576–588.

Gordon, Y. J., Gordon, R. Y., Romanowski, E. G., and Araullo-Cruz, T. P. (1993). Prolonged recovery of desiccated adenoviral serotypes 5, 8, and 19 from plastic and metal surfaces in vitro. *Opthalmology* 100(12), 1839–1840.

Graman, P. S., and Hall, C. B. (1989). Nosocomial viral respiratory infections. *Semin Respir Infect* 4, 253–260.

Hall, C. B., Douglas, R. G., and Geiman, J. M. (1980). Possible transmission by fomites of respiratory syncytial virus. *J Infect Dis* 141(1), 98–102.

He, Y., Jiang, Y., Xing, Y. B., Zhong, G. L., Wang, L., Sun, Z. J., Jia, H., Chang, Q., Wang, Y., Ni, B., and Chen, S. P. (2003). Preliminary result on the nosocomial infection of severe acute respiratory syndrome in one hospital of Beijing. *Zhonghua Liu Xing Bing Xue Za Zhi* 24(7), 554–556.

Henle, W., Henle, G., Stokes, J., and Maris, E. P. (1946). Experimental exposure of human subjects to viruses of influenza. *J Immunol* 52, 145–165.

Ho, P. L., Tang, X. P., and Seto, W. H. (2003). SARS: Hospital infection control and admission strategies. *Respirology* 8(Suppl), S41–S45.

HWFB (2003). SARS Bulletin (24 April 2003). Health, Welfare, and Food Bureau, Government of the Hong Kong Special Administrative Region, Hong Kong.

Iseman, M. D. (1992). A leap of faith: What can we do to curtail intrainstitutional transmission of tuberculosis? *Ann Intern Med* 117(3):251–253.

Kundsin, R. B. (1988). *Architectural Design and Indoor Microbial Pollution*. Oxford University Press, New York.

Liu, H. (2000). *Science and Engineering of Droplets: Fundamentals and Applications*. William Andrew Publishers, Norwich, NY.

Maki, D. G. (1978). Control of colonization and transmission of pathogenic bacteria in the hospital. *Ann Intern Med* 89, 777–780.

Maki, D. G., Alvarado, C. J., and Hassemer, C. A. (1982). Relation of the inanimate hospital environment to endemic nosocomial infection. *N Engl J Med* 307, 1562.

Marier, R. L., and Nelson, T. (1993). A ventilation-filtration unit for respiratory isolation. *Inf Contr Hosp Epidemiol* 14, 700–705.

McLean, R. (1961). The effect of ultraviolet radiation upon the transmission of epidemic influenza in long-term hospital patients. *Am Rev Resp Dis* 83, 36–38.

Menzies, D., Fanning, A., Yuan, L., and Fitzgerald, J. M. (2000). Hospital ventilation and risk for tuberculosis infection in Canadian health care workers. *An Intern Med* 133(10), 779–789.

Miller-Leiden, S., Lobascio, C. and Nazaroff, W. W. (1996). Effectiveness of in-room air filtration and dilution ventilation for tuberculosis infection control. *J Air and Waste Mgt Assoc* 46(9), 869.

MMWR (1997). Guidelines for prevention of nosocomial pneumonia. *Morb Mort Weekly* 46[RR-1 (01/03/1997)], 1–79.

Moser, M. R., Bender, T. R., Margolis, H. S., Noble, G. R., Kendal, A. P., and Ritter, D. G. (1979). An outbreak of influenza aboard a commercial airliner. *Am J Epidemiol* 110(1), 1–6.

Nardell, E. A. (1990). Dodging droplet nuclei. *American Review of Respiratory Disease* 142, 501–503.

Pippin, D. J., Verderame, R. A., and Weber, K. K. (1987). Efficacy of face masks in preventing inhalation of airborne contaminants. *J Oral Maxillofac Surg* 45, 319–323.

Riley, R. L., and O'Grady, F. (1961). *Airborne Infection*. Macmillan, New York.

Riley, R., Wells, W., Mills, C., Nyka, W., and McLean, R. (1957). Air hygiene in tuberculosis: Quantitative studies of infectivity and control in a pilot ward. *Am Rev Tuberc Pulmon* 75, 420–431.

Riley, R. L. (1994). Ultraviolet air disinfection: Rationale for whole building irradiation. *Inf Contr Hosp Epidemiol* 15, 324–328.

Riley, R. L., Mills, C. C., O'Grady, F., Sultan, L. U., Wittstadt, F., and Shivpuri, D. N. (1962). Infectiousness of air from a tuberculosis ward. *Am Rev Respir Dis* 85, 511–525.

Riley, R. L., and Nardell, E. A. (1989). Clearing the air: The theory and application of ultraviolet disinfection. *Am Rev Resp Dis* 139, 1286–1294.

Rutala, W. A., Jones, S. M., Worthington, J. M., Reist, P. C., and Weber, D. J. (1995). Efficacy of portable filtration units in reducing aerosolized particles in the size range of *Mycobacterium tuberculosis*. *Infect Contr Hosp Epidemiol* 16, 391–398.

Ryan, K. J. (1994). *Sherris Medical Microbiology*. Appleton & Lange, Norwalk.

Sherertz, R. J., Belani, A., Kramer, B. S., Elfenbein, G. J., Weiner, R. S., Sullivan, M. L., Thomas, R. G., and Samsa, G. P. (1987). Impact of air filtration on nosocomial *Aspergillus* infections. *Amer J Medicine* 83, 709–718.

Stead, W. W., Yeung, C., and Hartnett, C. (1996). Probable role of ultraviolet irradiation in preventing transmission of tuberculosis: A case study. *Inf Contr Hosp Epidemiol* 17, 11–13.

Turner, R. B. (1999). Nosocomial viral respiratory infections in pediatric patients; in *Hospital Epidemiology and Infection Control*, C. G. Mayhall, ed., Lippincott Williams & Wilkins, Philadelphia, 607–614.

USACE (2000). Guidelines on the Design and Operation of HVAC Systems in Disease Isolation Areas. *TG252*, U.S. Army Center for Health Promotion and Preventive Medicine.

Valenti, W. M. (1999). Influenza viruses; in *Hospital Epidemiology and Infection Control*, C. G. Mayhall, ed., Lippincott Williams & Wilkins, Philadelphia, 535–542.

Vickers, R. M., Yu, V. L., Hanna, S. S., Muraca, P., Diven, W., Carmen, N., and Taylor, F. B. (1991). Determinants of *Legionella pneumophila* contamination of water distribution systems: 15-hospital prospective study. *Infect Control* 8(9), 357–363.

Walsh, T. J., and Dixon, D. M. (1989). Nosocomial aspergillosis: Environmental micro-biology, hospital epidemiology, diagnosis and treatment. *Eur J Epidemiol* 5(2), 131–142.

Wells, W. F. (1934). On air-borne infection. Study II. Droplets and droplet nuclei. *Am J Hyg* 20, 611–618.

Wells, W. F. (1955). *Airborne Contagion.* Annals of the National Academy of Sciences, New York Academy of Sciences, New York.

WHO (1999). Guidelines for the Prevention of Tuberculosis in Health Care Facilities in Resource Limited Settings. *WHO/CDS/TB/99.269*, World Health Organization, Geneva.

Xu, P., Peccia, J., Fabian, P., Martyny, J. W., Fennelly, K. P., Hernandez, M., and Miller, S. L. (2003). Efficacy of ultraviolet germicidal irradiation of upper-room air in inactivating airborne bacterial spores and mycobacteria in full-scale studies. *Atmos Environ* 37, 405–419.

12

Surgical Site Infections

Introduction

Surgical site infections are often the most problematic and fatal type of infection that can occur in hospital settings. Many SSIs are thought to be transmitted only by direct contact, either hand to patient or hand to equipment, or equipment to patient, but this view overlooks the fact that both hands and equipment may pick up contamination from elsewhere, and that may include settling from the air. Microbes, especially bacteria in the size range of nosocomial agents, are relatively large and will settle within seconds or minutes. Once settled, they may be resuspended by activity. Levels of airborne bacteria are invariably highest near the floor and lowest near the ceiling. This may not be true for viruses, which can remain suspended indefinitely due to their size, but viruses are not normally SSIs.

Although direct contact is regarded as being the primary mode of contamination of surgical sites, airborne transmission can occur (Mangram et al. 1999). SSIs are nonrespiratory but may be partly airborne in origin, such as when common microbes like *Staphylococcus* and *Streptococcus* settle on open wounds, burns, or medical equipment (Fletcher et al. 2004). *Aspergillus* is undeniably airborne when it enters the hospital environment from outdoor air or from being disturbed by construction activities. Contamination of equipment and surfaces in the OR with fomites may produce direct contact infections as the end result of a process that includes aerosolization.

SSI Microbes

Microorganisms isolated from operating rooms are usually considered nonpathogenic or are commensals rarely associated with SSIs. Pathogens that cause SSIs are either acquired endogenously from the patient's own flora or exogenously from HCWs and the hospital environment. Bacteria account for the majority of SSIs with *Staphylococcus aureus* and coagulase-negative staphylococci

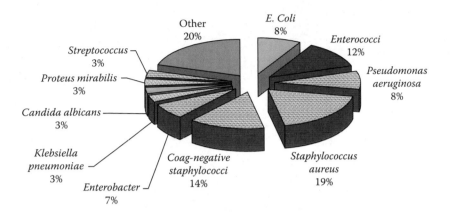

FIGURE 12.1
Breakdown of pathogens causing SSIs. Dotted patterns indicate potentially airborne microbes. Based on NNIS data for 1990–1992 (CDC 1993).

being the two most common pathogens isolated from clean wound sites. Figure 12.1 shows a breakdown of the most common SSI microbes.

Most SSI pathogens are believed implanted at the time of surgery (Mandell et al. 2000). *S. aureus* and other endogenous flora are believed to be directly inoculated onto the operative site during incision or subsequent manipulation. During contaminated or dirty procedures, the normal internal body flora can directly contaminate the operative site. The patient's endogenous flora may travel from distant sites to the open wound. The majority of *S. aureus* strains recovered from wound infections are endogenous.

In implant surgery, only a fraction of the coagulase-negative staphylococci that cause infections can be traced to the patient's skin. Coag-negative staphylococci can be shed from the skin and contaminate environmental surfaces, which may then serve as a secondary reservoir for further spread within a hospital (Ayliffe 1991). Blakemore et al. (1971) recovered *S. epidermis* from both air and blood samples obtained from a cardiopulmonary bypass machine during surgery. Because local air is drawn into the suction line of such equipment, numerous airborne bacteria can be aspirated from operating room air into the blood supply. Bacteria have been recovered from operating room air and such airborne microorganisms may contaminate the operative site during surgery (Duhaime et al. 1991).

The types and proportion of microbes causing wound infections depends on the nature of the operation. The type and concentration of skin microorganisms is thought to affect infections of intravascular devices. The skin of the arm has about 10 cfu of bacteria per 10 cm^2, while the skin of the neck and the upper thorax may have several thousand cfu per the same unit area (Wilson and Garrison 1995). Microbes cultured from intravascular devices include *Staphylococcus aureus, S. epidermis, Klebsiella, Enterobacter, Pseudomonas, Serratia,* and *Enterococcus.*

TABLE 12.1

Airborne SSI Microbes

Pathogen	Type	Airborne Class	Source
Aspergillus	Fungal spore	1	Environmental
Enterobacter	Bacteria	2	Humans and environmental
Enterococcus	Bacteria	2	Humans
Klebsiella pneumoniae	Bacteria	2	Humans and environmental
Proteus	Bacteria	2	Humans
Pseudomonas aeruginosa	Bacteria	1	Environmental
Serratia marcescens	Bacteria	2	Environmental
Staphylococcus aureus	Bacteria	1	Humans
Staphylococcus epidermis	Bacteria	2	Humans
Streptococcus pyogenes	Bacteria	1	Humans

Table 12.1 summarizes the most common potential airborne pathogens that have or are suspected of having been transmitted as SSIs.

Surgical site infections occur in cardiac surgery with an infection rate of about 2%. In sternal surgical site infection, almost 50% of infections are ascribed to *S. aureus* and *S. epidermis*, with the remainder due to *E. coli*, *Klebsiella*, *Enterobacter*, *Proteus*, and *Pseudomonas*. The frequency with which endogenous *Staphylococcus* species are implicated suggests that direct inoculation of overlying skin microorganisms into the operative site is an important route of acquisition (Wong 1999). Other exogenous sources of infection have included *Burkholderia cepacia*, *Enterobacter cloacae*, *Mycobacterium fortuitum*, *Candida*, and *Legionella*.

Perl, Chotani, and Agawala (1999) reviewed 18 outbreaks of *Aspergillus* in health care environments involving bone transplants and other operations, and many of these involved construction activities that raised dust. In some of the nonconstruction cases, the ventilation systems were found to have faults, such as lack of negative pressure or filtration problems. Infection control interventions designed to prevent airborne transmission have been effective in reducing airborne concentration and in reducing cases of aspergillosis among high-risk patients (Cooper, O'Reilly, and Guest 2003).

Aerobiology of Operating Rooms

The aerobiology of operating rooms depends primarily on the microbial flora of the occupants, with common skin microbes like *Staphylococcus* and *Streptococcus*, and intestinal flora like *Enterobacter* contributing to air and surface contamination. Environmental contaminants like *Pseudomonas aeruginosa*

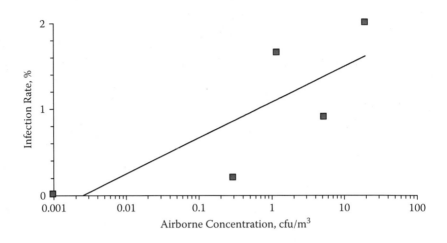

FIGURE 12.2
Plot of *Aspergillus* airborne levels versus infection rates, assuming 0% infections to be 0.001%
for the sake of plotting the data.

can also infiltrate hospitals and contaminate ORs. Airborne microorganisms
that settle on and contaminate equipment and surfaces in the OR may result
in direct contact infections.

The number of personnel in an operating suite influences the total counts
of airborne bacteria (Hambreaus, Bengtsson, and Laurell 1977). There is a
direct relationship between the number of personnel in a surgical suite
and the airborne bacteria (Hart 1938; Duvlis and Drescher 1980; Boyce et
al. 1990; Moggio et al. 1979; Kundsin 1976). Conversation among personnel can increase the bacterial load of the air, and contaminated facemasks
(measured postoperatively) occur in 9–10% of surgeons and nurses (Ritter
1984; Dubuc, Guimont and Ferland 1973). Streifel (1996) reviewed the literature on aspergillosis infection rates and summarized the associated levels (Sherertz et al. 1987; Arnow et al. 1991; Rhame et al. 1984; Petersen et
al. 1983). Figure 12.2 graphically illustrates the relation between infection
rates as the contamination level in the air increases. Data suggest that a
threshold may exist at about 0.1 cfu/m³ below which infection rates may
approach zero. Figure 12.3 shows an operation in progress in which ten
people, counting the patient, were in the OR.

Because microbes aerosolize and settle continuously during occupation
and activity, the amount of surface contamination is also likely to be related
to the level of airborne contamination. The majority of airborne bacteria in
ORs is primarily endogenous human microbes that cannot survive indefinitely on OR surfaces, and results suggest a direct relationship between airborne bacterial levels and occupancy. Various studies indicate that surgeons,
nurses, and patients are the source of such microbial contamination (Alford
et al. 1973; Bitkover, Marcusson, and Ransjo 2000).

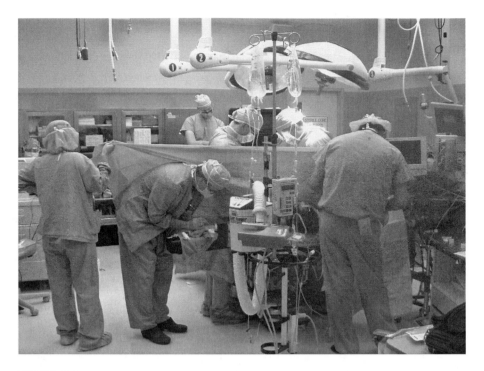

FIGURE 12.3
The number of people in an OR will directly impact the airborne bacteria levels. There are ten people in this room, counting the patient and the author (as an observer).

An individual may shed from 3,000 to 50,000 microorganisms per minute depending on activity and the effectiveness of protective clothing. The microbes most frequently identified in operating room air include *Staphylococcus epidermis* and *S. aureus* (Nelson 1978). *Streptococcus pyogenes* was found in about 12–18% of preoperative throat swabs from patients (Dubuc, Guimont, and Ferland 1973).

Foreign guidelines suggest that airborne microbial levels in an operating room should be kept below 10 cfu/m^3 during procedures (Wilson 2001; Gruendemann and Stonehocker 2001). Levels of airborne contaminants in ORs are typically lower than in the general wards, but not significantly so. For hospital air, WHO recommends the limits of 100 cfu/m^3 for bacteria and 50 cfu/m^3 for fungi (WHO 1988). There are currently no standards for OR aerobiology in the United States, but the current standard used in China is 200 cfu/m^3, while the EU has suggested a limit of 10 cfu/m^3, based on the ISO Class 7 cleanroom limit (EU Grade B) used in the pharmaceutical industry and as a target for ultraclean ventilation systems (Durmaz et al. 2005; Kowalski 2007). For high-risk procedures such as hip arthroplasty, air purity of less than 1 cfu/ft^3 (0.028 cfu/m^3) has been suggested (Hardin and Nichols 1995).

Airborne SSIs

The degree to which airborne concentrations are associated with SSIs has not been well studied and no estimates can be made as to the percentage of SSIs that result from airborne transmission, but it is certainly not zero. Walter, Kundsin, and Brubaker (1963) demonstrated airborne infection in a surgical suite due to *S. aureus*. Other evidence for the role of airborne transmission comes from studies of laminar and other airflow systems (Charnley 1972; Bradley, Enneking, and Franco 1975; Sanderson and Bentley 1976; Salvati et al. 1982). The use of ultraclean air and ultraviolet irradiation in the OR was found to be of benefit in orthopedic surgery for refined clean sites only, but not for the other wound classes (NAS 1964; Lidwell et al. 1983). Clearly, if a wound is clean to begin with it may become contaminated by airborne microbes in the OR, but if a wound is already partly or completely contaminated then increased OR air cleanliness will be unable to make a difference, unless the air could simultaneously disinfect deep tissue. The laminar airflow system of Charnley-Howorth provides direct downward flow of air that spreads out upon reaching the operating table and is considered to be the most effective ventilation system for operating rooms (Howorth 1985; Persson and van der Linden 2004). It has been noted that when a surgeon leans over a wound in the downward airflow of a Charnley-Howorth system, wound contamination increases significantly, apparently from the airstream stripping bacteria from the surgeon's face and neck and then depositing them on the wound when the air currents change.

Lidwell et al. (1983) found that infection rates in joint replacement surgery correlated with the airborne concentrations of bacteria near the wound. The relationship can be quantified from field data. Figure 12.4 plots the rate of joint sepsis versus the airborne bacterial count in operating rooms from six hospitals based on data from Lidwell et al. (1983). The data have been fit to a logarithmic equation as shown.

Friberg, Fribert, and Burman (1999) found that airborne concentrations of bacteria were strongly correlated with surface concentration in the wound area. Bacterial contamination of the surgical field occurs by one of two primary routes, direct contact and airborne (Schonholtz 1976). It has been suggested by Howorth (1985) that 80–90% of bacterial contaminants found in operative wounds gain access to the wound by the airborne route, although other researchers found evidence that airborne contamination plays a much smaller role, especially for contaminated wounds (Lowell et al. 1980; Moggio et al. 1979; Lowbury and Lidwell 1978). The exact contribution to infections in surgery is difficult to isolate from other factors and can require extensive multiyear studies. Based on the results of ultraviolet air disinfection reported by Hart et al. (1968), the contribution of airborne infections would appear to be about 6–7% maximum, including both clean and contaminated surgical wounds. Hoffman, Bennett, and Scott (1999) suggest that 10% of

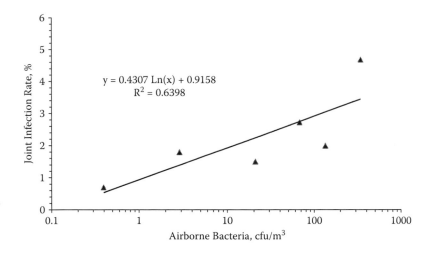

FIGURE 12.4
The relationship between airborne bacteria and the incidence of joint sepsis. Data adapted from Lidwell et al. (1983).

staphylococci are transmitted by the aerosol route, based on the results of Mortimer et al. (1966).

The danger of airborne bacteria causing SSIs is thought to be especially great for orthopedic implant procedures due to the fact that infections can arise from a small number of bacteria. A survey by Miner et al. (2005) of 411 US hospitals performing total knee replacement procedures found that 30% use laminar airflow, 42% use total body exhaust, and 5% use ultraviolet lights, sometimes in combination.

A system of classifying operative sites by the degree of contamination was introduced by the National Research Council during its cooperative study on the effects of ultraviolet irradiation in operating rooms on surgical site infections (NAS 1964). The basic classification scheme is shown in Table 12.2. The surgical site infection rates in the National Research Council cooperative study were 3.3% for clean sites, 7.4% for clean contaminated sites, 16.4% for contaminated sites, and 28.6% for dirty sites. As mentioned previously, only the clean sites (Refined Clean) are amenable to air disinfection (i.e., with UVGI), because the other classes are already contaminated, and sterile air cannot decontaminate surgical sites.

Airborne SSI Etiology

Exogenous pathogens may contaminate wound sites from the hands of HCWs by direct inoculation, and other body sites (i.e., hair, head, neck, nares,

TABLE 12.2

Surgical Site Classification

Class	Description	Airborne Transmission
Clean sites (refined clean)	Surgical sites that are primarily closed, have no inflammation, genital and urinary tracts are not entered.	Plays a role
Clean-contaminated sites	Surgical sites in which the respiratory, alimentary, genital, or urinary tract is entered under controlled conditions and without unusual contamination.	No impact
Contaminated sites	These include open, fresh accidental wounds or operations with major breaks in sterile technique, and sites in which acute, nonpurulent inflammation is encountered.	No impact
Dirty and infected sites	These include old traumatic wounds with retained devitalized tissue, foreign bodies, fecal contamination, and where a perforated viscus or pus is encountered during the operation.	No impact

oropharynges) may be sources for exogenous contamination. *Staphylococcus* and *Streptococcus* were found in infected wound sites and were also isolated from the nasopharynges of operating room personnel (Ha'eri and Wiley 1980). Holton et al. (1991) demonstrated a significant decrease in infection rates in hemodialysis patients by eradicating nasal carriage using mupirocin. *Streptococcus pyogenes* has been found in about 15% of preoperative throat swabs from patients (Dubuc, Guimont, and Ferland 1973).

The largest source of airborne microorganisms in the OR is the staff, but the floor also serves as a reservoir for settled microbes (Walter and Kundsin 1960). Conversation among operating room personnel also increases airborne concentrations of bacteria (Letts & Doermer 1983; Ritter 1984). Talking can aerosolize microbes, and microbes are also shed continuously from personnel at rates that vary individually. Breaking wind in the OR can also aerosolize microbes from the intestine, and anal carriers of Group A *Streptococcus* have been implicated (Schaffner et al. 1969; Sula 2002). Vaginal carriers have also been associated with OR SSIs (Stamm et al. 1978).

Most of the bacteria in surgical site infections are shed from the skin or are attached to squames or particulate matter less than 5 microns in size (Hardin and Nichols 1995; Noble, Lidwell, and Kingston 1963; Noble 1975). These particles readily remain airborne and float easily on turbulent air currents. A single disperser is potentially able to infect a considerable number of other patients, and staphylococci are routinely inhaled or transferred by direct contact (Williams 1966).

It is likely that pathogens are introduced during surgery, as well as during the postoperative period. Brown et al. (1996) examined the routes by which bacteria might enter a wound indirectly by contamination of instruments during skin preparation and draping, and found that airborne concentrations of bacteria were two to four times higher during preparations for hip or knee arthroplasty than during surgery. Bacteria can also contaminate the surgical knife (Ritter et al. 1975).

Spores of *Clostridium perfringens* have been isolated from the ventilation system and floors of operating rooms (Fredette 1958). Contaminated floors from tracked dirt and accumulated debris could become an internal source for *C. perfringens* spores or other spores (Clark 1985). On rare occasions where the inanimate environment of the OR has been implicated in SSIs, the sources have been contaminated antiseptics, solutions, or dressings. In these latter cases the suspect microbes have included *Rhizopus, C. perfringens, Pseudomonas,* and *Serratia marcescens* (Wong 1999). It is widely considered that wet mopping of the operating room floor with a disinfectant between cases is sufficient to reduce the risk of the OR surfaces causing SSIs. In wounds that have been contaminated by soil, *Clostridium* species are the only anaerobic bacteria that affect humans (Davis and Shires 1991).

Dust particles, skin squames, and respiratory secretions from operating room personnel can increase bacterial counts in the air. These particles may settle quickly and contaminate operative sites located a short distance from the contaminant source (Wong 1999). Some 10% of skin squames carry bacteria (Noble 1975).

More than 90% of bacteria contaminating clean surgical wounds come from the ambient air, and a substantial part of these bacteria contaminate the wound directly during clean surgery in conventionally ventilated operating rooms (Whyte, Hodgson, and Tinkler 1982). Direct airborne contamination results from deposition of airborne microbes on the wound, while indirect contamination results from airborne bacteria settling on surfaces outside the wound, which are then transferred via hands or surgical instruments to the wound (Persson and van der Linden 2004).

Control of Airborne SSIs

The effect of clean air delivery has been studied by Aglietti et al. (1974). The air flowing over the wounds of patients operated on in an enclosure with horizontal filtered airflow contained five times fewer microorganisms than air near the wound in an enclosure without laminar airflow. Wound cultures showed significantly less contamination when the filtered laminar airflow was used. One of the drawbacks in horizontal flow systems is that if the surgical team is not covered appropriately, endogenous microbes from the team may shed onto the open wound.

Airborne transmission is thought to be a primary transmission mechanism among patients undergoing bone marrow transplants (Perl, Chotani, and

Agawala 1999). In a study on the distribution of air documenting increased airflow to rooms with a higher infection rate, Leclair et al. (1982) suggested that increased quantities of viruses and droplets were expelled from the exhaust loop of a ventilator. Positive pressure in rooms with patients shedding VZV can also disseminate airborne VZV.

Procedural methods to reduce airborne concentrations of bacteria include reducing the number of personnel in the OR, using proper attire, and minimizing activity and conversation. Engineering controls for reduction of airborne bacteria include increasing outside airflow and increasing total filtered airflow and air distribution (ASHRAE 2003). Laminar airflow systems and UVGI will further decrease airborne contamination to very low levels, sometimes called ultraclean air (Wong 1999). Ultraclean air may only be of direct benefit for clean wounds, as noted earlier.

Infections during surgery are affected by the size of the wound and by the duration of the procedure (Mangram et al. 1999). Both these factors may be related to the degree of airborne contamination. Increased duration of surgery tends to increase the infection rate, as shown in Figure 12.5, which is based on data from Mead, Pories, and Hall (1986).

Bone marrow transplant patients are at similar risk for infections as other surgical patients, and these can be classified as endogenous factors and exogenous factors. Endogenous microorganisms are likely transmitted by direct routes such as self-inoculation, but exogenous sources of transmission depend on the surrounding environment.

Fungal spores are environmental, and if they contaminate the hospital environment they can cause severe infections. Infections caused by *Aspergillus*

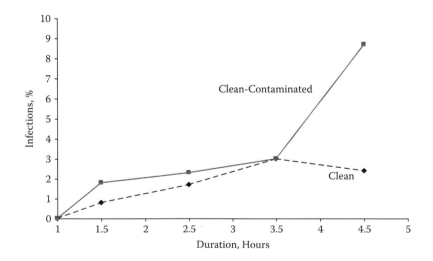

FIGURE 12.5
Duration of surgery versus infection rate for Clean and Clean-Contaminated procedures. Based on data from Mead, Pories, and Hall (1986).

flavus and *A. fumigatus* are related to dispersion of mold spores via dust (Arnow et al. 1991). Rhame et al. (1984) reported that 5.4% of bone marrow transplant patients developed invasive *Aspergillus* infections when the mean concentration in the air was 0.9 cfu/m^3. It would require a failure of the ventilation system filters to allow such spores to enter the operating room and their presence should be monitored to ensure system performance. Placing settle plates in empty ORs would be sufficient monitoring. Entry of environmental spores into an operating room may be by other routes, such as the floors or on the shoes and clothes of personnel. Sterilization or disinfection of floors outside ORs, and disinfection of footwear, may be one means of controlling environmental pathogen intrusion into ORs. Barnes and Rogers (1989) report on an outbreak of aspergillosis that was brought under control with laminar flow isolation. Invasive *Aspergillus* infections in an ICU have been attributed to spores accumulating in fibrous insulation material in the ceiling cavity and others to growth on air filters (Vesley and Streifel 1999; Arnow et al. 1991). Environmental fungal spores have been found in hospital systems including in the ventilation system, air conditioners, insulation, fireproofing material, blankets, and in demolition and construction areas. Once fungal spores contaminate a site, pathogenic fungi characteristically grow as yeast.

The impact of airborne contamination on the success of hip arthroplasty was studied in depth by Charnley (1972) who developed increasingly sophisticated means for isolating the HCWs from the air bathing the operative site. He achieved some success with laminar flow ventilation systems and the use of total body exhaust suits. An extensive prospective study of over 8000 hip and knee replacements concluded that laminar airflow, body exhaust gowns, and prophylactic antibiotic administration each contributed to a reduction in infection rates (Lidwell et al. 1982, 1987). Current infection rates for hip arthroplasty approach those in other types of clean surgical procedures, or about 1.5%. Ultraviolet light systems have also been used to control infections during hip arthroplasty and have been shown to reduce infection in clean surgical operations (Carlsson et al. 1986; Lowell and Kundsin 1980). Some of the microbes isolated in infections of total knee arthroplasty include *Staphylococcus aureus*, *S. epidermis*, *Streptococcus*, *Klebsiella*, *Enterobacter*, *Pseudomonas*, and *Enterococcus* (Rivero 1995). Some of the unusual microbes isolated in infections of total joint replacements include *Brucella*, *Listeria monocytogenes*, *M. fortuitum*, *Moraxella*, and *Yersinia*.

If the ventilation system is working per design, the exposed skin of the patient and HCWs becomes the next most important source of contaminants in the surgical setting (Whyte et al. 1992). Allo et al. (1987) reports on a contaminated ventilation system in an operating room that resulted in contamination of intravenous catheters. In such cases airborne spores may settle on catheters at some point from the air or else contact a contaminated surface or a hand. In any event, the normal requirements for operating room air filtration should certainly be sufficient to keep out *Aspergillus* spores.

TABLE 12.3

Typical Airborne Concentrations in Operating Rooms

Mean Level (cfu/m³)	Ventilation Type	Contaminant	Reference
7	Laminar	Bacteria	Ritter et al. 1975
7.7	Laminar	Bacteria	Berg, Bergman, and Hoborn 1991
19	Jourbert system	Bacteria	Luciano 1984
22	Laminar	Bacteria	Friberg and Friberg 2005
23	Conventional	Bacteria	Bergeron et al. 2007
24	Conventional	Bacteria	Berg, Bergman, and Hoborn 1989
28	Conventional	Bacteria	Nelson 1978
29	Ultraclean	Bacteria	Brown et al. 1996
35	Conventional	Bacteria	Lidwell 1994
46	OR zonal ventilation	Bacteria	Hambraeus, Bengtsson, and Laurell 1977
65	Conventional/curtain	Bacteria	Lowbury and Lidwell 1978
74	Conventional	Bacteria	Hambraeus, Bengtsson, and Laurell 1977
74	Conventional	Bacteria	Tighe and Warden 1995
52	Conventional	Fungi	Tighe and Warden 1995

Many typical ORs have a sterile core outside the ORs from which equipment is staged. Filtered air is supplied directly to both the sterile core and to the ORs. Air is exhausted from the ORs and from the surrounding hallways. Pressure relationships are typically monitored continuously through the use of wall-mounted pressure sensors. Supply air is filtered, often with a HEPA filter, but this does not guarantee sterility of the air in the OR. Published data on airborne levels in ORs indicates that the air is far from sterile, and is often no cleaner than the air in the general wards. Simply conforming to existing guidelines for ventilation rates and filter recommendations is no guarantee that air will be sterile. In fact, few hospitals ever monitor the air quality in ORs and so hospitals do not, in general, know whether the air in their ORs is contaminated or not.

Table 12.3 lists a sampling of data on airborne concentrations in ORs, broken down by the type of ventilation system used. The laminar and ultraclean type ventilation systems generally perform better on average than conventional systems in reducing mean airborne levels of bacteria (Neson et al. 1973). A study of ultraclean protective environments, ORs with exceptionally clean air, coupled with sterile garb and other measures showed a statistically significant reduction in overall infection rates in five of ten trials (Bennett, Jarvis, and Brachmen 2007).

Air supplied to ORs, especially through HEPA filters, may be highly disinfected, but most of the airborne bacteria come from room occupants, including the patient, and so increasing the rates of supply air above design guidelines brings only diminishing returns, often at high economic cost. High air exchange rates do not necessarily translate into lower levels of airborne bacteria.

Modeling airflow in ORs can be developed based on the dimensions, the design airflows, and the measured airborne concentrations of microbes. Such a model can compute the airborne concentrations minute by minute based on two types of contamination sources. The first source is the ambient environment and it produces airborne contamination during unoccupied periods. The source may be already present in the room, or may come from infiltration, or it may be delivered with the air supply. The occupants are the second source of contaminants.

Modeling a specific operating room using a completely mixed single-zone model is fairly straightforward given the dimensions and the design airflows. This model is based on an actual OR with a volume of 1528 ft^3 and 1650 cfm of HEPA-filtered outside air, making for almost 20 ACH. The ambient level proved to be 25 cfu/m^3 unoccupied with the ventilation system operating, which implies an ambient release rate of 1170 cfu/min. It could be assumed that this contamination comes from infiltration, room contaminants being aerosolized by air currents and activity, and imperfectly filtered supply air. If the latter is true then it can be seen that the high flow rate, 20 ACH, could be counterproductive. Data from 13 operations in the same or similar ORs led to an estimated release rate of 1521 cfu/min per occupant (Kowalski 2009). This empirical release rate is relatively low based on studies of the number of bacteria produced by humans (Sherertz et al. 2001). The model presented here computes the airborne concentrations minute by minute over the course of a two-hour operation in which personnel entered and exited the OR. Per this model, the airborne levels of bacteria will constantly rise and fall during the course of an operation, as shown in Figure 12.6.

Overhead surgical site UV systems have been in use in some operating rooms since at least 1936 (Hart and Sanger 1939; Brown et al. 1996). Duke University has successfully used overhead UVGI systems since 1940 to maintain a low level of orthopedic infections (Lowell et al. 1980; Goldner and Allen 1973). One source suggests that overhead UV can reduce airborne microbial concentrations to below 10 cfu/m^3 in the operating room (Berg, Bergman, and Hoborn 1991; Berg-Perier, Cederblad, and Persson 1992). A 49% decrease in airborne bacteria with an overhead UV system was demonstrated by Moggio et al. (1979). Lowell and Kundsin (1980) reports on an overhead UV system that produced a 99–100% decrease in aerosolized *E. coli*, which produced a 54% decrease in airborne bacteria during surgical procedures.

The overhead UV surgical system implemented by Ritter, Olberling, and Malinzak (2007) was able to reduce the surgical site infection rate from 1.77% to 0.5%. Overhead UV systems inhibit both airborne transport and survival of bacteria on surfaces, including microbes that settle on equipment, on personnel, and on floors. These systems will be addressed in more detail in Chapter 20.

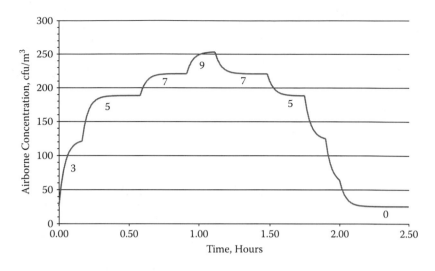

FIGURE 12.6
Airborne levels in an OR with 20 ACH computed based on release rates for occupants. Number below line indicates the number of occupants.

Postoperative wound infections may be limited to some degree by over-head UV exposure of the wound (Wright and Burke 1969). UV systems were tried as an adjunct at Duke University Hospital in the control of postoperative surgical infections from 1938 to 1948, and maintained an incidence rate of 1.39% in this period (Woodhall, Neill, and Dratz 1949). The postoperative infections largely involved *S. aureus* and *S. albus*.

References

Aglietti, P., Salvati, E. A., Wilson, P. D., and Kutner, L. J. (1974). Effect of a surgical horizontal unidirectional filtered air flow unit on wound bacterial contamination and wound healing. *Clin Orthop* 101, 99.

Alford, D. J., Ritter, M. A., French, M. L., and Hart, J. B. (1973). The operating room gown as a barrier to bacterial shedding. *Am J Surg* 125, 589–591.

Allo, M. D., Miller, J., Townsend, T., and Tan, C. (1987). Primary cutaneous aspergillosis associated with Hickman intravenous catheters. *N Engl J Med* 317(18), 1105–1108.

Arnow, P. M., Sadigh, M., Costas, C., Weil, D., and Chudy, R. (1991). Endemic and epidemic aspergillosis associated with in-hospital replication of *Aspergillus* organisms. *J Infect Dis* 164, 998–1002.

ASHRAE (2003). *HVAC Design Manual for Hospitals and Clinics.* American Society of Heating, Ventilating, and Air Conditioning Engineers, Atlanta, GA.

Ayliffe, G. A. J. (1991). Role of the environment of the operating suite in surgical wound infection. *Rev Infect Dis* 13(Suppl 10), 800–804.

Barnes, R. A., and Rogers, T. R. (1989). Control of an outbreak of nosocomial aspergillosis by laminar air-flow isolation. *J Hosp Infect* 14(2), 89–94.

Bennett, J. V., Jarvis, W. R., and Brachman, P. S. (2007). *Bennett & Brachman's Hospital Infections.* Lippincott Williams & Wilkins, Philadelphia.

Berg, M., Bergman, B. R., and Hoborn, J. (1989). Shortwave ultraviolet radiation in operating rooms. *J Bone Joint Surg* 71-B(3), 483–485.

_____. (1991). Ultraviolet radiation compared to an ultra-clean air enclosure. Comparison of air bacteria counts in operating rooms. *JBJS* 73(5), 811–815.

Berg-Perier, M., Cederblad, A., and Persson, U. (1992). Ultraviolet radiation and ultra-clean air enclosures in operating rooms. *J Arthroplasty* 7(4), 457–463.

Bergeron, V., Reboux, G., Poirot, J. L., and Laudinet, N. (2007). Decreasing airborne contamination levels in high-risk hospital areas using a novel mobile air-treatment unit. *Inf Contr Hosp Epidemiol* 28(10), 1181–1186.

Bitkover, C. Y., Marcusson, E., and Ransjo, U. (2000). Spread of coagulase-negative staphylococci during cardiac operations in a modern operating room. *Ann Thorac Surg* 69(4), 1110–1115.

Blakemore, W. S., McGarrity, G. J., Thurer, R. J., Wallace, H. W., MacVaugh, H. I., and Coriell, L. L. (1971). Infection by air-borne bacteria with cardiopulmonary bypass. *Surgery* 70, 830–838.

Boyce, J. M., Potter-Bynoe, G., Opal, S. M., Dziobek, L., and Medeiros, A. A. (1990). A common source outbreak of *Staphylococcus epidermis* infections among patients undergoing cardiac surgery. *J Infect Dis* 161, 493–499.

Bradley, L. P., Enneking, W. F., and Franco, J. A. (1975). The effect of operating-room environment on the infection rate after Charnley low-friction hip replacement. *J Bone Joint Surg* 57, 80–83.

Brown, I. W., Moor, G. F., Hummel, B. W., Marshall, W. G., and Collins, J. P. (1996). Toward further reducing wound infections in cardiac operations. *Ann Thorac Surg* 62(6), 1783–1789.

Carlsson, A. S., Nilsson, B., Walder, M. H., and Osterberg, K. (1986). Ultraviolet radiation and air contamination during total hip replacement. *J Hosp Infect* 7, 176–184.

CDC (1993). National Nosocomial Infections Surveillance (NNIS) Report, data summary from 1990–1992. Centers for Disease Control, Atlanta, GA.

Charnley, J. (1972). Postoperative infection after total hip replacement with special reference to air contamination in the operating room. *Clin Orthop* 87, 167–187.

Clark, R. P. (1985). Ventilation conditions and air-borne bacteria and particles in operating theatres: Proposed safe economies. *J Hyg* 95, 325–335.

Cooper, E. E., O'Reilly, M. A., and Guest, D. I. (2003). Influence of building construction work on *Aspergillus* infection in a hospital setting. *Inf Contr Hosp Epidemiol* 24, 472–476.

Davis, J. M., and Shires, G. T. (1991). *Principles and Management of Surgical Infections.* J. B. Lippincott Company, New York.

Dubuc, F., Guimont, A., Roy, L., and Ferland, J. J. (1973). A study of some factors which contribute to surgical wound contamination. *Clin Ortho* 96, 176–178.

Duhaime, A. C., Bonner, K., McGowan, K. T.., Schut, L., Sutton, L. N., and Plotkin, S. (1991). Distribution of bacteria in the operating room environment and its relation to ventricular shunt infections: A prospective study. *Childs Nerv Syst* 7, 211–214.

Durmaz, G., Kiremitci, A., Akgun, Y., Oz, Y., Kasifoglu, N., Aybey, A., and Kiraz, N. (2005). The relationship between airborne colonization and nosocomial infections in intensive care units. *Mikrobiyol Bul* 39(4), 465–471.

Duvlis, Z., and Drescher, J. (1980). Investigations on the concentration of air-borne germs in conventionally air-conditioned operating theaters. *Zentralbl Bakteriol B* 170(1-2), 185–198.

Fletcher, L. A., Noakes, C. J., Beggs, C. B., and Sleigh, P. A. (2004). The importance of bioaerosols in hospital infections and the potential for control using germicidal ultraviolet radiation. *Proceedings of the 1st Seminar on Applied Aerobiology*, Murcia, Spain.

Fredette, V. (1958). The bacteriologic efficiency of air-conditioning systems in operating-rooms. *Can J Surg* 1, 226–229.

Friberg, B., and Friberg, S. (2005). Aerobiology in the operating room and its implications for working standard. *Proc Inst Mech Eng* 219(2), 153–160.

Friberg, B., Friberg, S., and Burman, L. G. (1999). Correlation between surface and air counts of particles carrying aerobic bacteria in operating rooms with turbulent ventilation: An experimental study. *J Hosp Infect* 42, 61–68.

Goldner, J. L., and Allen, B. L. (1973). Ultraviolet light in orthopedic operating rooms at Duke University. *Clin Ortho* 96, 195–205.

Gruendemann, B. J., and Stonehocker, S. (2001). *Infection Prevention in Surgical Settings.* W. B. Saunders, Philadelphia.

Ha'eri, G. B., and Wiley, A. M. (1980). The efficacy of standard surgical face masks: An investigation using tracer particles. *Clin Orthop* 148, 160–162.

Hambraeus, A., Bengtsson, S., and Laurell, G. (1977). Bacterial contamination in a modern operating suite. *J Hyg* 79, 121–132.

Hardin, W. D. J., and Nichols, R. L. (1995). Aseptic technique in the operating room; in *Surgical Infections*, D. E. Fry, ed., Little, Brown and Company, Boston, 109–117.

Hart, D. (1938). Pathogenic bacteria in the air of operating rooms. *Arch Surg* 38(5), 521.

Hart, D., and Sanger, P. W. (1939). Effect on wound healing of bactericidal ultraviolet radiation from a special unit: Experimental study. *Arch Surg* 38(5), 797–815.

Hart, D., Postelthwait, R., Brown, I., Smith, W., and Johnson, P. (1968). Postoperative wound infections: A further report on ultraviolet irradiation with comments on the recent 1964 National Research Council Cooperative report. *Ann Surg* 167(5), 728–743.

Hoffman, P. N., Bennett, A. M., and Scott, G. M. (1999). Controlling airborne infections. *J Hosp Infect* 43(Suppl), 203–210.

Holton, D. L., Nicholle, L. E., Diley, D., and Bernstein, K. (1991). Efficacy of mupirocin nasal ointment in eradicating *Staphylococcus aureus* nasal carriage in chronic hemodialysis patients. *J Hosp Infect* 17, 133–137.

Howorth, F. H. (1985). Prevention of airborne infection during surgery. *Lancet* 1, 386.

Kowalski, W. J. (2007). Air-treatment systems for controlling hospital-acquired infections. *HPAC Engineering* 79(1), 28–48.

Kowalski, W. J. (2009). *Ultraviolet Germicidal Irradiation Handbook: UVGI for Air and Surface Disinfection.* Springer, New York.

Kundsin, R. (1976). Operating Room as a Source of Wound Contamination and Infection. *Workshop on Control of Operating Room Airborne Bacteria, Committee on Prosthetic Research and Development, Committee on Prosthetic Orthotic Education, Assembly of Life Sciences*, 167–172.

Leclair, J. M., Zaia, J. A., Levin, M. J., Congdon, R. G., and Goldman, D. A. (1982). Airborne transmission of chickenpox in a hospital. *N Engl J Med* 302, 450–453.

Letts, R. M., and Doermer, E. (1983). Conversation in the operating theater as a cause of airborne bacterial contamination. *JBJS* 65(3), 357–362.

Lidwell, O. M. (1994). Ultraviolet radiation and the control of airborne contamination in the operating room. *J Hosp Infect* 28, 245–248.

Lidwell, O. M., Elson, R.A., Lowbury, E. J. L., Blowers, R., Stanley, S. J., and Lowe, D. (1987). Ultraclean air and antibiotics for prevention of postoperative infection. *Acta Orthop Scand* 58, 4–13.

Lidwell, O. M., Lowbury, E. J., Whyte, W., Blowers, R., Stanley, S. J., and Lowe, D. (1982). The effect of ultraclean air in operating rooms on deep sepsis in the joint after total hip and total knee replacement: A randomized study. *Br Med J (Clin Red Ed)* 285(6334), 10–14.

———. (1983). Airborne contamination of wounds in joint replacement operations: The relationship to sepsis rates. *J Hosp Infect* 4, 111–131.

Lowbury, E. J., and Lidwell, O. M. (1978). Multi-hospital trial on the use of ultraclean air systems in orthopaedic operating rooms to reduce infection: Preliminary communication. *J Royal Soc Med* 71, 800–806.

Lowell, J., and Kundsin, R. (1980). Ultraviolet Radiation: Its Beneficial Effect on the Operating Room Environment and the Incidence of Deep Wound Infection Following Total Hip and Total Knee Arthroplasty. *A-810*, American Ultraviolet Company, Murray Hill, NJ.

Lowell, J. D., Kundsin, R. B., Schwartz, C. M., and Pozin, D. (1980). Ultraviolet radiation and reduction of deep wound infection following hip and knee arthroplasty; in *Airborne Contagion*, Annals of the New York Academy of Sciences, R. B. Kundsin, ed., NYAS, New York, 285–293.

Luciano, J. R. (1984). New concept in French hospital operating room HVAC systems. *ASHRAE Journal* Feb., 30–34.

Mandell, G. L., Gerald, L., Bennett, J. E., and Dolin, R. (2000). *Principles and Practice of Infectious Diseases.* Churchill Livingstone, Philadelphia.

Mangram, A. J., Horan, T. C., Pearson, M. L., Silver, L. C., Jarvis, W. R., and HICPAC (1999). Guideline for prevention of surgical site infection. *Am J Infect Control* 27(2):97–134.

Mead, P. B., Pories, S. E., and Hall, P. (1986). Decreasing the incidence of surgical wound infections. *Arch Surg* 121, 458.

Miner, A. L., Losina, E., Katz, J. N., Fossel, A. H., and Platt, R. (2005). Infection control practices to reduce airborne bacteria during total knee replacement: A hospital survey in four states. *Inf Contr Hosp Epidemiol* 26(12), 910–915.

Moggio, M., Goldner, J. L., McCollum, D. E., and Beissinger, S. F. (1979). Wound infections in patients undergoing total hip arthroplasty. Ultraviolet light for the control of airborne bacteria. *Arch Surg* 114(7), 815–823.

Mortimer, E. A., Wolinsky, E., Gonzaga, A. J., and Rammelkamp, C. H. (1966). Role of airborne transmission in staphylococcal infections. *Br Med J* 5483, 319–322.

NAS (1964). Postoperative wound infections: The influence of ultraviolet irradiation of the operating room and various other factors. *Ann Surg* 160(Suppl), 1.

Nelson, P. J. (1978). Clinical use of facilities with special air handling equipment. *Hosp Topics* 57(5), 32–39.

Neson, J. P., Glassburn, A. R., Talbott, R. D., and McElhinney, J. P. (1973). Clean room operating rooms. *Clin Ortho* 96, 179–187.

Noble, W. C. (1975). Dispersal of skin microorganisms. *Br J Dermatol* 93, 477–485.

Noble, W. C., Lidwell, O. M., and Kingston, D. (1963). The size distribution of airborne particles carrying micro-organisms. *J Hyg* 61, 385–391.

Perl, T. M., Chotani, R., and Agawala, R. (1999). Infection control and prevention in bone marrow transplant patients; in *Hospital Epidemiology and Infection Control*, C. G. Mayhall, ed., Lippincott Williams & Wilkins, Philadelphia, 803–844.

Persson, M., and van der Linden, J. (2004). Wound ventilation with ultraclean air for prevention of direct contamination during surgery. *Inf Contr Hosp Epidemiol* 25(4), 297–301.

Petersen, P. K., McGlave, P., Ramsay, N. K., Rhame, F. S., Cohen, E., Perry, G. S., Goldman, A. I., and Kersey, J. (1983). A prospective study of infectious diseases following bone marrow transplantation: Emergence of *Aspergillus* and *Cytomegalovirus* as the major causes of mortality. *Infect Control* 42(2), 81–89.

Rhame, F., Streifel, A., Kersey, J., and McGlave, P. (1984). Extrinsic risk factors for pneumonia in the patient at high risk of infection. *Am J Med*, 42–52.

Ritter, M. A. (1984). Conversation in the operating theatre as a cause of airborne bacterial contamination. *J Bone Joint Surg Am* 66(3), 472.

Ritter, M. A., Eitzen, H. E., French, M. L. V., and Hart, J. B. (1975). The operating room environment as affected by people and the surgical face mask. *Clin Ortho* 111, 147–150.

Ritter, M. A., French, M. L., and Eitzen, H. E. (1975). Bacterial contamination of the surgical knife. *Clin Ortho* 108, 158–160.

Ritter, M., Olberding, E., and Malinzak, R. (2007). Ultraviolet lighting during orthopaedic surgery and the rate of infection. *J Bone Joint Surg* 89, 1935–1940.

Rivero (1995). Infections of total joint prostheses; in *Surgical Infections*, D. E. Fry, ed., Little, Brown and Company, Boston, 407–414.

Salvati, E. A., Robinson, R. P., Zeno, S. M., Koslin, B. L., Brause, B. D., and Wilson, P. D. (1982). Infection rates after 3175 total hip and total knee replacements performed with and without a horizontal unidirectional filtered air-flow system. *Br J Surg* 64(4), 525–535.

Sanderson, M. C., and Bentley, G. (1976). Assessment of wound contamination during surgery: A preliminary report comparing vertical laminar flow and conventional theatre systems. *Br J Surg* 63, 431–432.

Schaffner, W., Lefkowicz, L. B., Goodman, J. S., and Koenig, M. G. (1969). Hospital outbreak of infections with Group A streptococci traced to an asymptomatic anal carrier. *N Engl J Med* 280, 1224–1225.

Schonholtz, G. H. (1976). Maintenance of aseptic barriers in the conventional operating room. *J Bone Joint Surg Am* 58(Suppl A), 439.

Sherertz, R. J., Belani, A., Kramer, B. S., Elfenbein, G. J., Weiner, R. S., Sullivan, M. L., Thomas, R. G., and Samsa, G. P. (1987). Impact of air filtration on nosocomial *Aspergillus* infections. *Amer J Medicine* 83, 709–718.

Sherertz, R. J., Bassetti, S., and Bassetti-Wyss, B. (2001). "Cloud" Health Care Workers. *Emerg Inf Dis* 7(2), 241–244.

Stamm, W. E., Feely, J. C., and Facklam, R. R. (1978). Wound infections due to group A *Streptococcus* traced to a vaginal carrier. *J Infect Dis* 138, 287–292.

Streifel, A. J. (1996). Controlling aspergillosis and *Legionella* in hospitals; in *Indoor Air and Human Health*, R. A. Gammage, CRC Press, Boca Raton, FL.

Sula, M. (2002). Killer on the loose: A gruesome outbreak of flesh-eating bacteria at Evanston Hospital sends doctors scrambling to find the source before other patients get infected. *Reader* 31(33), 1.

Tighe, S. W., and Warden, P. S. (1995). An investigation of microbials in hospital air environments. *Indoor Air Rev* May, 20–22.

Vesley, D., and Streifel, A. J. (1999). Environmental services; in *Hospital Epidemiology and Infection Control*, C. G. Mayhall, ed., Lippincott Williams & Wilkins, Philadelphia, 1047–1053.

Walter, C. W., and Kundsin, R. B. (1960). The floor as a reservoir of hospital infections. *Surg Gynecol Obstet* 111, 1–7.

Walter, C. W., Kundsin, R. B., and Brubaker, M. M. (1963). The incidence of airborne wound infection during operation. *JAMA* 186, 908–913.

WHO (1988). Indoor air quality: Biological contaminants. *European Series 31*, World Health Organization Copenhagen, Denmark.

Whyte, W., Hambreus, A., Laurell, G., and Hoborn, J. (1992). The relative importance of the routes and sources of wound contamination during general surgery. II. Airborne. *J Hosp Infect* 22, 41–44.

Whyte, W., Hodgson, R., and Tinkler, J. (1982). The importance of airborne bacterial contamination of wounds. *J Hosp Infect* 3, 123–135.

Williams, R. E. O. (1966). Epidemiology of airborne staphylococcal infection. *Bact Rev* 30(3), 660–672.

Wilson, J. (2001). *Infection Control in Clinical Practice*. Balliere Tindall, Edinburgh.

Wilson, M. A., and Garrison, R. N. (1995). Infections of intravascular devices; in *Surgical Infections*, D. E. Fry, ed., Little, Brown and Company, Boston, 541–549.

Wong, S. (1999). Surgical site infections; in *Hospital Epidemiology and Infection Control*, C. G. Mayhall, ed., Lippincott Williams & Wilkins, Philadelphia, 189–210.

Woodhall, B., Neill, R., and Dratz, H. (1949). Ultraviolet radiation as an adjunct in the control of post-operative neurosurgical infection. Clinical experience 1938-1948. *Ann Surg* 129, 820–825.

Wright, R., and Burke, J. (1969). Effect of ultraviolet radiation on post-operative neurosurgical sepsis. *J Neurosurgery* 31, 533–537.

13

ICUs and Patient Rooms

Introduction

The aerobiological hazards in intensive care units, patient rooms in the general wards, hospital laboratories, and other areas are addressed in this chapter. Topics addressed here include gastrointestinal illnesses, sinusitis, UTIs, and ocular infections. ICUs may have unique problems and are subject to infections from particular pathogens. Patient rooms and the general wards are often subject to respiratory and other infections. Hospital laboratories create special hazards for workers. Table 13.1 summarizes the various airborne pathogens that may occur in general wards, patient rooms, and ICUs, excluding the respiratory pathogens that were addressed in Chapter 11. One new potential airborne pathogen is included here, in addition to those in Table 4.1: *Listeria monocytogenes*. *Listeria* has been previously suspected of airborne transport and is herein assigned Airborne Class 2.

Intensive Care Units

Intensive care units (ICUs) are similar in design to patient rooms with 6 ACH total and 2 ACH of outdoor air. In ICUs almost a third of nosocomial infections are respiratory in nature, but not all of these are necessarily airborne because many are transmitted through direct contact (Wenzel 1981; Wilson 2001). Pneumonia accounts for approximately one-quarter of infections acquired in ICUs (Richards et al. 1999). The infections that occur most commonly in ICUs occur frequently in general wards as well. Durmaz et al. (2005) found that airborne-related nosocomial infections were higher in anesthesia ICUs than in surgical ICUs, and the microbes isolated most frequently from the air were *S. aureus* and *Acinetobacter*. Wilson, Huang, and McLean (2004) found that airborne MRSA in a hospital was strongly correlated with the presence and number of MRSA colonized patients in the ward.

TABLE 13.1

Airborne Nosocomial Pathogens in General Wards and ICUs

Pathogen	Type	Airborne Class	Source	Infection
Acinetobacter	Bacteria	2	Humans/ environmental	Sinusitis
Adenovirus	Virus	2	Humans	Ocular
Aspergillus	Fungal spore	1	Humans	Ocular
Bacillus cereus	Bacterial spore	2	Environmental	Ocular, FI
Clostridium botulinum	Bacterial spore	2	Environmental	FI
Clostridium difficile	Bacterial spore	1	Humans/ environmental	GI
Clostridium perfringens	Bacterial spore	2	Environmental	FI
Haemophilus influenzae	Bacteria	2	Humans	UTI, ocular
Histoplasma	Fungal spore	1	Environmental	Ocular
Klebsiella	Bacteria	2	Humans	GI, sinusitis
Listeria monocytogenes	Bacteria	2	Environmental	FI
Mycobacterium tuberculosis	Bacteria	1	Humans	Ocular
Neisseria meningitidis	Bacteria	2	Humans	UTI
Norwalk virus	Virus	1	Environmental	GI
Proteus	Bacteria	2	Humans	Sinusitis
Pseudomonas aeruginosa	Bacteria	1	Humans/ environmental	GI, sinusitis, UTI, ocular
Rotavirus	Virus	2	Humans	GI
Staphylococcus aureus	Bacteria	1	Humans	Sinusitis, UTI, ocular, FI
Streptococcus spp.	Bacteria	1	Humans	Sinusitis, UTI, ocular

Note: UTI = urinary tract infections, GI = gastrointestinal infections, FI = food-borne illness.

Sinusitis refers to infections of the sinus that may result in local pain, nasal congestion, and purulent nasal drainage. Nosocomial sinusitis may cause fever and sepsis in mechanically ventilated patients (Bonten 1999). The etiology of sinusitis resembles the spectrum of pathogens that cause other respiratory infections. In community-acquired sinusitis the bacterial isolates include *Haemophilus influenzae, Moraxella catarrhalis, Streptococcus, Poststreptococcus, Staphylococcus, Bacteroides,* and *Enterobacter* (Segal 1995). Bacterial isolates in nosocomial sinusitis include *Pseudomonas aeruginosa, Staphylococcus aureus, Klebsiella, Streptococcus, Bacteroides, E. coli, Proteus, Acinetobacter, Candida, Morganella,* and *Poststreptococcus*. Nasotracheal intubation is the most important risk factor for nosocomial sinusitis and colonization of the upper respiratory tract is universal in mechanically ventilated patients. Because the microbes involved are predominantly endogenous microbes that have been brought into direct contact with the sinus through intubation, nosocomial sinusitis must be considered a nonairborne infection.

Nosocomial gastrointestinal (GI) tract infections are distinguished from community-acquired infections by the incubation period (Farr 1999). Typically, if gastroenteritis occurs more than three days after admission it is considered nosocomial, although this distinction may be somewhat arbitrary. Hospital employees may import gastrointestinal pathogens from the community, or they may occur as a result of food contamination. Bacterial isolates include *Clostridium difficile, Salmonella, Shigella,* and *Campylobacter.* The normal intestinal microflora resist colonization by pathogens, but prior antibiotic use can disrupt the microflora and leave a patient open to infection by noncommensal and opportunistic bacteria such as *Pseudomonas, Klebsiella, Clostridium,* and *Candida* species. Transmission of nosocomial GI tract infections can occur from (1) direct patient-to-patient contact, (2) dissemination among patients on the hands of HCWs, (3) environmental contamination and direct or indirect spread, and (4) spread by contaminated medical equipment. Nosocomial gastroenteritis spreads by contact, primarily by the hands of HCWs.

Environmental contamination contributes to transmission of gastrointestinal illness in hospitals, particularly in the case of *Clostridium difficile* (McFarland et al. 1989). *C. difficile* is a spore and can survive on surfaces for as long as five months (Fekety et al. 1981). It is invariably found in proximity to infected patients and is commonly cultured from their rooms. Aerosolization of *C. difficile* spores occurs commonly but sporadically in patients with symptomatic infection (Best et al. 2010). With an aerodynamic mean diameter of about 2 microns, *Clostridium* spores, including *C. perfringens* and *C. botulinum,* are capable of airborne transport. The mean settling time of *Clostridium* spores, based on their size, is about 8 minutes, and because they can easily be stirred up by foot traffic and air currents, these spores will, in time, spread throughout the hospital environment. One study of an air cleaning system in a hospital greatly reduced *Clostridium difficile* infections (Nielsen 2008). Although *Clostridium difficile* is likely ingested or brought into direct contact with a patient to cause an infection, the evidence suggests it is subject to airborne transport in hospital environments. Estimates of terminal velocities of *Clostridium* spores by Snelling et al. (2011) indicate a mean settling time of about 0.4 m/hr. Measurements from Wilcox et al. (2011) suggest the terminal velocity of *Clostridium* spores is about ten times higher, or about 4 m/hr for the average spore. In either case it is clear the spores can remain suspended in air for great lengths of time.

Rotavirus is an enteric pathogen that may cause nosocomial infections in hospitals and is suspected of aerosol transport (Farr 1999; Santosham et al. 1983; Stals, Walther, and Bruggeman 1984). The frequency of simultaneous respiratory symptoms and the persistence of the virus in fomites are also suggestive of airborne spread. It can survive up to 10 days in fomites on inanimate surfaces (Sattar, Lloyd-Evans, and Springthorpe 1986).

A number of reports have speculated that nosocomial viral gastroenteritis may be transmitted by aerosol and these early reports likely involve

Norwalk virus (or Norwalk-like viruses), which have been suspected of airborne transmission (Marks et al. 2000). Norwalk virus has caused outbreaks of gastroenteritis on cruise ships and is considered capable of being aerosolized during projectile vomiting and possibly in the spray of cruise ships ploughing warm waters (Gunn et al. 1980; MMWR 2002). In one outbreak it was suggested that the moving of contaminated laundry contributed to the dissemination of airborne viral particles (Sawyer et al. 1988).

Urinary tract infections are generally the result of contaminated equipment and direct contact transmission, but some of the contaminants may arrive via airborne transport. Some of the bacterial pathogens that cause UTIs may be present in the pharynx as commensals or transient endogenous microbes, including *Pseudomonas* species, *Streptococcus pneumoniae*, *Staphylococcus aureus*, *Haemophilus influenzae*, and *Neisseria meningitidis* (Ryan 1994). These agents are most often merely colonizers and not common etiologic agents of UTIs.

Environmental spores may spread in ICUs just as they do in general areas of the hospital. Airborne concentrations of *Aspergillus* spores in an ICU were measured by Falvey and Streifel (2007), who found the mean levels were 8 cfu/m^3, while the mean levels in patient care areas ranged from 6 to 16 cfu/m^3.

Ocular infections in hospitals are uncommon and represent no more than 0.5% of all nosocomial infections (Weber, Durand, and Rutala 1999). Common infecting pathogens include *Staphylococcus*, *Pseudomonas*, *Streptococcus*, Adenovirus, *Aspergillus*, *Bacillus cereus*, *Moraxella*, *Haemophilus*, *Histoplasma*, and *M. tuberculosis* (Weber, Durand, and Rutala 1999). Outbreaks of fungal infections have been reported. The inoculation of the eye most likely always occurs as a result of hand contact or rubbing, but the possibility of airborne microbes that are common in hospital air settling on the eyes of prone patients should not be discounted altogether.

Patient Rooms and General Wards

Patients in general wards are subject to outbreaks of various nosocomial infections of which most are nonairborne. They are also subject to outbreaks of community-acquired infections that often result in nosocomial outbreaks that follow seasonal patterns in the community. Respiratory infections are dealt with in Chapter 11. This section addresses nonrespiratory infections that affect patients in general wards.

Food-borne outbreaks in hospitals have involved a variety of microorganisms including *Salmonella*, *Staphylococcus aureus*, *Clostridium perfringens* spores, *Shigella*, *Bacillus cereus* spores, *Campylobacter jejuni*, and other microbes (Farr 1999). All spores are capable of airborne transport and therefore *Clostridium perfringens* and *Bacillus cereus* could be considered airborne in hospital environments whenever they are present. *Clostridium botulinum* spores have also been implicated as an enteral pathogen causing nosocomial infections, and it too may be capable of airborne transport in hospitals.

Listeria monocytogenes is another enteral pathogen that has caused infections in the hospital and some evidence exists from the food industry that it may settle on foods from the air (Dobeic, Ivan, and Edward 2005). *Listeria* can survive aerosolization for up to 3.5 hours and can settle on and contaminate floors and other surfaces (Spurlock and Zottola 1991). One report from a current study indicates *Listeria* may survive in cold storage for up to 28 days (Dickson et al. 2011). Germicidal air cleaning systems have proven effective for controlling *Listeria* contamination on floors and surfaces in the meat packing industry (Cundith et al. 2002). In experimental airborne infection *Listeria* causes acute pneumonitis (Lefford, Amell, and Warner 1978).

Various kinds of UV systems have been applied successfully in general wards, patient rooms, and hallways to reduce infection rates, including in-duct airstream disinfection systems, upper- and lower-room systems, and UV barrier systems (Kowalski 2007; Dumyahn and First 1999). UV barrier systems installed in doorways between isolation wards were effective in preventing the spread of chickenpox (Wells 1938). Barrier systems were found to reduce cross-infections between patient cubicles (DelMundo and McKhann 1941; Sommer and Stokes 1942; Robertson et al. 1943). Schneider et al. (1969) applied in-duct UV for the supply air of an isolation ward and simultaneously irradiated the surrounding corridors, effectively controlling airborne pathogens. Upper-room UVGI systems have been used successfully to control disease transmission in hospitals and have been implemented at The Cradle in Evanston, The Home for Hebrew Infants, The Livermore CA Veteran's Hospital, the North Central Bronx Hospital, and St. Luke's Hospital in New York for the control of respiratory infections (Sauer, Minsk, and Rosenstern 1942; Higgons and Hyde 1947; Wells 1955; McLean 1961; EPRI 1997). The results of field trials of upper-room systems show a net average reduction of infections of 70% (see Chapter 19).

Hospital Laboratories

Any hospital laboratories that deal with biological agents face potential inhalation hazards from handling mishaps and casual exposure (Kowalski 2006). Hospital laboratories pose special risks of infection to workers from pathogenic microorganisms brought in with infected patients (McGowan 1999). These risks may be unknown until samples are analyzed and identified. Biological laboratories normally have a variety of systems and protocols to protect health care workers from such laboratory hazards, including laboratory hoods, air cleaning systems, pressurization zones, sterilization equipment, biohazard-rated facilities, personnel protective suits, and strict procedures for handling hazardous agents.

Existing procedures are considered adequate to protect workers and these are typically adhered to diligently. All of the existing guidelines and standards offer similar guidance concerning the design and operation of the ventilation or air cleaning systems (see Chapter 17). Typically these guidelines recommend about 6–15 air changes per hour (ACH). The use of filtered 100% outside air is generally specified as an option and this is the most common approach taken in laboratories today. Air is typically exhausted to outside, and certain codes may require HEPA filtration of the exhaust air, although the necessity for this is debatable. For systems that recirculate air, a minimum of 50% outside air (or maximum 50% return air) is suggested by some of the guidelines. HEPA filtration is also recommended for recirculated or exhaust air from biosafety cabinets (ASHRAE 1999).

There are four levels for categorized containment laboratories: Biosafety Level 1, 2, 3, and 4 (DHHS 1993; CDC 2003). The basic characteristics of these laboratories are summarized in Table 13.2. There are no specific requirements for the use of UV in any biosafety laboratories, but UV systems are often used in biosafety cabinets and for equipment disinfection.

The incidence of tuberculosis in laboratory personnel is many times higher than that in other occupations. The greatest risk to laboratory workers is the aerosolization of liquid specimens during handling (McGowan 1999). Airborne fungal spores have caused problems in laboratories with *Coccidioides* and *Histoplasma* considered the greatest threats that may be transmitted by aerosols. Arthroconidia from *C. immitis* easily become airborne, as do the

TABLE 13.2

Basic Characteristics of BSL Containment Laboratories

BSL	Requirements	Recommendations	Application
1	No specific HVAC requirements	3–4 ACH, slight negative pressure	Microbial agents of no known hazard or minimal hazard
2	No specific HVAC requirements	100% OA, 6–15 ACH, slight negative pressure, use of safety cabinets	Microbial agents of moderate potential hazard
3	Physical barrier, double doors, no recirculation, maintain negative pressure	Exhaust may require HEPA filtration	Microbial agents that pose a serious hazard via inhalation
4	Physical barrier, double doors, no recirculation, maintain negative pressure, etc.	Requirements determined by biological safety officer	Microbial agents that pose a high risk of lethality via inhalation

Note: ACH = air changes per hour.

infective conidia from *H. capsulatum*. Pulmonary infection with *Blastomyces dermatidis* has followed inhalation of the mold form by laboratory workers.

Yersinia pestis has been spread by direct inoculation and inhalation of aerosols or droplets in laboratories (McGowan 1999). Aerosol generation of *Bordetella pertussis* has also caused infections in laboratories, and *Francisella tularemia* is also capable of producing infectious aerosols from cultures (Sewell 1995). Laboratory-generated aerosols can provide a transmission mechanism for *Bacillus anthracis*, *Corynebacterium diphtheria*, *Coxiella burnetti*, *M. bovis*, and *Pseudomonas pseudomallei*. Aerosols can produce droplets that contaminate counters or floor surfaces and these may result in hand contamination and self-inoculation by workers. Microorganisms in droplets that dry on surfaces can remain viable for several days.

Other Hospital Areas

Hallways and storage areas surrounding operating rooms and ICUs can be a source of contaminants that may be tracked into the ORs and isolation rooms by foot traffic or may wash into ORs via opening of doors. Hallways are also often used as storage areas and both types of areas can accumulate microbiological contamination due to the greater surface area, which can act as both a microbial substrate and as protection from sunlight or desiccation. The greater the total surface area in any given environment, the greater the potential for accumulation of microbial contamination.

Contamination of hospital lobbies with *Aspergillus* spores may be one means by which these spores enter and contaminate other areas of the hospital. Falvey and Streifel (2007) showed that the lobby of a hospital had a higher mean recovery of *Aspergillus* spores than other areas, and attributed this to the fact that one set of double doors was often held open and higher traffic levels induced more turbulence. The lobby had mean levels of 21 cfu/m^3 even though it was supplied with air from 90–95% filters, which should theoretically remove all *Aspergillus* spores. Cleaning the environment of fungal spores should be performed in a manner that does not generate dust. Surfaces should be wiped with wet cloths and wet mops. Anderson et al. (1996) measured airborne *Aspergillus* levels rise from 24 cfu/m^3 to 62 cfu/m^3 when vacuum cleaners were put to use. Vacuum cleaners that filter air and do not blow dust around are available for hospital applications. Carpets can be a source for spores that accumulate over time, and spills on rugs can grow mold.

Storage areas for supplies and equipment are often located adjacent to ORs and near ICUs so that materials may be expediently delivered. Such storage areas can provide large amounts of surface area on which microbial contamination may accumulate over time. Because such areas are only transiently occupied, UV area decontamination systems may be applied to provide high levels of disinfection. In such areas it is also feasible to apply upper-room UV systems that will not only disinfect the air but will tend to disinfect the lower-room surfaces over time. Although the stray irradiance from upper-room UV systems is typically below ACGIH/NIOSH limits for human safety (NIOSH 1972), the accumulated dose to surfaces over extended time periods will provide fairly high levels of disinfection.

Bacteria and spores tend to settle downward over time and accumulate near the floor, and are resuspended by traffic or tracked into hallways by foot. The placement of lower-room UV units along the walls in several places will maintain the hallway floors in a decontaminated state.

References

Anderson, K., Morris, G., Kennedy, H., Croall, J., Michie, J., Richardson, M. D., and Gibson, B. (1996). Aspergillosis in immunocompromised paediatric patients: Associations with building hygiene, design, and indoor air. *Thorax* 51, 256–261.

ASHRAE (1999). Chapter 7: Health care facilities; in *ASHRAE Handbook of Applications*, ASHRAE, ed., American Society of Heating, Refrigerating and Air Conditioning Engineers, Atlanta, GA, 7.1–7.13.

Best, E. L., Fawley, W. N., Parnell, P., and Wilcox, M. H. (2010). The potential for airborne dispersal of *Clostridium difficile* from symptomatic patients. *Clin Inf Dis* 50, 1450–1457.

Bonten, M. J. M. (1999). Nosocomial sinusitis; in *Hospital Epidemiology and Infection Control*, C. G. Mayhall, ed., Lippincott Williams & Wilkins, Philadelphia, 239–246.

CDC (2003). Guidelines for environmental infection control in health-care facilities. *MMWR* 52(RR-10), 1–48.

Cundith, C. J., Kerth, C. R., Jones, W. R., McCaskey, T. A., and Kuhlers, D. L. (2002). Germicidal air-cleaning system effectiveness for control of airborne microbes in a meat processing plant. *J Food Microbiol Safety* 67(3), 1170–1174.

DelMundo, F., and McKhann, C. F. (1941). Effect of ultra-violet irradiation of air on incidence of infections in an infant's hospital. *Am J Dis Child* 61, 213–225.

DHHS (1993). Biosafety in Microbiological and Biomedical Laboratories. U.S. Department of Health and Human Services, Cincinnati, OH.

Dickson, J. S., Zimmerman, J. J., Bryden, K. M., and Hoff, S. J. (2011). Aerosol and airborne transmission of *Listeria monocytogenes*: A potential source of cross-contamination. USDA/Iowa State University. www.reeis.usda.gov/web/cris-projectpages/207482.html

Dobeic, M., Ivan, G., and Edward, K. (2005). Airborne microorganisms *Brochothrix thermosphacta*, *Listeria monocytogenes*, and *Lactobacillus alimentarius* in meat industry as a risk in food safety. *ISAH 2005*, Warsaw, Poland.

Dumyahn, T., and First, M. (1999). Characterization of ultraviolet upper room air disinfection devices. *Am Ind Hyg Assoc J* 60(2), 219–227.

Durmaz, G., Kiremitci, A., Akgun, Y., Oz, Y., Kasifoglu, N., Aybey, A., and Kiraz, N. (2005). The relationship between airborne colonization and nosocomial infections in intensive care units. *Mikrobiyol Bul* 39(4), 465–471.

EPRI (1997). UVGI for TB Infection Control in a Hospital. *TA-107885*, Electric Power Research Institute, Palo Alto, CA.

Falvey, D. G., and Streifel, A. J. (2007). Ten-year air sample analysis of *Aspergillus* prevalence in a university hospital. *J Hosp Infect* 67, 35–41.

Farr, B. M. (1999). Nosocomial gastrointestinal tract infections; in *Hospital Epidemiology and Infection Control*, C. G. Mayhall, ed., Lippincott Williams & Wilkins, Philadelphia, 247–274.

Fekety, R., Kim, K.-H., Brown, D., Batts, D. H., Cudmore, M., and J. Silva. (1981). Epidemiology of antibiotic-associated colitis. Isolation of *Clostridium difficile* from the hospital environment. *Am J Med* 70, 906–908.

Gunn, R. A., Terranova, W. A., Greenberg, H. B., Yashuk, J., Gary, G. W., Wells, J. G., Taylor, P. R., and Feldman, R. A. (1980). Norwalk virus gastroenteritis aboard a cruise ship: An outbreak on five consecutive cruises. *Am J Epidemiol* 112(6), 820–827.

Higgons, R. A., and Hyde, G. M. (1947). Effect of ultra-violet air sterilization upon incidence of respiratory infections in a children's institution. *New York State J Med* 47(7):15–27.

Kowalski, W. J. (2006). *Aerobiological Engineering Handbook: A Guide to Airborne Disease Control Technologies*. McGraw-Hill, New York.

_____. (2007). Air-treatment systems for controlling hospital-acquired infections. *HPAC Engineering* 79(1), 28–48.

Lefford, M. J., Amell, L., and Warner, S. (1978). Listeria pneumonitis: Induction of immunity after airborne infection with *Listeria monocytogenes*. *Infect Immun* 22(3), 746–751.

Marks, P. J., Vipond, I. B., Carlisle, D., Deakin, D., Fey, R. E., and Caul, E. O. (2000). Evidence for airborne transmission of Norwalk-like virus (NLV) in a hotel restaurant. *Epidemiol Infect* 124(3), 481–487.

McFarland, L. V., Mulligan, M. E., Kwok, R. Y. Y., and Stamm, W. E. (1989). Nosocomial acquisition of *Clostridium difficile* infection. *N Engl J Med* 320, 204–210.

McGowan, J. (1999). Nosocomial infections in diagnostic laboratories; in *Hospital Epidemiology and Infection Control*, C. G. Mayhall, ed., Lippincott Williams & Wilkins, Philadelphia, 1127–1136.

McLean, R. (1961). The effect of ultraviolet radiation upon the transmission of epidemic influenza in long-term hospital patients. *Am Rev Resp Dis* 83, 36–38.

MMWR (2002). Outbreaks of gastroenteritis associated with noroviruses on cruise ships—United States, 2002. *Morb Mortal Wkly Rep* 51(49), 1112–1115.

Nielsen, P. (2008). *Clostridium difficile* aerobiology and nosocomial transmission. Northwick Park Hospital, Harrow, Middlesex, UK.

NIOSH (1972). Occupational Exposure to Ultraviolet Radiation. *HSM 73-110009*, National Institute for Occupational Safety and Health, Cincinnati, OH.

Richards, M. J., Edwards, J. R., Culver, D. H., and Gaines, R. P. (1999). Nosocomial infections in medical intensive care units in the United States. *Crit Care Med* 27(5), 887–892.

Robertson, E. C., Doyle, M. E., Tisdall, F. F., Koller, L. R., and Ward, F. S. (1943). Use of ultra-violet radiation in reduction of respiratory cross-infections in a children's hospital. *JAMA* 121, 908–914.

Ryan, K. J. (1994). *Sherris Medical Microbiology.* Appleton & Lange, Norwalk, CT.

Santosham, M., Yolken, R. H., Quiroz, E., Dillman, L., Oro, G., Reeves, W. C., and Sack, R. B. (1983). Detection of rotavirus in respiratory secretions of children with pneumonia. *J Pediatr* 103, 583–585.

Sattar, S. A., Lloyd-Evans, N., and Springthorpe, V. S. (1986). Institutional outbreaks of rotavirus diarrhoea: Potential role of fomites and environmental surfaces as vehicles for virus transmission. *J Hyg (London)* 96, 277–289.

Sauer, L. W., Minsk, L. D., and Rosenstern, I. (1942). Control of cross infections of respiratory tract in nursery for young infants. *JAMA* 118, 1271–1274.

Sawyer, L. A., Murphy, J. J., Kaplan, J. E., Pinsky, P. F., Chacon, D., Walmsley, S., Schonberger, L. B., Phillips, A., Forward, K., Goldman, C., Brunton, J., Fralick, R. A., Carter, A. O., Gary, W. G., Glass, R. I., and Low, D. E. (1988). 25- to 30-nm virus particle associated with a hospital outbreak of acute gastroenteritis with evidence for airborne transmission. *Am J Epidemiol* 127, 1261–1271.

Schneider, M., Schwartenberg, L., Amiel, J. L., Cattan, A., Schlumberger, J. R., Hayat, M., deVassal, F., Jasmin, C. L., Rosenfeld, C. L., and Mathe, G. (1969). Pathogen-free isolation unit—Three years' experience. *Brit Med J* 29 March, 836–839.

Segal, M. N. (1995). Nosocomial sinusitis; in *Surgical Infections*, D. E. Fry, ed., Little, Brown and Company, Boston.

Sewell, D. L. (1995). Laboratory-associated infections and biosafety. *Clin Microbiol Rev* 8, 389–405.

Snelling, A. M., Beggs, C. B., Kerr, K. G., and Sheperd, S. J. (2011). Spores of *Clostridium difficile* in hospital air. *Clin Infect Dis* 51, 1104–1105.

Sommer, H. E., and Stokes, J. (1942). Studies on air-borne infection in a hospital ward. *J Pediat* 21, 569–576.

Spurlock, A. T., and Zottola, E. A. (1991). The survival of *Listeria monocytogenes* in aerosols. *J Food Prot* 54, 910–912.

Stals, F., Walther, F. J., and Bruggeman, C. A. (1984). Faecal and pharyngeal shedding of rotavirus and rotavirus IgA in children with diarrhea. *J Med Virol* 14, 333–339.

Weber, D. J., Durand, M., and Rutala, W. A. (1999). Nosocomial ocular infections; in *Hospital Epidemiology and Infection Control*, C. G. Mayhall, ed., Lippincott Williams & Wilkins, Philadelphia, 287–300.

Wells, W. F. (1938). Air-borne infections. *Mod Hosp* 51, 66–69.

――――. (1955). *Airborne Contagion and Air Hygiene.* Harvard University Press, Cambridge, MA.

Wenzel, R. P. (1981). *CRC Handbook of Hospital Acquired Infections.* CRC Press, Boca Raton, FL.

Wilcox, M. H., Bennett, A., Best, E. L., Fawley, W. N., and Parnell, P. (2011). Reply: Spores of *Clostridium difficile* in hospital air. *Clin Infect Dis* 51, 1105.

Wilson, J. (2001). *Infection Control in Clinical Practice.* Balliere Tindall, Edinburgh, Scotland.

Wilson, R. D., Huang, S. J., and McLean, A. S. (2004). The correlation between airborne methicillin-resistant *Staphylococcus aureus* with the presence of MRSA colonized patients in a general intensive care unit. *Anaesth Intensive Care* 32(2), 202–209.

14

Pediatric Nosocomial Infections

Introduction

The control of nosocomial infections in neonates is an area that increases in importance in direct relation to the number of new neonatal intensive care units (NICUs). Healthy babies are discharged expediently from new-born nurseries, but premature infants are kept in NICUs for extended periods in order to deal with the various health problems associated with low birth weight. The prolonged periods that infants are kept hospitalized leads to prolonged exposure to nosocomial pathogens, among which are a wide array of airborne nosocomial pathogens.

Pediatric nosocomial infection rates vary from 1% in community hospital well-baby nurseries to 22% in NICUs (Palmer, Giddens, and Palmer 1996). Distinguishing between neonatal infections acquired from a perinatal source at the time of delivery and infections that are hospital acquired is difficult, and therefore it is typical to count all infections in the first 28 days of life as nosocomial unless the source is clearly prenatal, transplacental acquisition. Congenital neonatal pathogens include rubella and cytomegalovirus. Vertically acquired pathogens include Group B streptococci, *Listeria monocytogenes*, and *Staphylococcus aureus*.

The types of infections that afflict neonates and premature infants are generally different from those that affect adults, and viruses can predominate. Skin infections, diarrhea, respiratory infections, and septicemia are the most common nosocomial infections in neonates. Pediatric patients are subject to central nervous system (CNS) infections, which account for almost 10% of infant infections in ICUs. Newborns can acquire infections from fomites, the hands of HCWs, invasive procedures, and by airborne transmission and inhalation. All newborns are immunologically immature at birth and this is especially true of premature neonates. The immune mechanisms at birth make only IgM. Breast milk contains IgG, which helps protect infants against gastrointestinal illnesses.

NICUs provide medical care for neonates and premature infants. They are often adjacent to and associated with pediatric ICUs (PICUs), and may be subject to the same aerobiological hazards. Most nosocomial upper respiratory tract infections in pediatric patients are nonbacterial and appear about two weeks after admission (Jarvis 1987). Table 14.1 summarizes potentially airborne pediatric infections, the sources, and the airborne class.

The types of facilities subject to nosocomial infections are not limited to NICUs and newborn nurseries but can include long-term and residential care facilities, chronic disease and specialty hospitals, and residential schools. Infection control policies and procedures developed for adult long-term care facilities are not applicable to pediatric extended care (PEC) facilities (Harris

TABLE 14.1

Airborne Pediatric Nosocomial Infections

Pathogen	Type	Airborne Class	Source	Infection
Acinetobacter	Bacteria	2	Environmental	Respiratory
Adenovirus	Virus	2	Human	Respiratory
Clostridium difficile	Bacteria	1	Human/ environmental	Gastrointestinal
Corynebacterium diphtheriae	Bacteria	2	Human	Respiratory
Coxsackievirus	Virus	2	Human	Respiratory
Enterobacter	Bacteria	2	Human/ environmental	Gastrointestinal
Enterococcus	Bacteria	2	Human/ environmental	Gastrointestinal
Haemophilus influenzae	Bacteria	2	Human	Respiratory
Influenza	Virus	1	Human	Respiratory
Klebsiella pneumoniae	Bacteria	2	Human/ environmental	Respiratory
Measles	Virus	1	Human	Measles
Moraxella catarrhalis	Bacteria	2	Human	Respiratory
Mycobacterium tuberculosis	Bacteria	1	Human	Respiratory
Mycoplasma hominis	Bacteria	2	Human	Respiratory
Neisseria menigitidis	Bacteria	2	Human	Respiratory
Parainfluenza	Virus	2	Human	Respiratory
Parvovirus	Virus	2	Human	Respiratory
Proteus	Bacteria	2	Human	Respiratory
Pseudomonas aeruginosa	Bacteria	1	Environmental	Respiratory
Reovirus	Virus	2	Human	Respiratory
Rhinovirus	Virus	2	Human	Respiratory
Rotavirus	Virus	2	Human	Gastrointestinal
RSV	Virus	1	Human	Respiratory
Serratia marcescens	Bacteria	2	Environmental	Respiratory
Staphylococcus aureus	Bacteria	1	Human	Respiratory
Staphylococcus epidermis	Bacteria	2	Human	Respiratory
Streptococcus pneumoniae	Bacteria	2	Human	Respiratory
VZV	Virus	1	Human	Chickenpox

2006). Outbreaks of *Haemophilus influenzae*, Coxsackievirus, influenza, RSV, adenovirus, VZV, and MRSA have all been reported in PEC facilities.

Pediatric Bacterial Infections

The most common bacterial pathogens in pediatric and newborn surgical site infections include *S. aureus, E. coli*, coag-negative staphylococci, *P. aeruginosa*, and *Enterococcus* (Wong 1999). Staphylococci account for some 70% of NICU infections (Baltimore 1984). *Streptococcus* species are also reported. These species are common endogenous microbes that are widely present in infants, mothers, visitors, and hospital personnel. The newborn is at risk for infection because of immaturity of the normal defenses and the immune system. Infection rates increase with overcrowding and understaffing (Moore 1999). NICU personnel may serve as the reservoir and source of resistant coag-negative staphylococci (i.e., *Staphylococcus epidermis*) that colonize neonates, whose skin is usually sterile prior to birth (Boyce 1999).

Streptococcus pyogenes (GAS) is a common cause of pharyngitis in the community but not in nosocomial settings. Bacterial tracheitis is most commonly due to *S. aureus* and *H. influenzae* and is thought to occur secondarily to viral infections, usually parainfluenza (Schutze and Yamauchi 1999). Diphtheria is due to *Corynebacterium diphtheriae* but has been uncommon in pediatric patients since a vaccine was developed. Sinusitis in children is typically due to *S. pneumoniae, H. influenzae*, and *Moraxella catarrhalis* (Wald 1995). Other pathogens involved in pediatric or nosocomial sinusitis include *S. aureus, Bacteroides, P. aeruginosa, Klebsiella, Enterobacter*, and *Proteus*. Otitis media is a common illness among children and is commonly caused by infection with *S. pneumoniae, M. catarrhalis*, or *H. influenzae*.

Lower respiratory tract infections account for some 15–20% of infections in children and the common etiologic agents include *S. aureus, S. epidermis, Klebsiella, Pseudomonas, Moraxella, Enterococcus*, and *E. coli*. The fecal-oral route can explain most infections except for *Pseudomonas* and *Acinetobacter*, which are not the usual inhabitants of the human gastrointestinal tract (Schutze and Yamauchi 1999). The Enterobacteriaceae may come from the patient's own endogenous flora but other microbes likely have an exogenous or environmental source, and the hands of HCWs may be important factors in transmission. In NICUs, *S. epidermis* is a major cause of nosocomial infections while *Chlamydia trachomatis, Ureaplasma urealyticum*, and *Mycoplasma hominis* may cause infections in premature infants.

Nosocomial gastrointestinal tract infections in pediatric patients are largely spread by person-to-person transmission or point-source infection through food and water (Mitchell and Pickering 1999). Enteropathogens that may conceivably also spread through pediatric wards by the airborne route include *Clostridium difficile*, rotavirus, adenovirus, parvovirus, *Klebsiella*, and *Pseudomonas*. Isolation rates of *Clostridium difficile* as high as 90% have occurred in NICUs, and this microbe also causes outbreaks in child care centers.

In pediatric patients meningitis is usually the result of bacteremia, and in children older than 3 months, meningitis is caused by pathogens such as *Haemophilus influenzae, Streptococcus pneumoniae, Neisseria meningitidis, E. coli, Klebsiella, Enterobacter,* and *Proteus* (Schutze and Yamaguchi 1999). Spread of *N. meningitis* and *H. influenzae* occurs by the respiratory route. Outbreaks of meningitis in NICUs have been attributed to other microbes also, including *Staphylococcus aureus, S. epidermis,* and *Serratia. Serratia marcescens* can cause serious endemic and epidemic infections in NICUs. Outbreaks of *Serratia* have caused widespread newborn gastrointestinal colonization with the infants serving as reservoirs (Moore 1999).

Bacterial strains that occur in the NICU are frequently antibiotic resistant (Moore 1999). The usual mode of transmission in the nursery is by contact, either direct physical contact or transfer from one infant to another on the hands of personnel. The hands of HCWs are usually transiently contaminated, and handwashing can interrupt transmission. Transmission by indirect contact with contaminated equipment can also occur. Water sources may be a source for *Pseudomonas, Serratia,* and *Flavobacterium.* Infected personnel and visitors, including parents, may introduce pathogens into nurseries.

Newborns are colonized with *S. aureus* within the first few days after birth, and the microbe may be acquired either from the mother or the hospital environs. MRSA infections are an increasing problem in newborn nurseries, although most outbreaks have been described in NICUs (Moore 1999). The major reservoir of MRSA is the infected newborn, with HCWs being an occasional source. Neonates are also at an elevated risk for surgical site infections and risk factors include increased incision length, increased duration of surgery, and contamination of the operating site (Davenport and Doig 1993).

Surfaces and equipment can harbor microbes that settle from the air or attach to surfaces due to natural forces. These fomites can be re-aerosolized at any time from activity. Figure 14.1 shows a NICU procedure room with a vast amount of surface area. Carpets and rugs in NICUs may harbor a wide variety of bacteria and these may be stirred up by activity. In one NICU investigated by the author, pathogenic bacteria surviving in a rug in a lunchroom were apparently tracked into the NICU by HCWs (Kowalski and Bahnfleth 2002).

NICUs are typically kept under negative pressure relative to the hallways. Rugs and carpeting in hallways outside the NICU may contain ground-in dust, fungal spores, and possibly harbor bacteria. It has been noted in previous studies that the sweeping of floors and vacuuming of rugs can aerosolize bacteria-containing dust and possibly spores. Such aerosolization of contaminants would cause them to be drawn into the NICU if it is under negative pressure relative to the hallway.

FIGURE 14.1
A typical NICU procedure room, with many possible surfaces where pathogens may settle and exist as transient fomites.

In large baby-washing sinks, splashing water might cause aerosolization of water droplets containing *Legionella*. Sinks should have faucets but not nozzles of the type that might aerosolize water droplets. *Legionella* is a common contaminant in hospital potable water supplies and may be found in 30–68% of hospitals (Rutala and Weber 1997). However, sinks are far more likely to contain Gram-negative bacteria such as *Pseudomonas, Burkholderia, Serratia, Acinetobacter,* and *Mycobacteria* than they are to contain *Legionella*.

Acute otitis media is primarily a disease of children in which pathogenic bacteria may infect the mucoperiosteum of the middle ear cleft (Nicklaus 1995). These bacteria may arrive via airborne transport or by direct contact. The most common infecting microbes include *Streptococcus pneumoniae, Haemophilus influenzae,* and *Moraxella catarrhalis*. Infections of the external auditory canal can include *Pseudomonas aeruginosa, Staphylococcus aureus,* and *Proteus*.

The incubators in which premature infants are kept come in many different types. These units are normally enclosed and air is recirculated through integral filters. They include air filters but there is no standard for the filter type. Infections sometimes result from catheters and intubation components and from the medical tape used to attach the components. Infections produced under such conditions may result from endogenous microbes or direct contact contamination. These conditions may not be controllable by disinfection of air supplies, but alternatives include antimicrobial materials and surface disinfection equipment. Antimicrobial materials are currently available and these include fabrics that can be used for coverings or clothing, and possibly in place of medical tape. Additional antimicrobial materials are available in the form of gels or foams that may be used for catheters or intubation equipment.

Some incubation units are supplied by patient air from compressors in a separate equipment room. The patient air system should be inspected as one potential source of microbial contamination, although due to the processing of this air it is highly unlikely the air could become contaminated. However, these units contain integral air filters that may accumulate biocontamination.

Airborne fungal spores are rare in NICUs as a cause of infection, but fungal infection can be severe in such settings (Goldman 1989). Newborns and patient care equipment should be protected from exposure to dust and debris that may contain fungal spores (Krasinski et al. 1985)

Pediatric Viral Infections

Viral infections in pediatric patients account for about 27% of all nosocomial infections. Most respiratory viruses are spread by large droplets expelled from the respiratory tract and travel distances of about 1 meter, settling on surfaces close to the infected person (Moore 1999). Newborns do not cough vigorously and do not generate aerosols efficiently. True airborne transmission may be infrequent in nurseries but pathogens like *M. tuberculosis*, influenza, varicella, and measles may transmit in this manner. Control of varicella-zoster virus (VZV) spread in pediatric hospitals can be accomplished by placing patients in negative-pressure isolation rooms (Anderson et al. 1985). Viruses have been detected in the air of pediatric facilities (Tseng, Chang, and Li 2010).

VZV is readily transmitted by the airborne route, but it is rare in nurseries because most adults are immune and most newborns are protected by maternal antibody (Moore 1999). VZV may be introduced into a nursery by mothers, HCWs, or visitors with asymptomatic infections, or by infants with perinatal varicella. Pediatric patients, especially those under 5 years, are highly susceptible to VZV (as chickenpox) and outbreaks can drag on for months (Zaia 1999). Airborne transmission of measles virus has been shown to occur in one pediatrician's office (CDC 1983). Isolation precautions intended to prevent the spread of infection by both air and direct contact would ideally consist of a private room under negative air pressure to protect those outside.

Respiratory virus outbreaks in NICUs reflect outbreaks in local communities and are important causes of infection in NICUs. Newborns shed viruses for prolonged periods after symptoms end and their surrounds are often contaminated. RSV, parainfluenza, and influenza viruses survive on hands and surfaces or equipment long enough to permit transfer between patients. HCWs frequently become infected and play a role in transmission. Concurrent outbreaks have occurred involving multiple species such as RSV, rhinovirus, echovirus, and parainfluenza (Moore 1999). During the respiratory virus season, mycoplasma is detected in a large percentage of pediatric patients admitted to the hospital, and shedding of virus and mycoplasma from the upper respiratory tract of patients is common (Turner 1999).

Rotavirus is endemic in many NICUs and nurseries, and epidemics have occurred. Transmission is by the fecal-oral route or via fomites on surfaces and equipment. The virus survives for over an hour on surfaces and toys (Keswick et al. 1983). Reovirus can cause URIs and pneumonia in infants (Feigin et al. 2009). Rhinovirus is not as serious an infection in neonates as influenza or RSV, but it can be a serious infection in premature infants and the immunocompromised (Goldmann 2001).

Measles virus has caused outbreaks in nosocomial settings, and is commonly transmitted in emergency departments and outpatient waiting areas, and approximately one-half of measles cases are transmitted in physicians' offices (Wainwright et al. 1999). Airborne transmission has been implicated in a nosocomial office setting (Bloch, Orenstein, and Ewing 1985; Remington et al. 1985; CDC 1983). The largest groups of HCWs who acquire measles are nurses, followed by physicians. Measles can be severe in immunocompromised patients. Secondary attack rates can exceed 90%.

Isolation precautions for airborne pathogens such as VZV, TB, or measles requires a single room with negative pressure ventilation. Forced air incubators are not a substitute for negative pressure ventilation because they recirculate air back into the local environment. Isolation areas can be used for nonairborne infections, and these may include separation by curtains or other partitions. Where droplet precautions are required, infants should be separated by at least 3 feet. Newborns are unlikely to generate large-droplet aerosols, but aerosolization may be a concern with infected infants on respirators (Moore 1999).

Nurseries and Childcare Facilities

Nurseries and child care centers, or day care centers, are subject to many of the same infections as other pediatric facilities. Respiratory infections are the most common infection and caregivers may inadvertently transmit infections between children, as well as being at increased risk from occupational exposure (Cordell and Solomon 1999). Infections in such settings are primarily transmitted by person-to-person spread from body substances including feces, saliva, nasal secretions, and urine. Children often have poor hygiene and handling by HCWs may result in hand contamination. Children also share secretions and excretions through fomites that may contaminate toys and local surfaces. Many of the microorganisms that cause pediatric infections can survive on surfaces for long periods. The concentrations of pathogens recovered from surfaces and air samples in child care center classrooms has been shown to be inversely related to the age of the children (Petersen and Bressler 1986). Transmission of infections within child care facilities reflects and amplifies the prevalence of pathogens in the community. Children attending child care may harbor drug-resistant strains of *Streptococcus pneumoniae*, *Haemophilus influenzae*, and *Staphylococcus aureus*. Child care attendance increases the risk of primary *Haemophilus influenzae* disease (Cordell and Solomon 1999).

Control of infections in child care centers depends on the same methods used in health care settings, with hand washing and environmental sanitation being the primary means of control. Caution in handling contaminated materials is also an appropriate measure. Selection of materials and surfaces that can be easily cleaned and disinfected is another approach. Cohorting, or separating children into infected and uninfected groups, has been tried but requires good surveillance techniques to be effective. Nurseries should be kept clean and dust-free because fungal spores from dust may cause serious infections. Humidifiers can serve as reservoirs of *Legionella*, *Pseudomonas*, and other pathogens, and should be regularly refilled with sterile water.

References

Anderson, J. D., Bonner, M., Scheifele, D. W., and Schneider, B. C. (1985). Lack of nosocomial spread of varicella in a pediatric hospital with negative pressure ventilated patient rooms. *Infect Control* 6, 120–121.

Baltimore, R. S. (1984). Nosocomial infections in the pediatric intensive care unit. *Yale J Biol Med* 57, 185–197.

Bloch, A. B., Orenstein, W. A., and Ewing, W. M. (1985). Measles outbreak in a pediatric practice: Airborne transmission in an office setting. 75, 676–683.

Boyce, J. M. (1999). Coagulase-negative staphylococci; in *Hospital Epidemiology and Infection Control*, C. G. Mayhall, ed., Lippincott Williams & Wilkins, Philadelphia, 365–384.

CDC (1983). Imported measles with subsequent airborne transmission in a pediatrician's office—Michigan. Centers for Disease Control. *MMWR* 32, 401–402.

Cordell, R. L., and Solomon, S. L. (1999). Infections acquired in child care centers; in *Hospital Epidemiology and Infection Control*, C. G. Mayhall, ed., Lippincott Williams & Wilkins, Philadelphia, 695–716.

Davenport, M., and Doig, C. M. (1993). Wound infection in pediatric surgery: A study of 1094 neonates. *J Pediatr Surg* 28, 26–30.

Feigin, R. D., Cherry, J., Demmler-Harrison, G. J., and Kaplan, S. L. (2009). *Feigin & Cherry's Textbook of Pediatric Infectious Diseases, Volume 1.* Saunders/Elsevier, Philadelphia.

Goldmann, D. A. (1989). Prevention and management of neonatal infections. *Infect Dis Clin North Am* 3, 779–813.

_____. (2001). Epidemiology and prevention of pediatric viral respiratory infections in health-care institutions. *Emerg Inf Dis* 7(2), 249–253.

Harris, J.-A. S. (2006). Infection control in pediatric extended care facilities. *Inf Contr Hosp Epidemiol* 27(6):598–603.

Jarvis, W. R. (1987). Epidemiology of nosocomial infections in pediatric patients. *Pediatr Infect Dis* 6, 344–351.

Keswick, B. H., Pickering, L. K., Dupont, H. L., and Woodward, W. E. (1983). Survival and detection of rotaviruses on environmental surfaces in day care centers. *Appl Environ Microbiol* 46, 813–816.

Kowalski, W. J., and Bahnfleth, W. P. (2002). Innovative strategies to protect hospitalized premature infants against airborne pathogens and toxins. Hershey Medical Center NICU, Hershey, PA.

Krasinski, K., Holman, R. S., Hanna, B., Greco, M. A., Graff, M., and Bhogal, M. (1985). Nosocomial fungal infection during hospital renovation. *Infect Control* 6(7), 278–282.

Mitchell, D. K., and Pickering, L. K. (1999). Nosocomial gastrointestinal tract infections in pediatric patients; in *Hospital Epidemiology and Infection Control*, C. G. Mayhall, ed., Lippincott Williams & Wilkins, Philadelphia, 629–648.

Moore, D. L. (1999). Nosocomial infections in newborn nurseries and neonatal intensive care clinics; in *Hospital Epidemiology and Infection Control*, C. G. Mayhall, ed., Lippincott Williams & Wilkins, Philadelphia, 665–694.

Nicklaus, P. J. (1995). Otitis externa and otitis media; in *Surgical Infections*, D. E. Fry, ed., Little, Brown and Company, Boston, 495–499.

Palmer, S., Giddens, J., and Palmer, D. (1996). *Infection Control.* Skidmore-Roth, El Paso, TX.

Petersen, N. J., and Bressler, G. K. (1986). Design and modification of the day care environment. *Rev Infect Dis* 8, 618–621.

Remington, P. L., Hall, W. N., Davis, I. H., Herald, A., and Gunn, R. A. (1985). Airborne transmission of measles in a physician's office. *JAMA* 253(11), 1574–1577.

Rutala, W. A., and Weber, D. J. (1997). Water as a reservoir of nosocomial pathogens. *Inf Contr Hosp Epidemiol* 18(9):609–616.

Schutze, G. E., and Yamaguchi, T. (1999). Nosocomial bacterial infections of the central nervous systems, upper and lower respiratory tracts, and skin in pediatric patients; in *Hospital Epidemiology and Infection Control*, C. G. Mayhall, ed., Lippincott Williams & Wilkins, Philadelphia, 615–628.

Tseng, C. C., Chang, L. Y., and Li, C. S. (2010). Detection of airborne viruses in a pediatrics department measured using real-time qPCR coupled to an air sampling filter method. *J Environ Health* 73(4), 22–28.

Turner, R. B. (1999). Nosocomial viral respiratory infections in pediatric patients; in *Hospital Epidemiology and Infection Control*, C. G. Mayhall, ed., Lippincott Williams & Wilkins, Philadelphia, 607–614.

Wainwright, S. H., Singleton, J. A., Torok, T. J., and Williams, W. W. (1999). Nosocomial measles, mumps, rubella, varicella, and human parvovirus B19; in *Hospital Epidemiology and Infection Control*, C. G. Mayhall, ed., Lippincott Williams & Wilkins, Philadelphia, 649–664.

Wald, E. R. (1995). Chronic sinusitis in children. *J Pediatr* 127, 339–347.

Wong, S. (1999). Surgical site infections; in *Hospital Epidemiology and Infection Control*, C. G. Mayhall, ed., Lippincott Williams & Wilkins, Philadelphia, 189–210.

Zaia, J. A. (1999). Varicella-zoster virus; in *Hospital Epidemiology and Infection Control*, C. G. Mayhall, ed., Lippincott Williams & Wilkins, Philadelphia, 543–553.

15

Immunocompromised and Burn Wound Infections

Introduction

Infections in the immunocompromised are mostly due to opportunistic pathogens and many of these are either endogenous and hail from humans or are environmental and hail from the outdoors. The majority of these pathogens are potentially airborne and are either inhaled or else they settle from the air. This chapter addresses immunocompromised infections, burn wound infections, and infections in cancer patients. Burn wounds are included here because they have a certain similarity to immunocompromised infections—burn wounds are inherently susceptible to opportunistic infections. Infections in cancer patients also bear some similarity because of the reduced immunity of patients undergoing cancer treatment.

Immunodeficiency Infections

AIDS patients and others with immunodeficiency are at greater risk both from nosocomial pathogens and from opportunistic microbes that normally pose little risk to healthy individuals. HIV patients suffer bloodstream infections and urinary tract infections but these are rarely the result of airborne transmission. Airborne transmission may be possible for infections that cause pneumonia (other than VAP) and skin infections with fungi, which are ubiquitous in the environment and routinely spread by the airborne route. Most immunocompromised infections are lung infections, which suggests inhalation as a route of infection.

Pneumonia occurs in some 15–22% of HIV patients. Some of the causes of pneumonia in HIV patients include *Staphylococcus aureus*, *Pseudomonas aeruginosa*, *Klebsiella pneumoniae*, *Enterobacter*, *Moraxella catarrhalis*, *Streptococcus pneumoniae*, *Serratia*, and *Pneumocystis jirovecii* (Craven, Steger, and Fleming 1999). *M. tuberculosis* is a major threat to HIV patients and incidence rates are high. *M. avium* and *M. intracellulare* infections also occur in HIV patients. *Clostridium difficile* infections in HIV patients are a leading cause of gastrointestinal illness. Other emerging fungal infections that affect AIDS patients include *Coccidioides immitis, Histoplasma capsulatum, Fugomyces cyanescens, Pneumocystis carinii, Cryptococcus neoformans, Penicillium marneffei,* and *Trichosporon beigelii* (Dixon et al. 1996).

Table 15.1 summarizes all potentially airborne pathogens that have been associated with infections in HIV patients. Some of these species are rare and have not been addressed previously, such as *M. intracellulare, Moraxella catarrhalis, Penicillium,* and *Paracoccidioides*. These microbes are therefore assigned to Airborne Class 2, being uncommon and suspected but not proven to be airborne in nosocomial settings. It can be observed that the majority of these infections have environmental sources, and therefore protecting the patient from outside air is a critical matter.

Infections involving fungi have become increasingly common, partly due to a growing number of immunocompromised patients undergoing surgery (Wong 1999). *Aspergillus* is the most common infectious fungal spore. A number of other fungal pathogens may be causes of bone marrow transplant infections, including *Trichosporon, Fusarium, Pseudallescheria boydii,* and *Scedosporium prolificans* (Perl, Chotani, and Agawala 1999). *Fusarium* produces skin infections and disseminates from these. The other fungi cause diseases similar to *Aspergillus*. Unusual fungal pathogens causing lung infections in surgical patients include *Blastomyces dermatidis, Coccidioides immitis, Cryptococcus neoformans* (yeast), *Histoplasma capsulatum, Paracoccidioides brasiliensis,* and *Fugomyces cyanescens* (Fry 1995a). Each of these fungal spores is transmitted in the air and inhalation may lead to acute pulmonary infection, which may be severe in AIDS patients.

Infections with *Nocardia asteroides, N. brasiliensis,* and *N. caviae* resemble fungal pulmonary infections and are equally as uncommon as postsurgical complications in immunodeficient patients. The common view is that the spores of these bacteria are airborne and that inhalation leads to infection (Fry 1995b). Because these infections characteristically occur in immunocompromised patients, they are often associated with simultaneous infections with other pathogens such as *Aspergillus, M. tuberculosis,* and *Pneumocystis carinii. P. carinii* exists as a saprophyte in the lungs and is communicable primarily by the airborne route (Hughes 1982). It is uncommon except in immunocompromised patients.

Aspergillus spores and other environmental opportunistic pathogens may blow in through the windows of wards containing immunocompromised patients (Vonberg and Gastmeier 2006). They may also enter through doorways and other routes from the outdoors, and may be carried indoors with visitors and supplies. The key to protecting any immunocompromised patient is to isolate him or her from the environment and to filter the outside air. Common filters rated MERV 12–15 should be adequate to remove most environmental spores and bacteria from the air. Positive pressure

TABLE 15.1

Airborne Pathogens in Immunocompromised Infections

Pathogen	Type	Airborne Class	Infection	Source
Adenovirus	Virus	2	Pneumonia	E
Alternaria	Fungal spore	2	Various	E
Aspergillus	Fungal spore	1	Pneumonia	E
Blastomyces dermatidis	Fungal spore	2	URI	E
Clostridium difficile	Bacterial spore	1	Gastrointestinal infection	H/E
Coccidioides immitis	Fungal spore	2	Coccidioidomycosis	E
Cryptococcus neoformans	Fungal yeast	2	Cryptococcosis	E
Enterobacter	Bacteria	2	BSI, UTI, pneumonia	H/E
Enterococcus	Bacteria	2	BSI, UTI	H
Fugomyces cyanescens	Fungal spore	2	Pneumonia	E
Fusarium	Fungal spore	2	Various	E
Histoplasma capsulatum	Fungal spore	1	Histoplasmosis	E
Influenza	Virus	1	Pneumonia	H
Klebsiella pneumoniae	Bacteria	2	BSI, UTI, pneumonia	H/E
Moraxella catarrhalis	Bacteria	2	Pneumonia	H
Mycobacterium avium	Bacteria	2	MAC disease	E
Mycobacterium intracellulare	Bacteria	2	MAC disease	E
Mycobacterium tuberculosis	Bacteria	1	TB	H
Paracoccidioides brasiliensis	Fungal spore	2	Mycosis	E
Penicillium marneffei	Fungal spore	2	Various	E
Pneumocystis carinii	Fungi	2	Pneumonia	E
Pseudallescheria boydii	Fungal spore	2	Pseudallescheriasis	E
Pseudomonas aeruginosa	Bacteria	1	BSI, UTI, pneumonia	E
Rotavirus	Virus	2	Gastrointestinal infection	H
RSV	Virus	1	Pneumonia	H
Scedosporium prolificans	Fungal spore	2	Various	E
Serratia marcescens	Bacteria	2	Pneumonia	E
Staphylococcus aureus	Bacteria	1	BSI, UTI, pneumonia	H
Streptococcus pneumoniae	Bacteria	2	Pneumonia	H
Trichosporon beigelii	Fungal yeast	2	Various	E
VZV	Virus	1	Varicella	H

Note: E = environmental; H = human.

isolation rooms, or positively pressurized houses or clinics, are one approach (Linscomb 1994). In the case of HIV patients with TB infections, multiple barriers to protect both the patient and personnel outside may be warranted (CDC 1990). UV air disinfection systems can also provide some protection.

Burn Wound Infections

Patients with burn wounds are among those with the highest risk for noso-comial infections (Mayhall 1999). A burn injury creates a breach in the integrity of the skin and impacts the ability of the skin and other burned tissues to carry out normal protective functions (Gilpin, Rutan, and Herndon 1995). Burn wounds become colonized with a variety of microbes soon after the injury occurs. Figure 15.1 illustrates the most common microorganisms colonizing burn wounds. Pneumonia in burn wound patients may result from either blood-borne or airborne organisms, but in either case the infecting microbes are usually present in the flora of the burn wound. Almost all burn wound infections, some 98%, are caused by aerobic bacteria.

The most common infecting microbes are those that normally live as commensals, including *Staphylococcus, Pseudomonas, Streptococcus, Proteus, Klebsiella, Serratia, Enterococcus, Enterobacter, Acinetobacter, Candida, E. coli,* and coag-negative staphylococcus such as *Staphylococcus epidermis*. Burn injuries are frequently accompanied by inhalation of toxic gases, and pulmonary infections, including pneumonia, are among the primary causes of death in

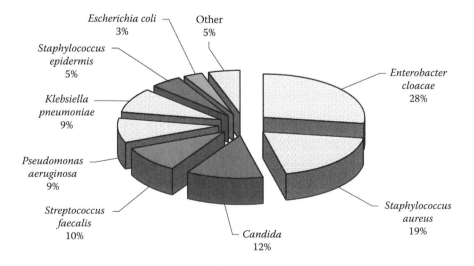

FIGURE 15.1
Microbes colonizing burn wounds. Based on data in Davis and Shires (1991).

burned patients. Immunodeficiency may accompany burn wounds and place the patients at high risk for infection. Microbes are present on the skin at the time of burning and are readily acquired from the patient's gastrointestinal tract. Pathogens can also be rapidly acquired from the hospital environment as well as from other patients in the facility. Early colonization of the burn wound takes place in the first 48 hours with Gram-positive commensals from within the sweat glands and hair follicles. Between 3 and 21 days, the wound becomes colonized with Gram-negative bacilli from the patient's own gastrointestinal tract or from other patients (Mooney and Gamelli 1989). Microorganisms that cause burn wound infections have been recovered from a number of sites in hospital environments including *Pseudomonas* from faucets, towel racks, and counter surfaces; *Enterobacter cloacae* from hydrotherapy equipment; *Acinetobacter* from mattresses; and *Providencia stuartii* from the air (Mayhall 1999; Wenzel et al. 1976). *Acinetobacter* has been suggested to have an airborne transmission component, as have several of the other most common infecting microbes (Allen and Green 1987).

Streptococcus pyogenes (GAS) have spread by the hands of HCWs and occasionally spread by the respiratory route in settings other than operating rooms (Crossley 1999). When patients are infected with large numbers of microbes, as in burn wounds, attention should be paid to the potential for airborne transmission and the role of fomites. GAS are resistant to desiccation and reports of infections in operating rooms adjacent to those in which colonized carriers have worked indicate that airborne transmission may occur (Richman, Breton, and Goldmann 1977; Berkelman et al. 1982). Table 15.2 summarizes the microbes among the most common burn wound infectious agents that may be transported by the airborne route.

The evidence indicates the primary transmission mechanism involves hand and surface contamination—the hands of HCWs become contaminated either from patients or else from local surfaces, and then transmit the microbes to other patients. Clearly such endogenous and ubiquitous microbes can transfer from hands to surfaces and vice versa, leaving the precise etiology confused. Burn wounds are certainly contaminated through self-inoculation, but these ubiquitous bacteria might also hail from other persons or they might settle out from the air. The actual cumulative dose of microbes that induce a staphylococcal infection, for example, may include staphylococci that came from the air. And of microbes that are inoculated onto a burn wound by hand or instrument contact, some may have also ultimately come from the air. For this reason all of the potentially airborne microbes mentioned as common infectious microbes above should be considered potentially airborne and may be amenable to appropriately designed air cleaning technologies. In this regard, barrier facilities and other techniques for limiting cross-contamination between patients in burn care wards have been effective in reducing infection rates, which is suggestive of an airborne component to cross-infection (McManus et al. 1985; Shirani et al. 1986).

TABLE 15.2

Airborne Pathogens in Burn Wounds

Pathogen	Type	Airborne Class	Source
Acinetobacter	Bacteria	2	Environmental
Alternaria	Fungal spore	2	Environmental
Aspergillus	Fungal spore	1	Environmental
Enterobacter	Bacteria	2	Human/environmental
Enterococcus	Bacteria	2	Human
Fusarium	Fungal spore	2	Environmental
Geotrichum	Fungal spore	2	Environmental
Klebsiella pneumoniae	Bacteria	2	Human/environmental
Proteus	Bacteria	2	Human
Mucor	Fungal spore	2	Environmental
Providencia stuartii	Bacteria	2	Human
Pseudomonas aeruginosa	Bacteria	1	Environmental
Rhizopus	Fungal spore	2	Environmental
Serratia marcescens	Bacteria	2	Environmental
Staphylococcus aureus	Bacteria	1	Human
Staphylococcus epidermis	Bacteria	2	Human
Streptococcus faecalis	Bacteria	2	Human
Streptococcus pyogenes	Bacteria	1	Human

Fungi account for some 7.5% of burn wound infections (Becker et al. 1991). Apart from *Candida*, which is a yeast, fungal spores that are commonly airborne and may settle on burn wounds include *Aspergillus*, *Mucor*, *Rhizopus*, *Fusarium*, and *Geotrichum*. Less commonly isolated fungi include *Drechslera*, *Alternaria*, and *Microspora*.

Microbes that cause burn wound infections have been recovered from a number of inanimate surfaces in burn wards (Mayhall 1999). Sites of environmental contamination in burn wards include sinks, bars of soap, towel racks, water supplies, counter surfaces, bed rails, and mattresses. Barrier techniques have been shown to be effective in diminishing infection rates in burn patients by isolating patients from each other. Local recirculation units using filtration and/or UV (see Figure 15.2) may be useful for controlling airborne microbial levels.

Cancer Clinics

Cancer patients are often at greater risk of infection due to immunosuppressive treatments. Nosocomial infections in cancer patients can be caused by a variety

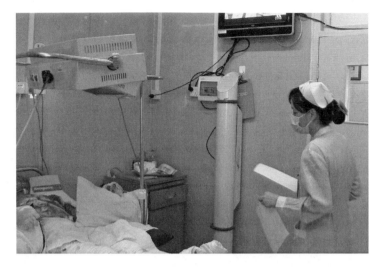

FIGURE 15.2
UVGI air disinfection unit (center) in use in a hospital burn ward. (Photo provided courtesy of Virobuster Electronic Air Sterilisation, Enschede, The Netherlands.)

of infectious microorganisms, with bacteria occurring in 75% of cases and viruses in 10% of cases (Hughes, Flynn, and Williams 1999). Fungal infections occurred from 2% to 22% of the time. Some of the bacterial causes of infection include species of *Corynebacterium, Enterococcus, Listeria, Staphylococcus, Streptococcus, Acinetobacter, Enterobacter, Haemophilus, Klebsiella, Moraxella, Legionella, Pseudomonas, Proteus, Neisseria,* and *Serratia.* Viral causes include adenovirus, influenza, parainfluenza, parvovirus, RSV, rotavirus, and VZV. Fungal causes include *Pseudallescheria boydii, Alternaria, Aspergillus, Coccidioides immitis, Fusarium, Geotrichum, Histoplasma capsulatum, Mucor, Penicillium,* and *Rhizopus.* The most common causes of respiratory infections include *S. aureus, P. aeruginosa, E. coli,* and *K. pneumoniae.* Precautions for cancer clinics include the same contact, droplet, and airborne precautions used in other health care facilities.

References

Allen, K., and Green, H. (1987). Hospital outbreak of multi-resistant *Acinetobacter anitratus*: An airborne mode of spread? *J Hosp Infect* 9, 110–119.

Becker, W. K., Cioffi, W. G., McManus, A. T., Kim, S. H., McManus, W. F., Mason, A. D., and Pruitt, B. A. (1991). Fungal burn wound infection. A 10-year experience. *Arch Surg* 126, 144–148.

Berkelman, R. L., Martin, D., Graham, D. R., Mowry, J., Freisem, R., Weber, J. A., Ho, J. L., and Allen, J. R. (1982). Streptococcal wound infections caused by a vaginal carrier. *JAMA* 247, 2680–2682.

CDC (1990). Guidelines for preventing the transmission of tuberculosis in health-care settings, with special focus on HIV-related issues. *MMWR* 39(RR-17), 1–29.

Craven, D. E., Steger, K. A., and Fleming, C. A. (1999). Nosocomial infections in adults infected with human immunodeficiency virus; in *Hospital Epidemiology and Infection Control*, C. G. Mayhall, ed., Lippincott Williams & Wilkins, Philadelphia, 745–766.

Crossley, K. B. (1999). Streptococci; in *Hospital Epidemiology and Infection Control*, C. G. Mayhall, ed., Lippincott Williams & Wilkins, Philadelphia, 477–503.

Davis, J. M., and Shires, G. T. (1991). *Principles and Management of Surgical Infections*. J. B. Lippincott Company, New York.

Dixon, D. M., McNeil, M. M., Cohen, M. L., Gellin, B. G., and LaMontagne, J. R. (1996). Fungal infections: A growing threat. *Pub Health Rep* 111, 226–235.

Fry, D. E. (1995a). Unusual fungal pathogens in surgical patients; in *Surgical Infections*, D. E. Fry, ed., Little, Brown and Company, Boston, 591–601.

———. (1995b). Actinomycosis and nocardiosis; in *Surgical Infections*, D. E. Fry, ed., Little, Brown and Company, Boston, 603–609.

Gilpin, D. A., Rutan, R. L., and Herndon, D. N. (1995). Burn wound infection; in *Surgical Infections*, D. Fry, ed., Little, Brown and Company, Boston, 169–177.

Hughes, W. T. (1982). Natural mode of acquisition for de novo infection with *Pneumocystis carinii*. *J Infect Dis* 145, 842.

Hughes, W. T., Flynn, P. M., and Williams, B. G. (1999). Nosocomial infections in patients with neoplastic diseases; in *Hospital Epidemiology and Infection Control*, C. G. Mayhall, ed., Lippincott Williams & Wilkins, Philadelphia, 767–780.

Linscomb, M. (1994). AIDS clinic HVAC system limits spread of TB. *HPAC* February.

Mayhall, C. G. (1999). Nosocomial burn wound infections; in *Hospital Epidemiology and Infection Control*, C. G. Mayhall, ed., Lippincott Williams & Wilkins, Philadelphia, 275–286.

McManus, A. T., McManus, W. F., Mason, A. D., Aitcheson, A. R., and Pruitt, B. A. (1985). Microbial colonization in a new intensive care burn unit. *Arch Surg* 120, 217–223.

Mooney, D. P., and Gamelli, R. L. (1989). Sepsis following thermal injury. *Compr Ther* 15, 22–29.

Perl, T. M., Chotani, R., and Agawala, R. (1999). Infection control and prevention in bone marrow transplant patients; in *Hospital Epidemiology and Infection Control*, C. G. Mayhall, ed., Lippincott Williams & Wilkins, Philadelphia, 803–844.

Richman, D. D., Breton, S. J., and Goldmann, D. A. (1977). Scarlet fever and Group A streptococcal wound infection traced to an anal carrier. *J Pediatr* 90, 387–390.

Shirani, K. Z., McManus, A. T., Pruitt, B. A., and Mason, A. D. (1986). Effects of environment on infection in burn patients. *Arch Surg* 121, 131–136.

Vonberg, R. P., and Gastmeier, P. (2006). Nosocomial aspergillosis in outbreak settings. *J Hosp Infect* 63, 246–254.

Wenzel, R. P., Hunting, K. J., Osterman, C. A., and Sande, M. A. (1976). *Providencia stuartii*, a hospital pathogen: Potential factors for its emergence and transmission. *Am J Epidemiol* 104, 170–180.

Wong, S. (1999). Surgical site infections; in *Hospital Epidemiology and Infection Control*, C. G. Mayhall, ed., Lippincott Williams & Wilkins, Philadelphia, 189–210.

16

Nursing Homes

Introduction

Nursing homes are defined as facilities licensed with a staff of professionals and inpatient beds that provide continuous nursing and other services to patients who are not in an acute phase of illness. Nursing homes house people who are unable to manage independently in the community. The terms nursing homes and long-term care facilities (LTCF) are substantially identical. The terms nosocomial and hospital-acquired infections (HAI) or health care–associated infections (HCAI) are also considered to apply to nursing homes for the sake of this text. These facilities share many of the same concerns about nosocomial or hospital-acquired infections but also have some unique infection problems. Table 16.1 summarizes all the potential airborne nosocomial pathogens that have caused infections in nursing homes.

The field of infection control in nursing homes is relatively new compared with hospital infection control, but detailed guidelines are already available from regulatory agencies (Mayon-White and Smith-Casey 2003; Crossley, Nelson, and Irvine 1992). Almost as many nosocomial infections occur annually in LTCFs as in hospitals in the United States (Haley et al. 1985). There are about 15,000 nursing homes in the United States and approximately 1.5 million residents with an average age of over 80 years (Smith et al. 2008). Residents of LTCFs are typically elderly and in declining health but may remain in residence for years. LTCFs often have acute care facilities and patients are regularly transferred between the two wings. The information in this chapter also applies to long-term acute care facilities (LTACs).

Nursing Home Bacterial Infections

Between 1.6 and 3.6 million infections occur annually in US LTCFs (Strausbaugh and Joseph 2000). Outbreaks of respiratory infections and gastroenteritis are common and estimated to result in thousands of outbreaks annually

TABLE 16.1

Potential Airborne Infections in Nursing Homes

Pathogen	Type	Airborne Class	Infection	Source
Adenovirus	Virus	2	Pneumonia	E
Bordetella pertussis	Bacteria	1	Whooping cough	H
Chlamydia pneumoniae	Bacteria	2	Pneumonia	H
Clostridium difficile	Bacterial spore	1	Gastrointestinal infection	H/E
Clostridium perfringens	Bacterial spore	2	Toxic reactions	E
Coronavirus	Virus	1	RTI	H
Coxsackievirus	Virus	2	RTI	H
Enterococcus (VRE)	Bacteria	2	BSI, UTI	H
Haemophilus influenzae	Bacteria	1	Pneumonia	H
Influenza	Virus	1	Pneumonia	H
Legionella	Bacteria	1	Pneumonia	E
Metapneumovirus	Virus	2	RTI	H
Mycobacterium tuberculosis	Bacteria	1	TB	H
Mycoplasma pneumoniae	Bacteria	2	Pneumonia	H
Norwalk virus	Virus	2	Gastrointestinal infection	E
Parainfluenza	Virus	1	Pneumonia	H
Rhinovirus	Virus	2	RTI	H
Rotavirus	Virus	2	Gastrointestinal infection	H
RSV	Virus	1	Pneumonia	H
Staphylococcus aureus (MRSA)	Bacteria	1	BSI, UTI, pneumonia	H
Streptococcus pneumoniae	Bacteria	1	Pneumonia	H
Streptococcus pyogenes	Bacteria	1	Fever, pharyngitis	E

Note: E = environmental; H = human.

(Strausbaugh et al. 2003). Elderly residents may have decreased immune system response and impaired health and nutrition, which makes them more susceptible to common infections (Castle 2000). Certain other factors common to nursing homes can impact the risk of infection such as family visitations, nurse turnover, staffing, and hospital transfer rates (Zimmerman et al. 2002).

Outbreaks in nursing homes have involved a number of potential airborne nosocomial pathogens or spores with airborne transport, including *Mycobacterium tuberculosis, Streptococcus pyogenes, S. pneumoniae, Enterococcus* (VRE)*, Bordetella pertussis, Haemophilus influenzae, Mycoplasma pneumoniae, Clostridium difficile, C. perfringens,* influenza, parainfluenza, RSV, adenovirus, rhinovirus, coronavirus, Coxsackievirus, rotavirus, and Norwalk or Norwalk-like viruses (Strausbaugh and Joseph 1999). Upper respiratory tract infections are common in nursing homes and follow seasonal community epidemics.

The most common infections in LTCFs are urinary tract infections (UTIs) and these may have little if any relation to airborne microbes. The primary aerobiological hazards are respiratory tract infections (RTIs) and pneumonia. Due to impaired immunity, RTIs may have greater impact on elderly residents than in the general population (Drinka et al. 1999). Most HCAI infections are sporadic in nature and are caused by colonizing microorganisms with relatively low virulence. Bacteria responsible for epidemics in nursing homes include *Mycobacterium tuberculosis, Streptococcus pneumoniae, Chlamydia pneumoniae, Legionella, Clostridium difficile, Salmonella, E. coli,* methicillin-resistant *Staphylococcus aureus,* vancomycin-resistant *Enterococcus,* and Group A *Streptococcus* (Smith et al. 2008).

Pneumonia is the second most common cause of infection and is the leading cause of mortality in nursing home residents (Mylotte 2002). Elderly residents are predisposed to pneumonia due to decreased clearance of bacteria from airways, underlying diseases, altered throat flora, inadequate oral care, and other factors (Marrie 2002). The most common etiologic agent of pneumonia is *Staphylococcus aureus,* representing about 35% of cases (Marrie 2002). *Streptococcus pneumoniae* accounts for about 13% of all cases (Gleich et al. 2000). The seasonal variation of pneumonia mirrors that of influenza and suggests that influenza virus infections play a role in pneumonia (CDC 2003). The case-fatality rate of pneumonia in LTCFs is approximately 6–23%. In addition, outbreaks have been associated with *Legionella, Chlamydia pneumoniae,* and RSV (CDC 2003).

Tuberculosis has been responsible for outbreaks in nursing homes, involving staff as well as residents, and attack rates are high (Garrett et al. 1999; Stead et al. 1985). Multidrug-resistant organisms (MDROs) have been associated with increased morbidity, mortality, and costs in nursing homes (Capitano and Nicolau 2003). MDROs are likely to pose a continuing problem to nursing homes, and as antibiotic resistance becomes pervasive, increased dependence on precautions, isolation, surveillance, and infection management is likely. Isolation in private rooms or isolation of entire wings is common for infections thought to be transmitted by direct contact or fomites.

MRSA is an important pathogen in nursing homes with prevalence rates from about 7% to 22% (Hartstein and Mulligan 1999; Barr et al. 2007). These institutions may serve as reservoirs for the introduction of MRSA into acute care hospitals. *Streptococcus pyogenes* (GAS) has been responsible for outbreaks in long-term care facilities, causing pneumonia and cutaneous infections (Crossley 1999). Nursing home residents have multiple risk factors for acquiring MRSA, including diabetes, hemodialysis, frequent hospitalization, and the sharing of rooms and other areas (Reynolds et al. 2011). In this latter study the overall point prevalence of MRSA was 31%, as compared to 6% in hospitals. Wendt et al. (2005) reports that 12–54% of nursing home residents in Germany were colonized with MRSA.

Clostridium difficile occurs in nursing homes at a rate of about 0.08 cases per 1000 resident days (Johnson and Gerding 1999). Nosocomial acquisition is the most important route of transmission, as opposed to endogenous activation. *Clostridium difficile* infections are especially problematic among elderly individuals, and this is the group in which the infection is most common (Campbell et al. 2009). As many as half of nursing home residents may have infections asymptomatically.

Chlamydia pneumoniae can cause large outbreaks of infection among elderly persons and its transmission is enhanced by close contact in nursing homes (Nakashima et al. 2006). The incubation period is thought to be from several weeks to several months and attack rates range from 44% to 68% with reinfection possible. Sharing of rooms is not a risk factor but close contact at meals may be a risk, suggesting this microbe is not airborne.

Nursing Home Viral Infections

Viruses that have been identified as causative agents of nursing home infections include influenza, parainfluenza, coronavirus, rhinoviruses, adenoviruses, and metapneumovirus (Boivin et al. 2007; Schlapbach et al. 2011). Influenza has caused outbreaks in nursing homes with clinical attack rates from 25% to 70% and with case fatality rates over 10% (Gravestein, Miller, and Drinka 1992). In one nursing home a typical outbreak began in November, peaked in February, and ended in April. The progression of the outbreak was complicated by other concurrent infections with RSV, parainfluenza, and *Mycoplasma pneumoniae* (Valenti 1999). Influenza is significantly more common in unvaccinated groups, with over twice the mortality rate than the vaccinated group. Coughing and sneezing by an influenza-infected patient generates clouds of aerosols and large droplets that can spray onto the surroundings and cause multiple secondary infections (Gross et al. 1988; Patriarca et al. 1987). These droplets may spray in a radius of up to 6 feet around the patient, although 3 feet is considered a safe minimum distance.

A study was conducted by Drinka et al. (2003) on the influenza infection risk between roommates in nursing homes. The introduction of influenza to one roommate in a double room puts the unaffected roommate at an almost 20% risk of being infected. Those who lived in rooms with an infected roommate had a relative risk of acquiring influenza of 3.07. However, the susceptibility of individuals was noted to vary greatly.

Norwalk virus or Norwalk-like viruses have been responsible for outbreaks in hospitals and nursing homes in which the agent is suspected of being transmitted by the airborne route (Chadwick and McCann 1994; Gellert, Waterman, and Ewert 1990). Aerosolization during projectile vomiting is believed to be capable of creating airborne virus particles as well as fomites. A norovirus outbreak in which 52% of residents and 46% of staff developed gastroenteritis was investigated by Wu et al. (2005). It was found that environmental contamination with norovirus was extensive, including rooms, dining room tables, and elevator buttons.

Outbreaks of severe respiratory tract infection due to human metapneumovirus have highlighted the emergence of this pathogen as an important cause of upper and lower respiratory tract infections in nursing homes (Boivin et al. 2007). No reports have yet implicated metapneumovirus as an airborne infection, but all respiratory viruses have this potential.

Nursing Home Options

Handwashing remains the best practice for the prevention of infections in LTCFs, but surveys indicate low adherence (about 30%) to appropriate handwashing, and that among nursing assistants an 82% frequency of potential microbial transmission had occurred after resident contact (Goldrick 1999). A series of guidelines and recommendations for preventing infection in LTCFs has been presented by the Association for Professionals in Infection Control (APIC) in a position paper (Smith and Rusnak 1997).

Because most nursing homes do not have isolation rooms or rooms with separate ventilation, they have no recourse to deal with potential airborne infections. Most residents in nursing homes tend to keep their doors wide open, or they are kept open by staff, and so mixing of air via hallways is inevitable. In facilities with central air handling units, filters and UVGI systems can be added to help control airborne infections. In facilities without central air, local recirculation units could be used to help reduce infection risks in individual rooms.

References

Barr, B., Wilcox, M. H., Brady, A., Parnell, P., Darby, B., and Tompkins, D. (2007). Prevalence of methicillin-resistant *Staphylococcus aureus* colonization among older residents of care homes in the United Kingdom. *Inf Contr Hosp Epidemiol* 28(7), 853–859.

Boivin, G., deSerres, G., Hamlin, M. E., Cote, S., Argouin, M., Tremblay, G., Maranda-Aubut, R., Sauvageau, C., Ouakki, M., Boulianne, N., and Couture, C. (2007). An outbreak of severe respiratory tract infection due to human metapneumovirus in a long-term care facility. *Clin Infect Dis* 44(9), 1152–1158.

Campbell, R. J., Giljahn, L., Machesky, K., Cibulskas-White, K., Lane, L. M., Porter, K., Paulson, J. O., Smith, F. W., and McDonald, C. (2009). *Clostridium difficile* infection in Ohio hospitals and nursing homes during 2006. *Inf Contr Hosp Epidemiol* 30(6), 526–533.

Capitano, B., and Nicolau, D. P. (2003). Evolving epidemiology and cost of resistance to antimicrobial agents in long-term care facilities. *J Am Med Assoc* 4(S), 90–99.

Castle, S. C. (2000). Clinical relevance of age-related immune dysfunction. *Clin Infect Dis* 31, 578–585.

CDC (2003). *Guidelines for Preventing Health-Care Associated Pneumonia*. Centers for Disease Control, Atlanta, GA.

Chadwick, P. R., and McCann, R. (1994). Transmission of a small round structured virus by vomiting during a hospital outbreak of gastroenteritis. *J Hosp Infect* 26, 251–259.

Crossley, K., Nelson, L., and Irvine, P. (1992). State regulations governing infection control issues in long-term care. *J Am Geriatr Soc* 40, 251–254.

Crossley, K. B. (1999). Streptococci; in *Hospital Epidemiology and Infection Control*, C. G. Mayhall, ed., Lippincott Williams & Wilkins, Philadelphia, 385–394.

Drinka, P. J., Gravenstein, S., Langer, E., Krause, P., and Shult, P. (1999). Mortality following isolation of various respiratory viruses in nursing home residents. *Infect Contr Hosp Epidemiol* 20, 812–815.

Drinka, P. J., Krause, P., Nest, L., Goodman, B. M., and Gravenstein, S. (2003). Risk of acquiring influenza A in a nursing home from a culture-positive roommate. *Inf Contr Hosp Epidemiol* 24(11), 872–874.

Garrett, D. O., Dooley, S. W., Snider, D. E., and Jarvis, W. R. (1999). *Mycobacterium tuberculosis*; in *Hospital Epidemiology and Infection Control*, C. G. Mayhall, ed., Lippincott Williams & Wilkins, Philadelphia, 477–503.

Gellert, G. A., Waterman, S. H., and Ewert, D. (1990). An outbreak of acute gastroenteritis caused by a small round structured virus in a geriatric convalescent facility. *Inf Contr Hosp Epidemiol* 11, 459–464.

Gleich, S., Morad, Y., Echague, R., Miller, J. R., Kornblum, J., Sampson, J. S., and Butler, J. C. (2000). *Streptococcus pneumoniae* serotype 4 outbreak in a home for the aged: Report and review of recent outbreaks. *Infect Contr Hosp Epidemiol* 21(11), 711–717.

Goldrick, B. A. (1999). Infection control programs in long-term-care facilities: Structure and process. *Inf Contr Hosp Epidemiol* 20(11), 764–769.

Gravestein, S., Miller, B. A., and Drinka, P. (1992). Prevention and control of influenza A outbreaks in long-term care facilities. *Infect Contr Hosp Epidemiol* 13, 49–54.

Gross, P. A., Rodstein, M., LaMontagne, J. R., Kaslow, R. A., Saah, A. J., Wallenstein, S., Neufeld, R., Denning, C., Gaerlan, P., and Quinnan, G. V. (1988). Epidemiology of acute respiratory illness during an influenza outbreak in a nursing home. *Arch Int Med* 148, 559–561.

Haley, R. W., Culver, D. H., White, J. W., Morgan, W. M., and Emori, T. G. (1985). The nationwide nosocomial infections rate: A new need for vital statistics. *Am J Epidemiol* 121, 159–167.

Hartstein, A. I., and Mulligan, M. E. (1999). Methicillin-resistant *Staphylococcus aureus*; in *Hospital Epidemiology and Infection Control*, C. G. Mayhall, ed., Lippincott Williams & Wilkins, Philadelphia, 347–364.

Johnson, S., and Gerding, D. N. (1999). *Clostridium difficile*; in *Hospital Epidemiology and Infection Control*, C. G. Mayhall, ed., Lippincott Williams & Wilkins, Philadelphia, 467–476.

Marrie, T. J. (2002). Pneumonia in the long-term-care facility. *Infect Contr Hosp Epidemiol* 23, 159–164.

Mayon-White, R., and Smith-Casey, J. (2003). Infection control in British nursing homes. *Inf Contr Hosp Epidemiol* 24(4), 296–298.

Mylotte, J. M. (2002). Nursing home-acquired pneumonia. *Clin Infect Dis* 35, 1205–1211.

Nakashima, K., Tanaka, T., Kramer, M. H., Takahashi, H., Ohyama, T., Kishimoto, T., Toshima, H., Miwa, S., Nomura, A., Tsumura, N., Ouchi, K., and Okabe, N. (2006). Outbreak of *Chlamydia pneumoniae* infection in a Japanese nursing home, 1999–2000. *Inf Contr Hosp Epidemiol* 27(11), 1171–1177.

Patriarca, P. A., Arden, N. H., Koplan, J. P., and Goodman, R. A. (1987). Prevention and control of type A influenza infections in nursing homes. *Ann Intern Med* 107, 732–740.

Reynolds, C., Quan, V., D, K., Peterson, E., Dunn, J., Whealon, M., Terpstra, L., Meyers, H., Cheung, M., Lee, B., and Huang, S. S. (2011). Methicillin-resistant *Staphylococcus aureus* (MRSA) carriage in 10 nursing homes in Orange County, Florida. *Inf Contr Hosp Epidemiol* 32(1), 91–93.

Schlapbach, L. J., Agyeman, P., Hutter, D., Aebi, C., Wagner, B. P., and Reidel, T. (2011). Human metapneumovirus infection as an emerging pathogen causing acute respiratory distress syndrome. *J Infect Dis* 203(2), 294–295.

Smith, P. W. (1984). *Infection Control in Long-Term Care Facilities.* John Wiley & Sons, New York.

Smith, P. W., Bennett, G., Bradley, S., Drinka, P., Lautenbach, E., Marx, J., Mody, L., Nicolle, L., and Stevenson, K. (2008). SHEA/APIC Guideline: Infection prevention and control in the long-term care facility. *AJIC* 36, 504–535.

Smith, P. W., and Rusnak, P. G. (1997). Infection prevention and control in the long-term care facility. *Inf Contr Hosp Epidemiol* 18, 831–849.

Stead, W. W., Lofgren, J. P., Warren, E., and Thomas, C. (1985). Tuberculosis as an endemic and nosocomial infection among the elderly in nursing homes. *N Engl J Med* 312, 1483–1487.

Strausbaugh, L. J., and Joseph, C. L. (1999). Epidemiology and prevention of infections in residents of long-term care facilities; in *Hospital Epidemiology and Infection Control*, C. G. Mayhall, ed., Lippincott Williams & Wilkins, Philadelphia, 1461–1482.

———. (2000). The burden of infection in long-term care. *Infect Control Hosp Epidemiol* 21, 674–679.

Valenti, W. M. (1999). Influenza viruses; in *Hospital Epidemiology and Infection Control*, C. G. Mayhall, ed., Lippincott Williams & Wilkins, Philadelphia, 535–542.

Strausbaugh, L. J., Sukumar, S. R., and Joseph, C. L. (2003). Infectious disease outbreaks in nursing homes: An unappreciated hazard for frail elderly persons. *Clin Infect Dis* 36, 870–876.

Wendt, C., Svoboda, D., Schmidt, C., Bock-Hensley, O., and Baum, H. V. (2005). Characteristics that promote transmission of *Staphylococcus aureus* in German nursing homes. *Inf Contr Hosp Epidemiol* 26, 816–821.

Wu, H. M., Fornek, M., Scwab, K. J., Chapin, A. R., Gibson, K., Schwab, E., Spencer, C., and Henning, K. (2005). A norovirus outbreak at a long-term-care facility: The role of environmental surface contamination. *Inf Contr Hosp Epidemiol* 26(10), 802–810.

Zimmerman, S., Gruber-Baldini, A. L., Hebel, J. R., Slone, P. D., and Magaziner, J. (2002). Nursing home facility risk factors for infection and hospitalization: Importance of registered nurse turnover, administration, and social factors. *J Am Geriatr Soc* 50, 1987–1995.

17

Procedural Controls and Guidelines

Introduction

A variety of guidelines and standards have been issued by various organizations that address issues related to air and surface disinfection and the technologies and procedures used in such applications. This chapter identifies and summarizes the various documents. Foremost among these documents are the guidelines issued by the Centers for Disease Control and Prevention (CDC), which provide extensive and relevant information. Disinfection procedures are often standards issued by organizations or else are in-house procedural standards unique to each facility. Handwashing and hand hygiene are an important part of infection management procedures and are also relevant to air and surface disinfection. The use of filtration, UVGI, and other disinfection methods for air disinfection are often addressed in part or in whole in many of these documents and it is noted whenever such mention is made.

A synopsis of the contents of these guidelines is provided and only aspects of these guidelines that relate to air and surface disinfection are reviewed here. Water disinfection issues addressed in the guidelines are not reviewed. Focus is provided on ventilation and air cleaning systems (i.e., filtration and UVGI) whenever the indicated document addresses these technologies. Focus is also provided on specific airborne microorganisms where they are discussed, while nonairborne microbes are omitted from the discussion.

Documents are classified according to the following three categories:

1. Disinfection Procedures and Protocols
2. Infection Management Procedures
3. Building and Facility Guidelines

Table 17.1 identifies the categories and lists the documents that are reviewed in this chapter. Additional relevant documents are summarized at the end of this chapter or are provided in the References.

TABLE 17.1

Guidelines Relevant to Air and Surface Disinfection

	Reference
Hygiene and Disinfection Protocols	
Guideline for Hand Hygiene in Health Care Setting	CDC 2002
Hand Washing, Cleaning, Disinfection, and Sterilization in Health Care	Health Canada 1998
Guideline for Disinfection and Sterilization in Health Care Facilities	CDC 2008
WHO Guidelines on Hand Hygiene in Health Care: A Summary	WHO 2009
APIC Guideline for Hand Washing and Antisepsis in Health Care Settings	APIC 1995
Infection Management Procedures and Protocols	
Guideline for Prevention of Surgical Site Infection	CDC 1999
Laboratory Infection Control Policy	Johns Hopkins Hospital 2007
Infection Control in the NICU: Recommended Standards	SGC 2001
Practical Guidelines for Infection Control in Health Care Facilities	WHO 2004
Prevention and Control of Tuberculosis in Correctional and Detention Facilities	CDC 2006
Management of Multidrug-Resistant Organisms In Health Care Settings	HICPAC 2006
Guidelines for Preventing Health Care–Associated Pneumonia	CDC 2003
Guidelines for Environmental Infection Control in Health Care Facilities	HICPAC 2003
Preventing the Transmission of *Mycobacterium tuberculosis* in Health Care Facilities	CDC 2005
Revised Hospital Interpretive Guidelines 42CFR482.42 Infection Control	CFR 2007
Prevention of Tuberculosis in Health Care Facilities in Resource-Limited Settings	WHO 1999
Guideline for Prevention of Catheter-Associated Urinary Tract Infections	HICPAC 2009
HA Infection Control Guidelines for Avian Influenza under Red Alert	CHP 2005
Infection Control in Anesthesia	AAGBI 2008
Protocol for Management of the Patient with H1N1 Influenza Virus	Sutter 2009
Guideline for Infection Control in Health Care Personnel	HICPAC 1998
Guideline for Isolation Precautions: Preventing Transmission of Infectious Agents	HICPAC 2007
SHEA/APIC Guideline: Infection Prevention and Control in the Long-Term Care Facility	Smith et al. 2008
APIC/CHICA-Canada Infection Prevention, Control, and Epidemiology	Friedman et al. 2008
Building and Facility Guidelines	
Guidelines for Construction and Equipment of Hospital and Medical Facilities	AIA 2006

TABLE 17.1 (*Continued*)

Guidelines Relevant to Air and Surface Disinfection

	Reference
HVAC Design Manual for Hospitals and Clinics	ASHRAE 2003
General Requirements: Purpose of the Facilities Standard for Public Buildings Service	GSA 2005
Guidelines on the Design and Operation of HVAC Systems in Disease Isolation Areas	USACHPPM 2000
Unified Facilities Criteria UFC 4-510-01 Design: Medical Military Facilities	UFC 2001
ASHRAE Position Document on Airborne Infectious Diseases	ASHRAE 2009
Manual of Recommended Practices for Industrial Ventilation	ACGIH 2010
Department of the Air Force Operating Room Ventilation Update	NRC 1998
Using Ultraviolet Radiation and Ventilation to Control Tuberculosis	DHS 1990
Environmental Control for Tuberculosis: Basic Upper-Room UVGI Guidelines for Health Care	NIOSH 2009

Hygiene and Disinfection Procedures and Controls

Disinfection procedures include surface disinfection, equipment disinfection, handwashing, and other related factors. Guidelines for hand hygiene in health care have been addressed by the Centers for Disease Control (CDC), the Association for Professionals in Infection Control (APIC), the World Health Organization (WHO), and the Healthcare Infection Control Practices/Advisory Committee (HICPAC).

> **Title:** Guideline for Hand Hygiene in Health Care Settings
>
> **Reference:** CDC 2002
>
> **Synopsis:** This guideline provides specific recommendations to promote improved hand hygiene in health care settings. It provides detailed background information on hand sepsis and its relation to the incidence of health care–associated infections. It describes methods used to evaluate the efficacy of various hand hygiene products and disinfectants, and identifies the log reduction rates that can be achieved with various products such as soaps, alcohols, chlorhexidine, chloroxylenol, hexachlorophene, iodine, quaternary ammonium compounds, triclosan, and other agents. Protocols for applying these products are reviewed and their effectiveness against individual species are discussed. The resistance of bacteria to certain antiseptic agents is discussed. The problem of nonadherence to procedures is discussed along with recommendations to promote improved hand hygiene. Gloving policies and their effectiveness are also addressed. Recommendations are

provided for improving hand hygiene practices and are classified in categories, from those that are strongly recommended to those that are suggested for implementation.

Title: Infection Control Guidelines: Hand Washing, Cleaning, Disinfection, and Sterilization in Health Care

Reference: Health Canada 1998

Synopsis: This guideline addresses hand washing, gloving, the cleaning and disinfection of patient care equipment, environmental sampling, housekeeping, laundry, and waste management. The characteristics of antiseptic agents are summarized and their relative effectiveness against bacteria, fungi, and viruses is described. Procedures for hand washing are provided along with specific recommendations graded according to a scale similar to but more detailed than the CDC categories. The disinfection of equipment is addressed in detail including listings of the type of hospital equipment, the products used for particular equipment, and the cleaning methods used. Also provided are manufacturer's recommendations for specific disinfectants including alcohols, chlorines, ethylene oxide, formaldehyde, glutaraldehydes, hydrogen peroxide, iodophores, peracetic acid, phenolics, and others. Ultraviolet radiation is briefly addressed as a means of disinfecting ventilation air but no mention is made of UVGI as a means of disinfecting surfaces or equipment. The various sterilization methods are discussed and tabulated with a description of their applications, advantages, and disadvantages. Recommendations for disinfection are graded according to the same categorization system used for hand washing. Microbiological sampling of the environment is addressed, and although routine sampling of the environment is not recommended due to a lack of evidence of its effectiveness in controlling infections, it is still recommended for sampling of water and certain fluids. Recommendations for housekeeping and laundry are provided and cleaning procedures are tabulated according to the application and categorized by the same scheme for hand washing and disinfection. Waste management is also addressed by category and application.

Title: Guideline for Disinfection and Sterilization in Healthcare Facilities

Reference: CDC 2008

Synopsis: This guideline provides a comprehensive review of the disinfection of health care equipment, factors affecting the efficacy of disinfection methods, cleaning methods, disinfectants and their applications, and sterilization methods and their applications, and provides extensive resources and references for source information. This guideline uses the Spaulding scheme, which distinguishes between critical items that confer a high risk of infection, semicritical items that contact mucous membranes or nonintact skin, and noncritical items for which virtually no infection risk has been documented. Detailed information is provided on the contamination of endoscopes, laparoscopes, arthroscopes, tonometers, dental equipment, surgical tools, and other equipment, and associated disinfection methods are discussed. Focus is placed on inactivation of *Clostridium difficile* and disinfection of other emerging health care pathogens. The necessity of surface disinfection for critical and noncritical surfaces in hospitals is addressed in detail including discussion of the effectiveness of various disinfectants and their efficacy in reducing the risk of disease transmission. Air disinfection is briefly covered and only in regard to disinfectant fog-spray techniques. Air disinfection by filtration and UVGI are deferred to another guideline (HICPAC 2003). Factors affecting disinfection and sterilization efficacy are addressed including the innate resistance of microorganisms, physical factors, duration of exposure to disinfectants, and biofilms. Cleaning methods and cleaning equipment are addressed along with an overview of all major disinfectants and their applications. Ultraviolet radiation is discussed as a disinfection method that is limited to the destruction of airborne organisms and inactivation of microorganisms on surfaces. The failure of UV to control postoperative wound infection is cited based on only a single study (NRC 1964). This document defines the sterility assurance level (SAL) as "the probability of a single viable microorganism occurring on a product after sterilization" and addresses the various sterilization technologies such as steam and ETO.

Title: WHO Guidelines on Hand Hygiene in Health Care: A Summary

Reference: WHO 2009

Synopsis: This guideline compiles the latest scientific evidence and lessons learned from the field on the use of hand hygiene practices for controlling health care–associated infections (HCAI). It provides evidence of the importance of hand hygiene and consensus recommendations for hand washing techniques. Worldwide statistics on HCAI are reviewed for both developed nations and developing countries. Transmission of HCAI through contaminated hands of HCWs is the most common pattern in most health care settings and requires five sequential steps: (1) microbes are present on the patient's skin or have been shed onto inanimate objects immediately surrounding the patient; (2) microbes must be transferred to the hands of HCWs; (3) microbes must be capable of surviving for several minutes on HCWs' hands; (4) hand washing or hand antisepsis of the HCWs must be inadequate or omitted entirely, or the agent used for hand hygiene is inappropriate; (5) the contaminated hands of the caregiver must come into direct contact with another patient or with an inanimate object that will come into direct contact with the patient. Risk factors for poor adherence to hand hygiene are discussed and strategies to improve hand hygiene are presented. The consensus recommendations for hand hygiene are based on the summarized evidence and are graded according to the system developed by the CDC/HICPAC. Specific detailed instructions for hand washing are provided including depictions of the step-by-step hand hygiene techniques for alcohol-based formulations and for soap and water. An overall implementation strategy is provided along with referenced sets of tools for HCWs to translate WHO recommendations into practice. Five essential elements of the implementation strategy are (1) system change—to ensure that the necessary infrastructure is in place; (2) training and education on the correct procedures for hand rubbing and hand washing; (3) evaluation and feedback to monitor good hand hygiene practices and infrastructure; (4) reminders in the workplace to prompt and remind HCWs about the importance of hand hygiene; (5) institutional safety climate—creating an environment for participation, awareness, and improvement at all levels.

Title: APIC Guideline for Handwashing and Antisepsis in Health Care Settings

Reference: APIC 1995

Synopsis: This guideline focuses on hand washing, surgical scrub techniques, and related topics, and provides information on the skin flora of hands and the characteristics of selected antimicrobial agents. Recommendations are made regarding (1) health care personnel hand washing, (2) personnel hand preparation for operative procedures, and (3) other aspects of hand care and protection. Transient flora are identified as microorganisms isolated from the skin but not demonstrated to be consistently present in the majority of persons. Resident flora or colonizing flora are persistently isolated from the skin of most persons, are considered permanent residents, and are not readily removed by mechanical friction. Three types of hand care are identified: (1) Hand washing for removing soil and transient flora, (2) hand antisepsis to remove and destroy transient flora, and (3) surgical hand scrub to remove or destroy transient flora and reduce resident flora. Some new technologies are described and those of future investigation are identified. Noncategorized recommendations are provided for hand washing and hand antisepsis.

Infection Management Procedures

Infection management forms an integral part of any infectious disease control program and the following documents provide guidance in this regard. Some redundancy exists between these guidelines and other guidelines that focus on hand washing and disinfection procedures.

Title: Guideline for Prevention of Surgical Site Infection

Reference: CDC 1999

Synopsis: This document presents recommendations for preventing surgical site infections (SSIs) and provides an overview of the epidemiology, microbiology, pathogenesis, and surveillance of SSIs. Recommendations are provided based on the consensus of HICPAC. Criteria are tabulated for defining an SSI and a distinction is made between superficial incisional SSIs, deep incisional SSIs, and organ/space SSIs. It is stated that microbial contamination of the surgical site is a precursor to an SSI and that the risk of an SSI is a function of the microbial dose, the virulence of the pathogen, and the host susceptibility. It is also stated that the source of most SSIs is the patient's endogenous flora but that exogenous sources include surgical personnel, the operating room and room air, and equipment brought into the room. *S. aureus* is discussed as a frequent SSI isolate often found in the patient's nares. Antiseptic agents are reviewed for their efficacy against classes of microorganisms and their preoperative applications are described. The operating room air is discussed as a source of microbes and some evidence for airborne transmission of Group A streptococci is cited. The use of laminar airflow is noted as having the potential to reduce SSI risk. It is noted that intraoperative UVGI has not been shown to decrease SSI risk according to two studies cited at the time this guideline was issued. Protocols for attire, asepsis, and surgical technique are reviewed. Prediction of SSI risk is discussed and four variables associated with SSI risk are identified: (1) abdominal operations, (2) operations over two hours, (3) wound classifications of contaminated or dirty/infected, and (4) operations on patients with three or more discharge diagnoses. Recommendations are provided according to categorical rankings and include preoperative practices, ventilation guidelines, sterilization of equipment, surgical attire, asepsis and surgical technique, postoperative care, and surveillance.

Title: Laboratory Infection Control Policy

Reference: Johns Hopkins Hospital 2007

Synopsis: These guidelines were issued by the Johns Hopkins Hospital (JHH) Department of Pathology to prevent and control infections in hospital laboratories. They address issues related to employee health, education, attire, procedures for handling patient specimens, general practices, cleaning and disinfection of equipment, housekeeping procedures, and maintenance, and direct users to other JHH procedures. Personnel exposed to communicable diseases are expected to report the exposure and follow appropriate procedures. Employee education consists of orientation and an annual update. Hand washing procedures are described. Six categories of isolation for patients are identified: (1) contact precautions, (2) droplet precautions, (3) pediatric droplet precautions, (4) airborne isolation, (5) special precautions, and (6) maximum precautions. Guidelines for cleaning and disinfecting equipment are summarized.

Title: Infection Control in the NICU: Recommended Standards

Reference: SGC 2001

Synopsis: These recommended standards were issued by the Study Group for the Control of Infections in NICUs for the Hospital Authority and are largely based on the "Guidelines for Perinatal Care" issued by the American Academy of Pediatrics and CDC guidelines. Topics addressed include the physical setup of the NICU, ventilation, scrub areas, and airborne isolation rooms. Ventilation requirements for NICUs and airborne isolation rooms are in accordance with ASHRAE guidelines. Administrative issues are addressed including surveillance for nosocomial infections, hand washing procedures, cleaning procedures, and procedures for cleaning and disinfecting patient care equipment and catheters.

Title: Practical Guidelines for Infection Control in Health Care Facilities

Reference: WHO 2004

Synopsis: These guidelines were issued by the World Health Organization to provide comprehensive information to health care workers in the prevention and control of transmissible diseases. They provide directions related to facilities, equipment, and procedures necessary for control of infections; cleaning and disinfection of equipment; waste management; and protection of health care workers from transmissible infections. They define standard precautions separately from "additional (transmission-based) precautions," which are used for diseases transmitted by air, droplets, and contact. They define the objectives of an infection control program and the responsibilities of an infection control committee and outline the education and training requirements of health care staff.

Title: Prevention and Control of Tuberculosis in Correctional and Detention Facilities: Recommended Practices from CDC

Reference: CDC 2006

Synopsis: This document addresses the problem of TB control in jails, prisons, and other correctional and detention facilities. Effective control is stated to be achievable through early identification, appropriate use of airborne precautions, comprehensive discharge planning and contact investigation, and education of staff and detainees. Recommendations are provided for screening inmates, procedures for handling inmates with suspected TB, diagnostic testing methods, environmental controls including ventilation and air cleaning, respiratory protection, treatment, management, discharge planning, education, and related issues. Three factors contribute to the high rate of TB in these facilities: (1) incarcerated persons have a high risk for TB, (2) facilities are often overcrowded and may have inadequate ventilation, and (3) movement of inmates into and out of overcrowded and inadequately ventilated facilities make implementation of TB-control measures difficult. Environmental controls should be implemented when the risk for TB transmission persists despite efforts to screen and treat infected inmates. Primary environmental controls consist of controlling the source of infection using local exhaust ventilation like hoods or booths and exhausting contaminated air using general ventilation. Source control techniques can prevent or reduce the spread of infectious droplet nuclei into the air. Secondary environmental controls consist of controlling the airflow to prevent contamination of air in areas adjacent to the source and cleaning the air using HEPA filtration and/or UVGI. Using UVGI as a control measure can involve portable UVGI recirculation units, upper-room UVGI, and in-duct UVGI. Unsuspected and undiagnosed cases of TB can be controlled with general ventilation and air cleaning. Increasing the removal efficiency of an existing filtration system is likely to lessen the potential for TB transmission when carriers are undiagnosed. Inmates with TB should be placed in an AII room that meets applicable design and operational guidelines. Temporary facilities or rooms may be used to house inmates suspected to be infectious, and the ventilation system may have to be modified with auxiliary exhaust fans that direct the air to the outdoors and appropriate levels of filtration. Secondary air cleaning techniques such as UVGI can also be used to increase air-cleaning effectiveness. Hoods and booths should have sufficient airflow to purge contaminants after departure of patients and before next use, and a table is provided for determining the length of time for purging at different air exchange rates. A table is also provided giving ventilation recommendations for selected areas. If discharging exhaust air to the outside is not feasible, HEPA filters should be used to clean the air before recirculating. For general areas that recirculate air the minimum recommended filtration is MERV 8, but higher efficiencies are recommended if possible. A combination of MERV-rated filters and UVGI may be used to increase effective air cleaning. When used, UVGI should be applied in-duct or in the upper room and the system should be designed, installed, and operated to ensure that irradiation levels are both sufficient for disinfection and safe for human occupation. Maintenance of environmental controls is critical and staff should be educated and trained accordingly.

Title: Management of Multidrug-Resistant Organisms in Health Care Settings

Reference: HICPAC 2006

Synopsis: This document addresses multidrug-resistant organisms (MDROs) with specific mention of MRSA, VRE, gram-negative bacilli (GNB), *E. coli*, *Klebsiella pneumoniae*, *Acinetobacter baumannii*, *Stenotrophomonas maltophilia*, *Burkholderia cepacia*, *Ralstonia pickettii*, and multidrug-resistant *Streptococcus pneumoniae* (MDRSP). The epidemiology of MDROs is reviewed with some statistics from the National Nosocomial Infection Surveillance (NNIS) system. It is stated that once MDROs are introduced into a health care setting, transmission and persistence of the strain is determined by the availability of vulnerable patients, selective pressure exerted by antimicrobial use, increased potential for transmission from expanding numbers of infected patients, and the effectiveness of prevention efforts. Evidence is cited that transmission occurs via the hands of health care workers. Appropriate clinical practices for the control of MDROs includes optimal management of catheters, prevention of lower respiratory tract infection in intubated patients, accurate diagnoses, and judicious use of antimicrobials. An overview is presented of MDRO control literature including identification of successful control strategies from an extensive list of published reports. Most of these studies applied multiple interventions. The types of successful interventions are grouped into seven categories: (1) administrative support, (2) judicious use of antimicrobials, (3) surveillance, (4) standard and contact precautions, (5) environmental measures, (6) education, and (7) decolonization. These interventions are discussed in detail along with supporting studies. General recommendations are provided according to the commonly used CDC/HICPAC categorization scheme.

Title: Guidelines for Preventing Health Care–Associated Pneumonia

Reference: CDC 2003

Synopsis: This guideline focuses on nosocomial pneumonia and provides recommendations from CDC/HICPAC to reduce the incidence of pneumonia and other lower respiratory tract infections in health care settings. It reviews the various bacterial, fungal, and viral agents that cause pneumonia along with their etiology and epidemiology. These include *Pseudomonas, Acinetobacter, Proteus, Staphylococcus aureus, E. coli, Klebsiella, Streptococcus pneumoniae, Haemophilus influenzae, Legionella, Chlamydia, Mycoplasma pneumoniae,* influenza, RSV, *Aspergillus, Flavobacterium, Bordetella pertussis,* adenovirus, measles, parainfluenza, rhinovirus, SARS virus, and varicella-zoster virus with special attention on the most common causes of nosocomial pneumonia. Focus is provided on the primary risk factors including mechanical ventilators, intubation, nebulizers, humidifiers, catheters, anesthesia equipment, and other devices that may become contaminated. Modes of transmission for each agent are discussed along with prophylactic measures, cleaning methods, and other control measures. Specific recommendations for the most common etiologic agents are provided for prevention of health care–associated pneumonia and these are categorized according to the standard CDC scheme.

Title: Guidelines for Environmental Infection Control in Health Care Facilities

Reference: HICPAC 2003

Synopsis: This document provides recommendations from the Centers for Disease Control and Prevention (CDC) and the Healthcare Infection Control Practices Advisory Committee (HICPAC) and reviews strategies for the prevention of environmentally mediated infections, especially as they relate to health care workers and immunocompromised patients. An extensive list of studies is provided including epidemiological studies and environmental studies, the results of which are used as the basis of recommendations to minimize the risk for transmission of pathogens. Modes of transmission of airborne diseases are discussed and a distinction is made between droplets that cause potential exposure within three feet of the source and droplet nuclei that may remain suspended in air indefinitely. Focus is provided on airborne *Aspergillus* spores, *Mycobacterium tuberculosis, S. aureus,* VZV, measles, influenza, and other microbes.

The impact of heating, ventilating, and air conditioning (HVAC) systems on the spread of health care–associated infections is addressed, including a review of AIA and ASHRAE guidelines. Pressurization, filtration, and UVGI are discussed as part of the engineering controls. UVGI in-duct air disinfection and upper-room systems are noted as supplemental technologies for air disinfection. Laminar airflow systems are described and it is stated that delivery of air at 0.5 m/s in protective environments (PEs) helps to minimize opportunities for microbial proliferation. The importance of HVAC system maintenance is stressed and a tabulation of associated hazards is presented. Air sampling is suggested during periods of construction and as a periodic check on indoor air quality, but it is noted that a lack of standards for air quality (in the United States) hinders the usefulness of such testing. A limit of *15 cfu/m³* for gross counts of fungi is mentioned. Environmental infection control measures for operating rooms, PE rooms, and airborne infection isolation (AII) rooms are reviewed. It is stated that the microbial level in operating rooms is proportional to the number of people in the room. Laminar airflow and UVGI systems are mentioned as adjuncts to operating room systems, but no conclusive evidence has yet been provided to show the SSI risk is decreased by these technologies. It is mentioned that newly cleaned floors become rapidly recontaminated from airborne microorganisms, and that carpets can harbor diverse microbial populations that may be re-aerosolized by vacuuming. Special focus is given to vancomycin-resistant enterococci (VRE), MRSA, *Clostridium difficile*, and other problematic pathogens. Environmental sampling of air and surfaces is discussed in terms of general principles and sampling is only recommended for four situations: (1) to support an investigation of an outbreak where environmental reservoirs are implicated, (2) research, (3) monitoring of a potentially hazardous situation, and (4) quality assurance of equipment, systems, or practices. Air sampling methods and their applications are tabulated. Other subjects addressed include laundry, animals in health care settings as a source of diseases and allergens, and medical waste. An extensive list of categorized recommendations is provided including some recommendations for UVGI systems

Title: Guidelines for Preventing the Transmission of *Mycobacterium tuberculosis* in Health Care Settings

Reference: CDC 2005

Synopsis: This guideline revises and updates earlier CDC guidelines and is intended to deal with the resurgence of TB and the appearance of multidrug-resistant strains. It presents recommendations based on a risk assessment process for all hospital and other facilities that deal with infectious TB patients. The pathogenesis and epidemiology of TB are reviewed and risk factors for patients are identified. Environmental factors that increase the probability of TB transmission are identified and these include exposure to TB in small enclosed spaces, inadequate ventilation that results in insufficient dilution and removal of infectious particles, recirculation of air containing droplet nuclei, and inadequate cleaning and disinfection of equipment. Administrative controls, environmental controls, and respiratory controls are discussed, and detailed recommendations are provided for preventing transmission of TB. Risk assessment includes three risk classifications: (1) low risk, (2) medium risk, and (3) potential ongoing transmission. Recommendations address general ventilation and air-cleaning methods. Airborne infection isolation (AII) rooms must have at least 6 ACH and this could be increased to 12 ACH (actual) by modifying the ventilation system, or can be effectively increased to 12 ACH through the use of internal recirculation HEPA filter units or UVGI systems. AII rooms should be checked for negative pressure. The general ventilation of waiting rooms and other areas can be improved with air-cleaning technologies such as (1) single-pass nonrecirculating systems with outside air exhaust, (2) recirculation systems with HEPA filters, and (3) room-air recirculation units employing HEPA filters and UVGI. HEPA filters must be used when (1) discharging air from local exhaust ventilation booths and (2) discharging air from an AII room when venting to outside is not possible. Air can be recirculated through HEPA filters in areas in which (1) no general ventilation system is present, (2) the existing system is incapable of providing the required ACH, or (3) when air cleaning without affecting the fresh-air supply or negative pressure system is desired. UVGI can be used in a room or corridor to irradiate the upper air in a room and can be installed in a duct to irradiate air or can be incorporated into room recirculation units. UVGI should not be used in place of HEPA filter units. Upperair UVGI systems should be properly installed and maintained, and irradiance levels should be checked with a radiometer to ensure that exposures within the work area are within safe levels. HCWs should be trained to understand the basic principles of UVGI systems, the potential hazards to HCWs and patients, and the importance of maintenance. Supplemental information is provided that addresses environmental controls, exhaust ventilation, booths, hoods, tents, general ventilation, dilution rates, directional airflow, airflow patterns in rooms, negative pressure, pressure differential, monitoring, and related topics. Supplemental reviews of HEPA filtration and UVGI systems are provided. UVGI has been demonstrated to be effective in killing *M. tuberculosis* and in reducing the transmission of other infectious agents in hospitals and is recommended as an adjunct technology to other control measures. A variety of UVGI studies are cited in support of the use of UVGI for duct irradiation, upper-air irradiation, and recirculation UVGI units, and health and safety issues are addressed.

Title: Revised Hospital Interpretive Guidelines 42CFR482.42 Infection Control

Reference: CFR 2007

Synopsis: This Code of Federal Regulations (CFR) section addresses special challenges in infection control including multidrug-resistant organisms (MDROs), ambulatory care, communicable disease outbreaks, and bioterrorism. It provides interpretive guidelines for hospital infection control officers including hospital epidemiologists (HEs) and infection control professionals (ICPs). A hospital infection control program should include appropriate control measures for patients who present a risk for the transmission of infectious agents by the airborne or droplet route. Communicable disease outbreaks present infection control issues that include (1) preventing transmission among patients, health care personnel, and visitors; (2) identifying persons who may be infected and exposed; (3) providing treatment or prophylaxis to large numbers of people; and (4) logistics issues. The infection control officer has the responsibility to (1) maintain a sanitary hospital environment, (2) develop and implement infection control measures for hospital personnel, (3) mitigate risks associated with patient infections, (4) mitigate risks that contribute to infections, (5) active surveillance, and (6) monitor compliance with all infection control program requirements, among others. Maintenance of the hospital physical environment includes ventilation issues and maintaining safe air-handling systems in areas that require special ventilation, and techniques for cleaning and disinfecting environmental surfaces. A variety of measures are identified for mitigation of risks that may contribute to health care–associated infections including hand washing, isolation procedures, disinfectants, adherence to CDC guidelines, and appropriate use of infection control technologies such as isolation rooms, portable air filtration equipment, UV lights, treatment booths, etc.

Title: Guidelines for the Prevention of Tuberculosis in Health Care Facilities in Resource-Limited Settings

Reference: WHO 1999

Synopsis: These guidelines were issued by the World Health Organization (WHO) to address the problem of *Mycobacterium tuberculosis* (TB) in developing countries. They provide discussion and recommendations based on three levels of infection control: (1) administrative, (2) environmental, and (3) personal respiratory protection. The first priority is administrative control measures to prevent the generation of infectious droplet nuclei to reduce the exposure of health care workers (HCWs) and patients. Next is environmental control methods used to reduce the concentration of droplet nuclei in the air in high-risk areas. Environmental control methods include maximizing natural ventilation or mechanical ventilation, HEPA filtration, and ultraviolet germicidal irradiation (UVGI). The third priority is to protect HCWs against inhalation hazards via personal respiratory protection. Environmental control measures are recommended for reducing the number of aerosolized infectious droplet nuclei in the work environment in facilities that have appropriate resources. The simplest and least expensive technique is to remove and dilute the air from TB patient areas by maximizing natural ventilation. Mechanical ventilation is used in isolation rooms to produce negative pressure and prevent the escape of contaminated air. The more costly methods of air filtration can be used to remove infectious particles and UVGI can be used to kill *M. tuberculosis* organisms. Recommendations are provided to maximize natural ventilation and control ventilation patterns. Where mechanical or natural ventilation is not feasible it is suggested that UVGI or portable HEPA filters may be useful in larger wards, TB clinics, and other areas where TB patients may congregate. UVGI may be applied in several forms: (1) in sputum collection booths, (2) continuous upper air irradiation, (3) portable UVGI floor units, and (4) the use of in-duct UVGI systems in air handling units. Continuous upper air irradiation is noted to be the most applicable in resource-limited countries.

Title: Guideline for Prevention of Catheter-Associated Urinary Tract Infections

Reference: HICPAC 2009

Synopsis: This guideline addresses the problem of catheter-associated urinary tract infections (CAUTI), which are the most common type of health care–associated infections, and it is intended for use by infection prevention staff and other health care workers. It provides categorized recommendations based on a systematic review of the best available evidence for appropriate urinary catheter use, proper techniques for urinary catheter insertion, proper catheter maintenance techniques, quality improvement programs, administrative guidelines, and surveillance. Recommendations for urinary catheterization are prioritized, and detailed background information is provided on the epidemiology and pathogenesis of CAUTI including an evidence review.

Title: HA Infection Control Guidelines for Avian Influenza under Red Alert

Reference: CHP 2005

Synopsis: This guideline addresses infection control measures for health care facilities during outbreaks of avian influenza. The particular antigenic strains of influenza that are of concern are identified as H5, H7, and H9. It is noted that the exact modes of transmission of avian influenza are not fully understood and there is no evidence to suggest airborne transmission, but that precautions against airborne transmission should be taken. Precautionary principles are summarized along with diagnostic guidelines, patient care practices, cleaning and disinfection, waste management, viral prophylaxis, and other infection control measures.

Title: Infection Control in Anesthesia

Reference: AAGBI 2008

Synopsis: This guideline, published by the Association of Anesthetists of Great Britain and Ireland, provides guidance for anesthetists for the control and management of health care–associated infections. Standard precautions and hand hygiene are reviewed with a focus on gloving, gowning, and face masks. It is recommended that movement within the operating theater and handling of patients and linens be limited to prevent airborne contamination. A review of the contamination potential of anesthetic equipment is provided along with detailed coverage of the decontamination process. The decontamination process is categorized according to four processes: (1) cleaning, (2) low-level disinfection, (3) high-level disinfection, and (4) sterilization. Risk assessment is subdivided into three categories: (1) high risk, (2) intermediate risk, and (3) low risk. Standard precautions of anesthetic procedures are reviewed and some special focus is provided for prion diseases.

Title: Protocol for Management of the Patient with H1N1 Influenza Virus

Reference: Sutter 2009

Synopsis: This protocol was issued by the Sutter Medical Center to provide a standardized approach for the management of patients with H1N1 influenza and to prevent the spread of H1N1 to health care workers and patients. The symptoms of H1N1 influenza are described and high-risk individuals are identified. Procedures are identified that may contribute to aerosolizing particles and diagnostic, testing, and treatment procedures are summarized. Control measures are reviewed and categorized, and these include patient placement, personal protection, environmental and equipment cleaning, and other factors. Additional relevant information is provided for screening, treating, and isolating pregnant women and newborn children who may be infected. Employees who develop flu symptoms are advised to stay home until they have been free of fever for 24 hours.

Title: Guideline for Infection Control in Health Care Personnel

Reference: HICPAC 1998

Synopsis: This guideline is intended to provide methods for reducing the transmission of infections from patients to health care personnel and from personnel to patients. The prevention strategies addressed in this document include immunization, isolation precautions, management of HCW exposure, and work restrictions for infected workers. It focuses on infections known to be transmitted in health care settings. A distinction is made between *droplet contact*, which refers to conjunctival, nasal, or oral mucosal contact with droplets containing microorganisms generated from an infective person that are propelled short distances, and airborne transmission, which refers to contact with droplet nuclei containing microorganisms that can remain suspended in the air for long periods or to contact with dust particles containing an infectious agent that can be widely disseminated by air currents. Recommendations on immunization for health care personnel and management of job-related illnesses are addressed in detail. Work restrictions are identified for each specific disease or pathogen to which HCWs may be exposed, and in most cases this means "exclude from duty." Specific diseases are addressed that represent the most frequently occurring infections transmitted by direct, indirect, or airborne routes. Norwalk-like gastrointestinal viruses are noted to have a postulated but unproved airborne route of transmission. Measles is noted to be transmissible by both close contact with infected patients and by the airborne route. *Bordetella pertussis* is transmitted by contact with respiratory secretions or large aerosol droplets. Group A *Streptococcus* is transmitted primarily through direct contact, but airborne transmission during outbreaks has been suggested and airborne dissemination of GAS from rectal and vaginal carriage has been noted. Varicella-zoster virus (VZV) has been transmitted in hospitals by airborne transmission. Influenza is transmitted by droplet contact, by droplet nuclei or small-particle aerosols. Recommendations are provided for the prevention of infections in health care personnel and are graded according to the scheme used in other CDC and HICPAC guidelines.

Title: Guideline for Isolation Precautions: Preventing Transmission of Infectious Agents in Health Care Settings

Reference: HICPAC 2007

Synopsis: The objectives of this guideline include providing infection control recommendations for all types of health care facilities, reaffirming Standard Precautions as the foundation for preventing transmission during patient care, reaffirming the importance of implementing Transmission-Based Precautions, and providing epidemiologically sound, evidence-based recommendations. Modes of transmission are subdivided into Contact transmission, Droplet transmission, and Airborne transmission. Contact transmission, the most common mode of transmission, is subdivided into Direct contact, where microbes are transferred directly from one infected person to another, and Indirect contact, in which transmission occurs via an intermediate object or person. Droplet transmission is distinguished from direct transmission by the fact that droplets travel directly from the respiratory tract of an infectious individual to susceptible mucosal surfaces of the recipient, generally over short distances (typically 3 feet or less). According to the guideline sources, organisms transmitted by the droplet route do not remain infective over long distances and therefore do not require special air handling and ventilation. Airborne transmission occurs by dissemination of either airborne droplet nuclei or small particles in the respirable size range containing infectious agents that remain infective over time and distance. Infectious agents to which this applies include *Mycobacterium tuberculosis*, *Aspergillus* spores, measles virus, and VZV. For such agents, precautions include the use of airborne infection isolation rooms (AIIRs). For other pathogens suspected to transmit via small-particle aerosols such as influenza, rhinovirus, norovirus, and rotavirus, AIIRs are not routinely required. In spite of several cited studies indicating airborne transmission, SARS virus is stated as having not been proved to transmit by this route, while it is also stated that opportunistic airborne transmission cannot be excluded. A new classification is introduced to break down routes of aerosol transmission: (1) obligate—under natural conditions disease occurs following transmission of the agent only through inhalation of small particle aerosols; (2) preferential—natural infection results from transmission through multiple routes, but small particle aerosols are the predominant route; and (3) opportunistic—agents that naturally cause disease through other routes, but under special circumstances may be transmitted via fine particle aerosols. Some airborne infectious agents derive from the environment without person-to-person transmission, such as anthrax, *Legionella*, and *Aspergillus* spores. Administrative measures for preventing transmission are detailed, including surveillance, education, hand hygiene, and personal protective equipment (PPE). Respiratory protection of N95 or higher is recommended to be worn for diseases that could be transmitted by the airborne route, including TB and SARS. Two tiers of precautions are defined: Standard Precautions and Transmission-Based Precautions. Standard Precautions are applied to all patients in all settings and constitute the primary strategy for preventing transmission. There are thee categories of Transmission-Based Precautions: Contact Precautions, Droplet Precautions, and Airborne Precautions. The preferred placement for patients who require airborne precautions is in an AIIR. Categorized recommendations are provided for all three precaution categories and precautions are detailed for specific infections.

Title: SHEA/APIC Guideline: Infection Prevention and Control in the Long-Term Care Facility

Reference: Smith et al. 2008

Synopsis: This position paper from the Society for Healthcare Epidemiology of America (SHEA) and the Association for Professionals in Infection Control and Epidemiology (APIC) provides basic infection control recommendations that could be widely applied to long-term care facilities (LTCFs) with the expectation that health care associated infections (HCAIs) could be minimized. Some specific hospital guidelines are cited and relevant portions are adapted for LTCFs. This paper addresses all levels of care in LTCFs, or nursing homes, which care for elderly or chronically ill residents. The various similarities and differences between LTCFs and hospitals are noted and statistics on LTCF infections are provided. The leading infection in LTCFs is UTI, while other common infections include URI, pressure ulcers, gastroenteritis, bacteremia, and epidemics of various diseases including influenza, tuberculosis, MRSA, VRE, and Group A *Streptococcus*. Infection control programs for LTCFs are reviewed, as are regulatory aspects and infection surveillance. The importance of isolation as an infection control strategy and a precaution is discussed in relation to MDROs, for which residents have a high colonization rate. Recommendations categorized according to the CDC system are provided for infection control programs, administrative structure, the ICP, surveillance, outbreak control, facilities, isolation and precautions, asepsis and hand hygiene, resident care, and other issues.

Title: APIC/CHICA-Canada Infection Prevention, Control, and Epidemiology: Professional and Practice Standards

Reference: Friedman et al. 2008

Synopsis: These standards are designed to be used in identifying areas for professional growth, developing job descriptions, and providing criteria for performance evaluations. These standards encompass a broad spectrum of practice settings and professional backgrounds and include key indicators that are intended for evaluating both the competency of individuals and their practice. They describe a level of individual competence in the professional role. Qualifications are described in terms of indicators, as is professional development and other factors. Practice standards are also described in terms of key indicators including infection prevention and control practice, surveillance, epidemiology, education, consultation, and other relevant topics.

Building and Facility Guidelines

Several guidelines are available to assist the design and construction of health care facilities and the Code of Federal Regulations also addresses many of the issues. The guidelines from AIA and ASHRAE are sufficiently similar and redundant in all their fundamental aspects that using either of them will meet the requirements of both.

Title: Guidelines for Construction and Equipment of Hospital and Medical Facilities

Reference: AIA 2006

Synopsis: These guidelines provide guidance for the design and construction of health care facilities and represent a consensus for such designs that have been developed over many decades. These guidelines are concise and virtually identical to ASHRAE guidelines in regard to ventilation requirements. Ventilation and filtration requirements are tabulated in terms of the required air exchange rates (ACH) for particular types of rooms and these requirements match those provided in ASHRAE guidelines (ASHRAE 1999, 2003). Recirculation units that use HEPA filters may be used in isolation rooms for airborne infection control. Airflow must be directed from the cleanest patient areas to less clean areas. In AII rooms, supplemental recirculating devices employing HEPA filters may be used in the patient room to increase the equivalent room air exchanges, but these do not provide the minimum outside air requirements. Air conditioning, heating, ventilating, and related equipment shall be installed in accordance with NFPA 90A, Standard for the Installation of Air Conditioning and Ventilation Systems.

Title: HVAC Design Manual for Hospitals and Clinics

Reference: ASHRAE 2003

Synopsis: This design manual reiterates ASHRAE and AIA guidelines and provides additional detailed design information for health care facilities. Two primary modes of disease transmission are identified: (1) direct contact transmission and (2) airborne transmission. Airborne transmission may result from sneezing, coughing, or talking by an infected person, resuspension of settled microbes, aerosolization of contaminated water droplets, carriage on human skin squames, and amplification or reproduction of microbes within HVAC equipment. Detailed discussion of the impact of ventilation on infection control addresses dilution, directional airflow, filtration, and UVGI. In-duct UVGI systems, recirculation UVGI systems, and upper-room UVGI systems are discussed with the caveat that UVGI can only be used as supplemental protection (to HEPA filtration systems). Duct cleaning is highlighted as a means of preventing microbial growth in ductwork. Operating parameters are provided for AII rooms and operating rooms including required airflow per floor area of room space and preferred design approaches for distributing air. Information is also provided on microbes that may grow indoors or in ventilation equipment like cooling coils, and removal rates that can be expected from air exchange rates using filters and UVGI systems.

Title: General Requirements: Purpose of the Facilities Standards for Public Buildings Service

Reference: GSA 2005

Synopsis: These Facilities Standards for the Public Buildings Service establish design standards and criteria for new buildings and modifications for the Public Buildings Service (PBS) of the General Services Administration (GSA). They address HVAC components including air-handling units (AHUs), fans, cooling and heating coils, drains, air filters, and ultraviolet lamps. Filter requirements are in accordance with other common standards (i.e., ASHRAE and AIA). Ultraviolet lamps (UVC emitters) are specified to be installed downstream of all cooling coils in AHUs and above drain pans to control airborne and surface microbial growth. Air distribution systems are to be designed in accordance with ASHRAE guidelines (ASHRAE 1985, 1999).

Title: Guidelines on the Design and Operation of HVAC Systems in Disease Isolation Areas: TG 252

Reference: CHPPM 2000

Synopsis: This document was issued by the US Army Center for Health Promotion and Preventive Medicine and includes guidelines for heating, ventilating, and air conditioning (HVAC) systems for new isolation rooms, existing isolation rooms, and other hospital and clinical areas. It is intended for architects, engineers, hospital facility managers, hospital safety managers, and other hospital staff, and provides guidance for the design and maintenance of areas used to treat disease isolation patients. HVAC guidelines address pressurization, comfort, airflow patterns, air exchange rates, and exhaust systems, and are largely in accordance with existing ASHRAE (2003) and AIA (2006) guidelines. Detailed recommendations are provided for HVAC system testing and maintenance. Some detailed recommendations for the use of UVGI systems are provided including using UVGI as a method of air disinfection to supplement other engineering controls. UVGI can only be used in conjunction with an engineered HVAC system that has been designed in accordance with other relevant sections of this guideline. Two types of UVGI systems are addressed: in-duct irradiation and upper-room air irradiation systems. Duct irradiation is recommended for isolation and treatment rooms where recirculated air is not HEPA filtered, and for recirculation systems in patient rooms, waiting rooms, emergency rooms, and other general use areas where there may be unrecognized infectious patients. UVGI may not be used as a substitute for HEPA filters. UV lamps can be installed in ducted systems perpendicular to the airflow in exhaust ductwork to decontaminate air prior to recirculation. Upper-room irradiation systems are recommended for isolation and treatment rooms as a supplemental method of air disinfection and can also be used in patient rooms, waiting rooms, and emergency rooms. Upper-room UVGI may be used to supplement the existing ventilation if the HVAC system does not meet the required number of air changes per hour required for existing facilities, but not for new or renovated facilities. Upper-room systems are not to be used when a room is connected to a recirculating HVAC system, except for return air systems where direct exhaust is not possible. Upper-room systems shall use louvers to direct irradiation to the upper room area and minimize exposure to patients and staff. Supply air shall be drawn through the radiation field from air registers and this air shall pass down into the room, over the patient, and out through the exhaust ducts. UVGI operation, maintenance, warning signs, and measurements are specified in terms typical of other guidelines or as specified by manufacturers. Monitoring of rooms with UV shall verify that the NIOSH relative exposure limit (REL) is not exceeded. UVGI shall not be used as a substitute for negative pressure rooms, and shall not be installed in series with HEPA filters, because they provide no significant additional benefit.

Title: Unified Facilities Criteria UFC 4-510-01 Design: Medical Military Facilities

Reference: UFC 2001

Synopsis: This document provides mandatory design and construction criteria for medical military facilities in the Department of Defense, including defense medical facilities and dental treatment facilities. The purpose of these criteria is to standardize and simplify the design of the health care facility process by concisely stating which design criteria is to be used. Design criteria are summarized for heating, ventilating, and air conditioning (HVAC) systems. The minimum HVAC design criteria must be in accordance with ASHRAE and AGCIH publications and guidelines. HVAC systems shall be designed to remove or reduce to acceptable levels airborne microbiological contaminants within the facility. Systems must control moisture and dust accumulation in air-handling units to avoid conditions permitting the growth of pathogenic and allergenic microorganisms. Filtration requirements are specified in accordance with ASHRAE guidelines. Air change rates shall be in accordance with ASHRAE Standard 62.1 (ASHRAE 2001). Disease isolation rooms shall be designed in accordance with CDC guidelines for TB (CDC 2005). In existing facilities, TB isolation rooms shall have 12 air exchanges per hour (ACH), or shall have a minimum of 6 air changes per hour supplemented with either HEPA filtration or UVGI systems specifically designed for TB rooms and providing an equivalent air exchange of at least 6 ACH.

Title: ASHRAE Position Document on Airborne Infectious Diseases

Reference: ASHRAE 2009

Synopsis: ASHRAE issued this document to state and explain its position on airborne infectious diseases and how the problem relates to heating, ventilating, and air conditioning (HVAC)

system design. ASHRAE states that many infectious diseases are transmitted through inhalation of airborne infectious particles termed droplet nuclei and that airborne infectious particles can be disseminated through buildings including ventilation systems. Airborne infectious disease transmission can be reduced using dilution ventilation, specific in-room flow regimes, room pressure differentials, personalized and source capture ventilation, filtration, and UVGI. Some diseases are transmitted through the airborne route when the mean aerodynamic diameter of a droplet or particle is less than 20 microns. These particles may be generated by coughing or sneezing, and to a lesser extent by talking and singing. These particles may remain airborne for hours and may be transported far from the source, and their distribution may be impacted by HVAC system operation. Increasing the amount of airflow in an indoor environment can dilute the concentrations of infectious particles and thereby lower the infection risk. Filtration and ultraviolet germicidal irradiation (UVGI) can also be used for engineering control. Three UVGI strategies are discussed by ASHRAE: in-duct air disinfection, upper-room UVGI, and whole-room disinfection. A tabulation of diseases spread by droplet or airborne transmission, even if airborne transmission is not the primary mode, is provided that identifies the health care workers at risk.

Title: Manual of Recommended Practices for Industrial Ventilation

Reference: ACGIH 2010

Synopsis: This manual is a new edition of a book first published in 1951 and is intended to be used by engineers and industrial hygienists to design and evaluate industrial ventilation systems, including hospital systems. It includes a chapter on air cleaning devices and provides various fundamental design information of use to designers of health care facilities.

Title: Department of the Air Force Operating Room Ventilation Update

Reference: NRC 1998

Synopsis: This memorandum provides guidance on recommended ventilation requirements in medical facilities. Heating, ventilating, and air-conditioning (HVAC) systems in hospitals provide for (1) the control of infectious agents released into the air during medical procedures or carried by infectious patients, (2) maintaining conditions of comfort for patients, and (3) inhibiting bacterial growth and virus activation. Examples of infectious agents of concern include TB, *Legionella*, varicella, rubella, and *Aspergillus*. Bacteria can generally be removed by filtration because they are typically larger than one micron. Airborne viruses that transmit infection are submicron in size; however, there is no current method to filter 100% of the viable particles; therefore, ventilation-pressure relationships are the primary means to prevent airborne transmission of infectious agents. The basics of operating ventilation systems and pressurizing rooms are covered, and means to evaluate ventilation system performance and calculate air exchange rates are discussed. Ultraviolet germicidal irradiation may be used to supplement any of the ventilation methods for air cleaning.

Title: Using Ultraviolet Radiation and Ventilation to Control Tuberculosis

Reference: DHS 1990

Synopsis: This booklet was issued to provide officials with a technically oriented, practical guide to the use of ultraviolet light and ventilation to control tuberculosis. Airborne infections such as TB can be prevented by killing the infectious microorganisms in the air with ultraviolet (UV) radiation. The spread of airborne infections can also be reduced with proper ventilation. Although environmental control measures such as air disinfection and ventilation can decrease the transmission of TB, they are only supplements to the usual control measures, which they cannot replace. When the identification and follow-up of TB cases is difficult, UV air disinfection and exhaust ventilation can provide an extra measure of protection. Design guidance is provided in this document for overhead or upper-room UVGI systems, and for UV lamps located inside ducts. Upper-room UV systems require good mixing between the upper-room air and the lower-room air. Overhead UV lamps are useful in crowded and poorly ventilated buildings where the conventional control methods are inadequate. Indoor air that might contain infectious particles cannot safely be recirculated back into a room unless the airborne microorganisms have been removed or inactivated. Filtering the air, installing UV lamps inside the air ducts, and other forms of air cleaning can effectively remove or kill bacteria. High-efficiency air cleaners are more expensive to install and maintain than overhead UV lamps. UV safety considerations are reviewed and a list of recommendations is provided for the proper use of UV upper air irradiation.

Title: Environmental Control for Tuberculosis: Basic Upper-Room Ultraviolet Germicidal Irradiation Guidelines for Health Care

Reference: NIOSH 2009

Synopsis: This guideline, issued in conjunction with the CDC and HICPAC, reviews the literature on applications of UV to the control of tuberculosis and provides guidelines for the installation and use of upper-room UVGI systems. Test results from some 21 studies on bacteria including TB surrogates are tabulated. The effectiveness varied from 40–98% for those studies where it was reported. Estimates for the most effective upper-room fluence rate range from 30–50 $\mu W/cm^2$. Factors that may impact effectiveness are discussed, including air mixing, location of supply diffusers and exhaust grilles, and humidity. Little effect was noted for humidity levels below 60%. Recommended temperature range is 68–75°F (20–24°C). A variety of practical guidelines are provided for installation and maintenance of UV systems and measurement of irradiance levels, and for safety considerations.

References

AAGBI (2008). Guidelines: Infection control in anaesthesia. *Anaesthesia* 63, 1027–1036.

ACGIH (2010). Manual of Recommended Practices for Industrial Ventilation. American Conference of Governmental Industrial Hygienists, Cincinnati, OH.

AIA (2006). Guidelines for construction and equipment of hospital and medical facilities; in *Mechanical Standards,* American Institute of Architects, ed., Washington, DC.

APIC (1995). APIC Guideline for handwashing and antisepsis in health care settings. *Am J Inf Control* 23, 251–269.

ASHRAE (1985). *Handbook of Fundamentals.* ASHRAE, Atlanta, GA.

_____. (1999). *Handbook of Systems and Equipment.* ASHRAE, Atlanta, GA.

_____. (2001). Standard 62: Ventilation for acceptable indoor air quality. ASHRAE, Atlanta, GA.

_____. (2003). *HVAC Design Manual for Hospitals and Clinics.* American Society of Heating, Ventilating, and Air Conditioning Engineers, Atlanta, GA.

_____. (2009). *ASHRAE Position Document on Airborne Infectious Diseases.* American Society of Heating, Refrigerating and Air-Conditioning Engineers, Atlanta, GA.

CDC (1999). *Guideline for Prevention of Surgical Site Infection.* Centers for Disease Control, Atlanta, GA.

_____. (2002). *Guideline for Hand Hygiene in Health-Care Settings.* Centers for Disease Control, Atlanta, GA.

_____. (2003). Guidelines for Environmental Infection Control in Health-Care Facilities. *MMWR* 52(RR-10), 1–48.

_____. (2005). *Guidelines for Preventing the Transmission of* Mycobacterium tuberculosis *in Health-Care Facilities.* Centers for Disease Control, Atlanta, GA.

_____. (2006). *Prevention and Control of Tuberculosis in Correctional and Detention Facilities: Recommendations from CDC.* Federal Register CDC, ed., US Government Printing Office, Washington, DC.

_____. (2008). *Guideline for Disinfection and Sterilization in Healthcare Facilities.* Centers for Disease Control, Atlanta, GA.

CFR (2007). Code of Federal Regulations Revised Hospital Interpretive Guidelines. 42CFR482.42 Infection Control, US GPO, Washington, DC.

CHP (2005). HA Infection Control Guidelines for Avian Influenza under Red Alert. Central Committee on Infectious Disease, Hospital Authority Infection Control Branch, Centre for Health Protection, Hong Kong.

CHPPM (2000). Guidelines on the Design and Operation of HVAC Systems in Disease Isolation Areas. *TG 252*, US Army Center for Health Promotion and Preventive Medicine, Aberdeen Proving Ground, MD.

DHS (1990). Using Ultraviolet Radiation and Ventilation to Control Tuberculosis. Department of Health Services, California.

Friedman, C., Curchoe, R., Foster, M., Hirji, Z., Krystofiak, S., Lark, R. L., Laxson, L., Ruppert, M. J., and Spaulding, L. (2008). APIC/CHICA-Canada infection prevention, control, and epidemiology: Professional and practice standards. *AJIC* 36, 385–389.

GSA (2005). *General Requirements: Purpose of the Facilities Standards for the Public Buildings Service.* Public Buildings Service of the General Services Administration, Washington, DC. http://hydra.gsa.gov/pbs/p100/

Health Canada (1998). Hand Washing, Cleaning, Disinfection and Sterilization in Health Care. *Canada Communicable Disease Report Volume 24SB*, Health Canada, Laboratory Centre for Disease Control, Ottawa, ON.

HICPAC (1998). Guideline for Infection Control in Health Care Personnel. Centers for Disease Control, Atlanta, GA.

_____. (2003). Guidelines for Environmental Infection Control in Health-Care Facilities. *MMWR* 52(RR-10), 1–48.

_____. (2006). *Management of Multidrug-Resistant Organisms in Healthcare Settings.* Centers for Disease Control, Atlanta, GA.

_____. (2007). Guideline for Isolation Precautions: Preventing Transmission of Infectious Agents in Healthcare Settings. Centers for Disease Control, Atlanta, GA.

_____. (2009). Guideline for Prevention of Catheter-Associated Urinary Tract Infections. Centers for Disease Control, Atlanta, GA.

Johns Hopkins Hospital (2007). Laboratory Infection Control Policy. Johns Hopkins Hospital Department of Pathology, Baltimore, MD.

NIOSH (2009). Environmental Control for Tuberculosis: Basic Upper-Room Ultraviolet Germicidal Irradiation Guidelines for Healthcare Settings. *NIOSH 2009-105*, Department of Health and Human Services, Centers for Disease Control and Prevention, National Institute for Occupational Safety and Health, Atlanta, GA.

NRC (1964). Postoperative wound infections—The influence of ultraviolet irradiation of the operating room and of various other factors. *Ann Surg* 160, 1–125.

NRC (1998). Department of the Air Force Operating Room Ventilation Update. National Resource Center, Brooks Air Force Base, San Antonio, TX.

SGC (2001). Infection Control in the NICU. Study Group for the Control of Infection in NICUs, Hospital Authority, Hong Kong.

Smith, P. W., Bennett, G., Bradley, S., Drinka, P., Lautenbach, E., Marx, J., Mody, L., Nicolle, L., and Stevenson, K. (2008). SHEA/APIC Guideline: Infection prevention and control in the long-term care facility. *AJIC* 36, 504–535.

Sutter (2009). Protocol for Management of the Patient with H1N1 Influenza Virus, Sutter Medical Center, Sacramento, CA.

UFC (2001). Unified Facilities Criteria, UFC 4-510-01 Design: Medical Military Facilities, U.S. Army Corps of Engineers, Department of Defense, Washington, DC.

WHO (1999). Guidelines for the Prevention of Tuberculosis in Health Care Facilities in Resource Limited Settings. *WHO/CDS/TB/99.269*, World Health Organization, Geneva.

_____. (2004). Practical Guidelines for Infection Control in Health Care Facilities. *SEARO Reg Pub 41*, World Health Organization, Geneva.

_____. (2009). WHO Guidelines on Hand Hygiene in Health Care: A Summary. World Health Organization, Geneva.

18

Air and Surface Sampling

Introduction

Microbiological testing methods are a simple and reliable means of identifying the microbial flora that inhabit or contaminate hospital air, surfaces, and equipment. Although sampling is only used on an as-needed basis in hospitals and regular sampling is generally not recommended by regulatory agencies, sampling provides the only means of investigating hospital aerobiology and surface contamination. Sampling methods for air and surfaces are fairly straightforward and the equipment needed for such testing is relatively inexpensive. Knowledge of the surface and airborne concentrations in different areas of the hospital and the species detected can be invaluable in identifying potential problems and determining the success of decontamination efforts. The microbiological air quality in operating rooms may be considered a mirror of the hygienic condition of the OR (Pasquarella et al. 2004). Three sampling methods are addressed in this chapter: (1) settle plate sampling, (2) volumetric air sampling, and (3) surface sampling. Table 18.1 lists some of the more common growth media used for bacteria and fungi.

Air Sampling with Settle Plates

Identification of bacteria or fungi in hospital environments can be performed with settle plate sampling. Settle plates are petri dishes placed in rooms for specified periods of time and on which airborne microbes will settle. Settle plate sampling can only be used to identify pathogens and does not provide data that are absolutely quantitative in terms of airborne cfu/m^3. Settle plates can be placed around a room in multiple locations, such as on the floor or at breathing height, and in the corners, sides, or center of a room. Separate settle plates are needed for bacteria and for fungi. Multiple plates are not needed if testing is for identifying species only, but if testing is comparative (i.e., Before vs. After airborne concentrations) then triplicate plates should be

TABLE 18.1

Typical Growth Media for Air and Surface Sampling

Medium	Microbe	Application Notes	Incubation Temperature
Buffered charcoal yeast extract agar	Bacteria	*Legionella*	35–37°C
MacConkey agar	Bacteria	Gram-negative bacteria	35–37°C
Heart infusion blood agar	Bacteria	Human commensal bacteria	35–37°C
R2A with cycloheximide	Bacteria	Environmental bacteria with fungal suppression	20–30°C
Soybean-casein digest agar	Bacteria	Environmental bacteria with fungal suppression	20–30°C
Soybean-casein digest agar	Bacteria	Thermophilic bacteria (*Actinomyces*)	50–55°C
Tripticase soy agar (TSA)	Bacteria	Bacteria	35–37°C
DG 18 (Dichloran 18%) agar	Fungi	Slows fungal growth, low water activity (Aw = 0.95)	20–25°C/35–37°C
Malt extract agar (MEA)	Fungi	Broad spectrum, saprophytic, allergenic, and pathogenic fungi	20–25°C/35–37°C
Inhibitory mold agar	Fungi	Same as MEA but suppresses bacterial colonies	20–25°C/35–37°C
Malt extract agar with NaCl, sucrose, or dichloranglycerol	Fungi	Xerophilic fungi	20–25°C/35–37°C
Rose bengal agar	Fungi	Broad spectrum like MEA but suppresses bacterial colonies	20–25°C/35–37°C

used. Testing three different locations in any room, with three plates at each test point, will require at least nine plates total for each test. Settle plates may be sufficient to test for large particles, but microbes in the 1–5 micron size range and lower may remain airborne indefinitely and may not be efficiently sampled with settle plates alone. See the suggested Test Protocol for Settle Plate Sampling that follows.

Air Sampling with an Air Sampler

Air sampling provides for both qualitative and quantitative measures of air-borne microbial levels and is easily performed using any one of the available air samplers on the market today. These samplers can vary in performance, and although these differences can be great it should be noted that they

TEST PROTOCOL FOR SETTLE PLATE SAMPLING

The following test protocol is suggested for use when the settle plate method is used for sampling hospital air for bacteria and/or fungi.

A.1: Preliminary Conditions and Materials

    A.1.1: Select and procure appropriate growth media in plates suitable for the intended application: Bacteria or Fungi. A list of typical growth media has been provided in Table 18.1. For comparative testing at least 18 plates will be needed, with three plates to be placed at each of three locations. Otherwise single plates can be used. Keep one to three plates unused as controls.

    A.1.2: Room conditions should be within normal indoor operating temperature and humidity ranges. Test locations should be selected and will preferably include room floor locations at the center, the side, and the corner. Occupancy can be kept as normal or unoccupied rooms can be tested for baseline conditions.

    A.1.3: Number the individual plates (either before they are placed at test locations or after they are removed from the test locations). A minimum of three plates (Bacteria and/or Fungi) is recommended for each test location for comparative testing.

B.1: Test Protocol

    B.1.1: Close any window blinds if there is direct sunlight in the room.

    B.1.2: Place the settle plates at the designated locations in the room. Each three-plate set of plates may be placed side by side, at each test location. Record each test location and the time of placement. Record room occupancy, including brief entrances or exits.

    B.1.3: After a period of 1–2 hours, cover and remove the settle plates. Make sure the plates are labeled or numbered before or at the time they are removed, and record the removal time.

    B.1.4: Incubate the plates at the required temperature or deliver them to a laboratory for incubation and counting.

    B.1.5: Repeat the above procedure for fungi (or bacteria) if necessary. Fungi and bacteria may be tested coincidentally, with six plates placed side by side at each test location, if desired.

C.1:   Evaluation of Results
    C.1.1:   Incubate the plates for 24–48 hours, or deliver to a laboratory for culturing.
    C.1.2:   Count the colonies on the plates and record them on appropriate forms. Evaluate the counts and determine the mean counts of the settle plates as compared with the control plates.
    C.1.3:   Speciate the samples and identify any hazardous pathogens.

are typically used for comparative purposes only, and therefore repeatability may be more relevant than absolute accuracy. Variations between air samplers of as much as an order of magnitude are not unusual (Straja and Leonard 1996; Ambroise et al. 1999; Jensen and Schafer 1998; Li et al. 1999; Lin et al. 1999; Griffiths, Upton, and Mark 1993). If the data collected are used for absolute comparisons (i.e., to meet an indoor guideline or limit) then the choice of sampler may be more critical than if they are used to assess the efficacy of decontamination measures.

Petri dishes are inserted in the air sampler and it is typically operated for a specified period of time (i.e., 20–30 minutes) during which a measured volume of air passes through the sampler. Afterward the plates are removed and incubated. For additional information on sampling human pathogens, see Artenstein et al. (1967). For examples of sampling fungi see Flannigan (1997) and Flannigan, McEvoy, and McGarry (1999). For more general sampling information see Boss and Day (2001) and Cox and Wathes (1995).

Evaluation of the air for fungal spore content will yield useful information about the effectiveness of the ventilation system and filters. Hospital air samples should draw at least 35 ft^3 or 1 m^3 to detect low levels of spores (Streifel 1999).

Figure 18.1 shows an example of a petri dish that was used for an air sample taken by the author in an occupied operating room. Culturing revealed the presence of species of *Staphylococcus*, *Streptococcus*, and *Neisseria* in the operating room air.

A suggested protocol for air sampling is shown above. This protocol can be adapted as necessary to suit local conditions or any applicable requirements. Room occupancy can greatly affect results and therefore it may be appropriate to sample room air when it is unoccupied (to provide a baseline) and to sample it again when occupied. Always be sure to record either the number of occupants or the time of occupancy for each occupant. This is especially true when monitoring the occupation of an OR during a surgical procedure because personnel may enter and exit several times in the course of an hour or two. The best location for an air sampler is on the floor either in the corners, the sides, or the center of the room, because this is where the highest

TEST PROTOCOL FOR AIR SAMPLING

The following is a suggested test protocol for air sampling in hospital environments.

A.1: Preliminary Materials and Conditions

    A.1.1: Select and procure an appropriate air sampler and growth media in plates suitable for the selected air sampler. For bacteria use 1–3 plates at each sample location, and sample at least three locations. Ditto for fungi. Keep at least one plate unused as a control.

    A.1.2: Room conditions should be within normal indoor ranges. Room occupancy should be kept as normal or room should be unoccupied to provide a baseline condition.

    A.1.3: Identify at least three locations for samples. Air sampler should preferably be placed on the floor, in the center, sides, or corners of the room.

B.1: Test Protocol

    B.1.1: Install the petri dish, filter, or other media in the air sampler. Place the air sampler in a preselected location and operate it for the time period specified by the manufacturer or the instructions.

    B.1.2: After the elapsed time period, shut down the air sampler and carefully remove the plate (or filter).

    B.1.3: Label the plate with an appropriate code or description of the location, the type of sample (Bacteria or Fungi), and date as necessary. Record the sample information on a separate log of all samples. Record room occupancy, including brief entrances or exits.

    B.1.4: Either deliver the sample to a designated laboratory or begin the incubation process by placing the sample in an incubator at the appropriate temperature.

    B.1.5: Repeat the above sequence of steps as necessary for the sample plates.

C.1: Evaluation of Air Sample Results

    C.1.1: If samples have been delivered to a laboratory then the results provided by the laboratory can be directly inspected. The plate counts should be tabulated with each sample taken and averaged for each specific location.

    C.1.2: If plates are digitally imaged for the record, the digital images may be used for counting and the plates disposed of.

C.1.3: The laboratory will identify all species and provide this information. The appearance of dangerous or unusual pathogens should be brought to the attention of the client or medical authorities (i.e., CDC) as necessary.

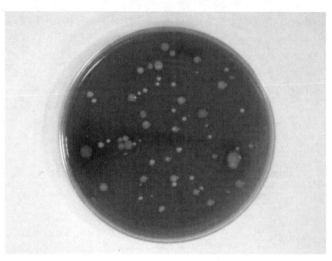

FIGURE 18.1
Cultured petri dish from an air sample showing colonies of *Streptococcus, Staphylococcus, Neisseria,* and *Bacillus subtilis* after 48 hours of incubation.

concentrations of bacteria and spores will be found. This may not be true for viruses because they will tend to remain airborne almost indefinitely. Air samplers can also be located at table height when investigating air quality for surgical sites or studying bacterial shedding from surgeons and nurses.

Surface Sampling

Surface sampling provides qualitative data in terms of contaminating species but does not provide quantitative data because there is no absolute reference for levels of surface contamination. It may, however, be used quantitatively to determine the efficacy of decontamination procedures or to compare different procedures and disinfectants. Any effective decontamination procedure should result in reduced levels of surface contamination, as in a Before versus After test. Surfaces to be sampled can include room floors and walls, doorknobs, faucet handles, medical equipment, pharmaceutical supplies, sterile containers or packaging, HVAC duct internal surfaces, cooling coils, drain pans, etc.

Surface sterilization is difficult to prove absolutely, as even the act of sampling a surface can add slight amounts of contamination. Attempting to verify sterilization may require large numbers of samples, high sensitivity, or large-area sampling. The sensitivity of detection for microbes can depend on the type of surface, area of the surface, sampling technique, microbial handling procedures, and sampling media. For small-area sampling the number of microbes recovered may be low, while for large-area sampling (i.e., a square meter) low levels of pathogens may be detected with greater sensitivity (Buttner et al. 2004). Only in highly controlled environments like the pharmaceutical industry is it possible to demonstrate sterilization through testing, and even so there will always be some level of error.

Surface sampling is generally performed by wiping swabs across the surface or by pressing materials, including petri dishes or contact plates, against the surface. Adhesive tape and adhesive sheets are also available for use in surface sampling (Yamaguchi et al. 2003). The swabs are drawn across a petri dish or will be inserted into sterile solutions that are plated on dishes in a laboratory. The actual area of the surface being sampled may not be critical in most cases because it is used for comparative purposes only, and typically about 2 in^2 or 5–10 cm^2 is sufficient. Speciation of surface samples will typically provide the most important information of interest to hospitals. Actual plate counts and concentrations per unit of surface area have limited absolute meaning unless they are used for comparative purposes. Comparing Before versus After surface samples can be useful for establishing the effectiveness of disinfection processes.

Often, the sampling of surfaces such as walls, floors, HVAC ducts, or cooling coils is conducted to verify that high levels of disinfection have been achieved. In such cases a relative level of sterilization, or virtual sterilization, may be defined as a matter of convenience. The common definition of sterilization in mathematical terms is six logs of reduction, or a 99.9999% reduction of surface microbial counts (assuming there are no survivors). This is a reasonable and practical approach and forms a workable goal for disinfection processes; however, it can depend to some degree on the area that is sampled, the level of contamination, and the effectiveness of the sampling process. The surface area sampled should be kept constant for any series of samples to be compared. Sampling devices capable of sampling larger areas may have the advantage of requiring fewer samples to be collected per site and greater detection sensitivity.

If a template is used to mask off the area to be sampled, as shown in Figure 18.2, it must be sterile. Prepackaged sterile paper templates are not reusable, but must be discarded after each use. Reusable templates (i.e., metal or plastic) must be disinfected before each use, *including before the first use*, or else they may contaminate all samples from the first to the last. Sterilization of reusable templates involves spraying or wiping a disinfectant such as

FIGURE 18.2
Surface swabbing with a reusable template. (Image provided courtesy of SKC Inc., Eighty Four, PA.)

alcohol on them and then drying them. *Failure to sterilize a reusable template can result in all samples becoming contaminated with the same microbe.* Such types of contamination will be obvious from the fact that the same contaminating bacteria or fungal species will occur at high levels even when all other microbes are effectively eliminated. Reusable templates are poor practice for microbiological sampling, and the preferred method is to use either disposable sterile templates or none whatsoever (i.e., simply estimating the sample area visually).

Gloves must be worn during the sampling process, and it should be recognized that the person doing the sampling likely carries various environmental or endogenous microbes. They should be careful not to touch the surface with any part of their clothing. Gloves should preferably be sterile or be sterilized before use. Gloves can be sterilized by spraying with alcohol.

Swabbing may be done with wet or dry sterile swabs, typically made of cotton. These come prepackaged and the paper or plastic package must be opened before each use. Both wet and dry sampling are possible, but one study reports that dry sampling is slightly more sensitive (Buttner et al. 2004). The area to be sampled should be swabbed with a back and forth motion, gently and with no more than a dozen or so strokes. Once a swab is taken from a surface it is then brushed across a petri dish, and the dish is covered up and stored away. Petri dishes should always be stored upside down so that no leakage will enter the plate to settle and cause contamination. Plates can also be swabbed in the vertical position so as to minimize contamination during the sampling process.

Some types of swabs come prepackaged with a sterile solution, and these are normally crushed to break them open before each use. Once the surface

TEST PROTOCOL FOR SURFACE SAMPLING

The following procedure applies generally to surface sampling, including cooling coils. See the text for special information if cooling coil surfaces are to be sampled.

A.1: Materials and Preliminary Conditions
    A.1.1: Procure surface sampling materials as necessary. These should include sterile swabs (wet or dry), sterile gloves, and petri dishes (if necessary). At least 3–9 plates (or swabs) will be needed for each condition, Before and After.
    A.1.2: Gloves should be sterilized through the use of a disinfectant such as alcohol. If a sterile template is used to mask off a surface area, then the template must be discarded after each use. Keep at least one plate unused as a control.

B.1: Test Protocol
    B.1.1: Disengage any UVGI systems and any operating equipment in the vicinity that may be a hazard to test personnel. If entry into an air-handling unit is required, the fan should be shut down.
    B.1.2: Identify and record the location of a suitable surface sampling point. The sample location should be approximately the same for subsequent samples. Either visually estimate an area of approximately 2 in^2 or use a sterile template for doing so.
    B.1.3: Using the wet or dry swab, draw it gently across the sample area with a back and forth motion, and either insert the swab back into the sterile container or draw the swab across the petri dish. Cover the plate, seal as necessary, and label the swabs and/or the plates with a code or description of the location sampled.
    B.1.4: Repeat the above process for all the preselected sample locations.
    B.1.5: Start up the UVGI system if any, operate it for specified time period (i.e., 24 hours to 2 weeks) and repeat the test for the After condition.

C.1: Evaluation of Results
    C.1.1: Deliver the samples (either swabs or plates) to a laboratory, or place the plates in an incubator as soon as possible, and incubate for 24–48 hours at the required temperature.

> C.1.2: After incubating for 24–48 hours, and before the plates become overgrown, remove them and count each plate. Alternatively, digital images of the plates can be made and used for counting.
>
> C.1.3: Tabulate and summarize the results as necessary on appropriate forms.
>
> C.1.4: If the client requires identification of the bacterial or fungal species, the laboratory should be directed to perform this function.

sample is taken, the swab is merely inserted back into the sterile solution. The sterile solution is then used for culturing on petri dishes in the laboratory.

The Test Protocol for Surface Sampling outlined above may be adapted to any hospital surface sampling application. For any surface that is to be sampled comparatively, the Before samples should correspond roughly to the location for the After samples, but not exactly because the sampling process may actually clean up the surface. For sampling cooling coils and ducts the ventilation system should be turned off, because the airstream air may carry microbes and create contamination problems.

Hospitals are often interested in sampling for Gram-positive versus Gram-negative bacteria (such as MRSA and VRE). One source reports that moistened swabs registered a sensitivity of 54% for Gram-positive cocci and 74.2% for Gram-negative bacteria while Rodac plates had 69.5% sensitivity for Gram-positive cocci and 42.7% for Gram-negative bacteria (Lemmena et al. 2001).

Cooling coil UV disinfection systems have proven popular in hospital facilities due to the economic benefits of keeping coils free of biofouling. Surface samples are often taken from the coils before and after a one- or two-week UV irradiation period to verify disinfection. Because the coil surfaces are actually fins and not smooth surfaces, the surface area being sampled may be difficult to quantify. The typical fin spacing is between 8 and 12 fins per inch, meaning the area sampled is not a perfectly smooth surface. Although the actual physical surface area sampled will be less than about 2 in² (6.5 cm²), due to the small spaces between the fins, this may be counterbalanced by the fact that spores and other debris tend to concentrate on the leading edges of the fins. In any event, the exact surface area to be sampled is not critical and what matters is that the area sampled is the same for all samples taken. Samples should be taken on the upstream side of the coil because this is where the spores will tend to accumulate. Before samples may prove to be so highly contaminated that they are uncountable. It is sufficient in such cases that the After samples show a significant reduction. High contamination levels may suggest, however, that manual coil cleaning might be warranted.

Virus Detection and Sampling

Viruses can be sampled from the air using filters, or liquid and solid impingement. Filters are the most efficient method because they can capture particles less than 5 microns in size. Sampling can affect viral sensitivity and therefore culturing may be inadequate for determining the true airborne concentration. Viruses can be identified using polymerase chain reaction (PCR) assay methods to detect the presence of viral DNA, but cannot at present distinguish viable viruses. PCR kits are available with specific primers for individual viruses. Airborne nosocomial viruses that have been sampled in air include influenza, measles, adenovirus, Coxsackievirus, parainfluenza, reovirus, rotavirus, coronavirus (SARS), rhinovirus, and RSV (see Verreault, Moineau, and Duchaine [2008] for source studies). Real-time PCR methods can be used to identify viruses much faster than traditional PCR methods. Agranovski et al. (2008) used an air sampler coupled with PCR technology to facilitate continuous long-term monitoring of airborne microbes including viruses. The combined device decreased the detection time from a few days to 2.5 hours.

Booth et al. (2005) detected airborne SARS coronavirus in various rooms of a hospital in Toronto during a SARS outbreak using a slit sampler and wet media as well as membrane filters. Samples were evaluated with reverse transcriptase (RT)-PCR and culture. Surface samples were taken from frequently touched surfaces using Dacron swabs premoistened with viral transport medium. Real-time PCR assaying was also used but was found to be less sensitive.

Myatt et al. (2004) used 37-mm Teflon filter cassettes to collect air samples on the floor and these were analyzed for the presence of rhinoviruses using an RT-PCR kit for picornaviruses. These were compared with nasal samples taken from occupants to determine if airborne viruses were the same as those infecting individuals. At least one match was found, suggesting the airborne rhinovirus had come from the individual coughing or sneezing.

Viruses can be detected in used ventilation filters. A study by Goyal et al. (2011) identified influenza and parainfluenza viruses in ventilation filters from two large public buildings using PCR methods.

Human cytomegalovirus (CMV), a major immunocompromised pathogen, was detected in the rooms of patients with CMV pneumonia and latent infection using filter sampling and a PCR assay (McCluskey, Sandina, and Greene 1996). Tseng, Chang, and Li (2010) used air filters and real-time quantitative PCR (qPCR) to detect influenza, adenovirus, and enteroviruses from the emergency room of a pediatrics department in a medical center. Influenza has been detected in 79% of air samples and RSV in 71% of air samples taken from the air of an urgent medical clinic using an air sampler coupled with a real-time PCR assay (Lindsley et al. 2010).

TABLE 18.2

ISO Air Cleanliness Classifications

Classification Number	Maximum Concentration, particles/m³			
	0.1 μm	0.2 μm	0.3 μm	0.5 μm
ISO Class 1	10	2	—	—
ISO Class 2	100	24	10	4
ISO Class 3	1000	237	102	35
ISO Class 4	10000	2370	1020	352
ISO Class 5	100000	23700	10200	3520
ISO Class 6	1000000	237000	102000	35200

Air and Surface Disinfection Standards

There are currently standards for airborne concentrations in the United States, although such standards are being adopted in Europe and Asia. For hospital air, WHO recommends limits of 100 cfu/m³ for bacteria and 50 cfu/m³ for fungi (WHO 1988). According to the criteria of Federal Standard 209E (FD 209E) on cleanrooms, conventionally ventilated operating rooms rank less than class 3.5 (Durmaz et al. 2005). ISO Cleanroom classifications (from ISO-14644-1, Classification by Airborne Particles) are shown in Table 18.2 for particles 0.5 microns and smaller. Because the smallest particles in this standard are 0.1 microns, viruses smaller than this are not covered, and therefore cleanroom classifications are not entirely suitable as a criteria for hospitals and operating rooms.

The definition of which pathogens constitute dangerous health hazards depends on the patient population. As has been shown in previous chapters, the spectrum of pathogens of concern can vary with the patients at risk, which may include the elderly, infants, surgical patients, or the immunocompromised. The limits of airborne concentrations may also be a function of the infectious dose. For some microbes, like TB bacilli, even a single cfu could be considered unacceptable. It has been suggested that no level of dangerous pathogens is acceptable in the indoor environment (AIHA 1989; Rao and Burge 1996).

Operating rooms must have the most stringent standards. To reduce the risk of airborne infection in ORs, Hardin and Nichols (1995) recommend that ventilation systems must be capable of reducing the level of airborne contamination to 15–20 cfu/ft³ (530–706 cfu/m³). For high-risk procedures the limit for air purity is suggested as 1 cfu/ft³ (35 cfu/m³). According to Audurier et al. (1985), the airborne concentration of bacteria in the operating room should not exceed 30 cfu/m³. Arrowsmith (1985) suggests that operating room air should not exceed 35 cfu/m³ when empty and when in use should not exceed 180 cfu/m³. For ultraclean operating theaters, there should be less than 10 cfu/m³ within 30 cm of the wound, and not more than 20 cfu/ m³ at the level of the operating table (Brown et al. 1996; Lidwell et al. 1983; Whyte et al. 1983; Holton and Ridgeway 1993). A limit of 15 cfu/m³ for gross counts of fungi is mentioned by HICPAC (2003). A limit of *10 cfu/m³*, used in the pharmaceutical industry and as a target for ultraclean ventilation (UCV) systems, would probably be an appropriate criterion for hospital ORs and ICUs. Sampling frequency for ORs should be more than once a year (Fox and Whyte 1996). Similar limits could be applied to isolation rooms and for housing immunocompromised patients.

Indoor airborne bacteria are typically either human commensals or environmental bacteria brought in with fresh air and are, in general, not considered dangerous pathogens except to the immunocompromised. Previously, Flannigan, McCabe, and McGarry (1991) suggested 4500 cfu/m³ as a limit for bacteria in nonhospital indoor environments while another source recommended a limiting range of 500–1000 cfu/m³ (EPD 1999). Such limits may be useful for general wards.

References

Agranovski, I. E., Safarov, A. S., Agafonov, A. P., Pyankov, O. V., and Sergeev, A. N. (2008). Monitoring of airborne mumps and measles viruses in a hospital. *Clean - Soil, Air, Water* 36(10-11), 845–849.

AIHA (1989). The practitioner's approach to indoor air quality investigations. *Indoor Air Quality International Symposium*, Fairfax, VA, 43–66.

Ambroise, D., Greff-Mirguet, G., Gorner, P., Fabries, J. F., and Hartemann, P. (1999). Measurement of indoor viable airborne bacteria with different bioaerosol samplers. *J Aerosol Sci* 30(Suppl 1), S699–S700.

Arrowsmith, L. W. M. (1985). Air sampling in operating theatres. *J Hosp Infect* 6, 352–353.

Artenstein, M. S., Miller, W. S., Rust, J. H., and Lamson, T. H. (1967). Large-volume air sampling of human respiratory disease pathogens. *American Journal of Epidemiology* 85(3), 479–485.

Audurier, A., Fenneteau, A., Rivere, R., and Raoult, A. (1985). Bacterial contamination of the air in different operating rooms. *Rev Epidemiol Sante Publique* 33(2), 134–141.

Booth, T. F., Kournikakis, B., Bastien, N., Ho, J., Kobasa, D., Stadnyk, L., Li, Y., Spence, M., Paton, S., Henry, B., Mederski, B., White, D., Low, D. E., McGeer, A., Simor, A., Vearncombe, M., Downey, J., Jamieson, F. B., Tang, P., and Plummer, F. (2005). Detection of airborne severe acute respiratory (SARS) coronavirus and environmental contamination in SARS outbreak units. *JID* 191, 1472–1477.

Boss, M. J., and Day, D. W. (2001). *Air Sampling and Industrial Hygiene Engineering*. Lewis Publishers, Boca Raton, FL.

Brown, I. W. J., Moor, G. F., Hummel, B. W., Marshall, W. G., and Collins, J. P. (1996). Toward further reducing wound infections in cardiac operations. *Ann Thorac Surg* 62(6), 1783–1789.

Buttner, M., Cruz, P., Stetzenbach, L., Klima-Comba, A., Stevens, V., and Emmanuel, P. (2004). Evaluation of the biological sampling kit (BiSKit) for large-area surface sampling. *Appl Environ Microbiol* 70(12), 7040–7045.

Cox, C. S., and Wathes, C. M. (1995). *Bioaerosols Handbook.* CRC/Lewis, Boca Raton, FL.

Durmaz, G., Kiremitci, A., Akgun, Y., Oz, Y., Kasifoglu, N., Aybey, A., and Kiraz, N. (2005). The relationship between airborne colonization and nosocomial infections in intensive care units. *Mikrobiyol Bul* 39(4), 465–471.

EPD (1999). Guidance notes for the management of indoor air quality in offices and public places. The Hong Kong Government of Special Administrative Region, Environmental Protection Department, Hong Kong.

Flannigan, B. (1997). Air sampling for fungi in indoor environments. *J Aerosol Sci* 28(3), 381–392.

Flannigan, B., McCabe, E. M., and McGarry, F. (1991). Allergenic and toxigenic microorganisms in houses; in *Pathogens in the Environment*, B. Austin, ed., Blackwell Scientific, Oxford, UK.

Flannigan, B., McEvoy, E. M., and McGarry, F. (1999). Investigation of airborne and surface bacteria in homes. *Indoor Air 99: Proceedings of the 8th International Conference on Indoor Air Quality and Climate*, Edinburgh, Scotland, 884–889.

Fox, C., and Whyte, A. (1996). Theatre air sampling: Once a year is not enough. *J Hosp Infect* 32, 319–320.

Goyal, S., Anantharaman, S., Ramakrishna, M., Sajja, S., Kim, S., Stanley, N., Farnsworth, J., Kuehn, T., and Raynor, P. (2011). Detection of airborne viruses in used ventilation filters from two large public buildings. *Am J Infect Contr* 39(7), e30–e38.

Griffiths, W. D., Upton, S. L., and Mark, D. (1993). An investigation into the collection efficiency and bioefficiencies of a number of aerosol samplers. *J Aerosol Sci* 24(S1), s541–s542.

Hardin, W. D. J., and Nichols, R. L. (1995). Aseptic technique in the operating room; in *Surgical Infections*, D. E. Fry, ed., Little, Brown and Company, Boston, 109–117.

HICPAC (2003). Guidelines for environmental infection control in health-care facilities. *MMWR* 52(RR-10), 1–48.

Holton, J., and Ridgeway, G. L. (1993). Commissioning operating theatres. *J Hosp Infect* 23, 153–160.

Jensen, P. A., and Schafer, M. P. (1998). Chapter J: Sampling and characterization of bioaerosols; in *NIOSH Manual of Analytical Methods*, NIOSH Publication 94-113, M. E. Cassinelli and P. F. O'Connor, eds, National Institute for Occupational Safety and Health, Atlanta, GA, 82–112.

Lemmena, S., Hafnera, H., Zolldanna, D., Amedicka, G., and Luttickenb, R. (2001). Comparison of two sampling methods for the detection of Gram-positive and Gram-negative bacteria in the environment. *Int J Hyg Environ Health* 203(3), 245–248.

Li, C.-S., Hao, M. L., Lin, W. H., Chang, C. W., and Wang, C. S. (1999). Evaluation of microbial samplers for bacterial microorganisms. *Aerosol Sci & Technol* 30, 100–108.

Lidwell, O. M., Lowbury, E. J. L., Whyte, W., Blowers, R., Stanley, S. J., and Lowe, D. (1983). Airborne contamination of wounds in joint replacement operations: The relationship to sepsis rates. *J Hosp Infect* 4, 111–131.

Lin, X., Reponen, T. A., Willeke, K., Grinshpun, S. A., Foarde, K. K., and Ensor, D. S. (1999). Long-term sampling of airborne bacteria and fungi into a non-evaporating liquid. *Atmos Environ* 33(26), 4291–4298.

Lindsley, W. G., Blachere, F. M., Davis, K. A., Pearce, T. A., Fisher, M. A., Khakoo, R., Davis, S. M., Rogers, M. E., Thewlis, R. E., Posada, J. A., Redrow, J. B., Celik, I. B., Chen, B. T., and Beezhold, D. H. (2010). Distribution of airborne influenza virus and respiratory syncytial virus in an urgent medical care clinic. *CID* 50, 693–698.

McCluskey, R., Sandina, R., and Greene, J. (1996). Detection of airborne cytomegalovirus in hospital rooms of immunocompromised patients. *J Virol* 56(1), 115–118.

Myatt, T. A., Johnston, S. L., Zuo, Z., Wand, M., Kebadze, T., and Rudnick, S. (2004). Detection of airborne rhinovirus and its relation to outdoor air supply in office environments. *Am J Respir Crit Care Med* 169, 1187–1190.

Pasquarella, C., Masia, M. D., Nnanga, N., Sansebastiano, G. E., Savino, A., Signorelli, C., and Veronesi, L. (2004). Microbial air monitoring in operating theatre: Active and passive samplings. *Ann Ig* 16(1-2), 375–386.

Rao, C. Y., and Burge, H. A. (1996). Review of quantitative standards and guidelines for fungi in indoor air. *J Air & Waste Mgt Assoc* 46(Sep), 899–908.

Straja, S., and Leonard, R. T. (1996). Statistical analysis of indoor bacterial air concentration and comparison of four RCS biotest samplers. *Environment International* 22(4), 389.

Streifel, A. J. (1999). Design and maintenance of hospital ventilation systems and the prevention of airborne nosocomial infections; in *Hospital Epidemiology and Infection Control*, C. G. Mayhall, ed., Lippincott Williams & Wilkins, Philadelphia, 1211–1221.

Tseng, C. C., Chang, L. Y., and Li, C. S. (2010). Detection of airborne viruses in a pediatrics department measured using real-time qPCR coupled to an air sampling filter method. *J Environ Health* 73(4), 22–28.

Verreault, D., Moineau, S., and Duchaine, C. (2008). Methods for sampling airborne viruses. *Microbiol Molec Biol Rev* 72(3), 413–444.

WHO (1988). Indoor air quality: Biological contaminants. *European Series 31*, World Health Organization, Copenhagen, Denmark.

Whyte, W., Lidwell, O. M., Lowbury, E. J. L., and Blowers, R. (1983). Suggested bacteriological standards for air in ultraclean operating rooms. *J Hosp Infect* 4, 133–139.

Yamaguchi, N., Yoshida, A., Saika, T., Senda, S., and Nasu, M. (2003). Development of an adhesive sheet for direct counting of bacteria on solid surfaces. *J Microbiol Meth* 53, 405–410.

19

UVGI Air Disinfection

Introduction

UVGI air disinfection systems have various applications in hospitals and health care facilities. There are four main types of UV air disinfection systems: in-duct UV systems, stand-alone recirculation units (also called unitary UV systems), upper-room systems, and UV barrier systems. Upper-room and barrier systems are passive disinfection systems that depend on local room air currents while in-duct UV systems and recirculation units are forced air systems. UV airstream disinfection systems are specifically designed to remove airborne microorganisms such as bacteria, viruses, and fungi. These systems require the use of filters to control dust accumulation on the lamp, and the filter plays a role in reducing airborne contamination, especially when it comes to spores. All in-duct systems utilize filters, as specified by codes and standards for hospital ventilation systems (ASHRAE 1999), and most but not all recirculating units use filters.

This chapter examines the four main types of UV air treatment systems and summarizes field trials of air disinfection systems. The degree to which these systems are effective depends on local conditions and it cannot be said that one type of system is better than another, especially as economics often plays a part in the selection process. UVGI is not a stand-alone solution to air contamination problems but is an adjunct that, when properly designed and applied, can provide benefits in terms of reduced infection rates and energy savings (Memarzadeh, Olmsted, and Bartley 2010). Other factors, such as careful design of the hospital built environment, installation and effective operation of the HVAC system, and a high degree of attention to traditional methods of cleaning and disinfection, should be assessed before any decision is made to apply UVGI to meet indoor air quality requirements in health care facilities.

Airstream Disinfection

The performance of a UV air disinfection system can be quantified in terms of the UV exposure dose it produces in the airstream. The UV dose is dependent on the mean irradiance in the duct and the exposure time through the irradiation chamber. The exposure time depends on the airflow and the dimensions of the duct (width, height, and length). The exposure time in seconds, E_t, is computed as follows:

$$E_t = \frac{Vol}{Q} = \frac{WHL}{Q} \tag{19.1}$$

where
  Vol = volume of the UV chamber, m³
  Q = airflow, m³/s
  W = width, m
  H = height, m
  L = length, m

Equation (19.1) depends on the airflow being completely mixed in the exposure zone, which is a reasonable assumption for any airstream in which turbulent conditions may exist. At the typical design air velocity of 2.54 m/s (500 fpm), Reynolds numbers will be on the order of 150,000 and complete mixing will be approached (Kowalski et al. 2000). The survival of any airborne population of microbes will be predicted by the standard exponential decay model and the UVGI removal rate (DR) can be computed as follows:

$$DR = 1 - e^{-kI_m E_t} \tag{19.2}$$

where
  DR = disinfection rate, fraction or %
  k = UV rate constant, m²/J
  I_m = mean irradiance, W/m²

The mean irradiance must be known in order to predict UV air disinfection performance. Methods for computing the irradiance around a UV lamp based on a view factor model have been developed by Kowalski et al. (2000). Irradiance can also be measured with a radiometer, but taking accurate readings inside an enclosed duct presents difficulties. Simplified charts and tables for determining the mean irradiance for various lengths and sizes of ducts based on the lamp UV wattage have been presented by Kowalski (2009).

The UVGI Rating Value (URV) scale is a convenient way to describe the UV dose produced by any air disinfection system (IUVA 2005). Table 19.1 shows

TABLE 19.1

UVGI Rating Values (URV) and Filter Recommendations

URV	Dose (J/m²)	Dose (µW-s/cm²)	Mean Dose (J/m²)	Filter MERV	Adenovirus	Influenza	TB	MRSA
					Air Disinfection Rates from UV Alone (%)			
1	0.01	1	0.055	6	0	0	0	1
2	0.10	10	0.15	6	1	1	5	6
3	0.20	20	0.25	6	1	2	9	11
4	0.30	30	0.4	6	2	4	13	16
5	0.50	50	0.63	6	3	6	21	26
6	0.75	75	0.88	6	4	9	30	36
7	1.0	100	1.25	7	5	11	38	45
8	1.5	150	2	8	8	16	51	59
9	2.5	250	3.75	9	13	26	69	77
10	**5**	**500**	**7.5**	**10**	**24**	**45**	**91**	**95**
11	**10**	**1000**	**12.5**	**11**	**42**	**70**	**99**	**100**
12	**15**	**1500**	**17.5**	**12**	**56**	**83**	**100**	**100**
13	**20**	**2000**	**25**	**13**	**66**	**91**	**100**	**100**
14	**30**	**3000**	**35**	**14**	**80**	**97**	**100**	**100**
15	**40**	**4000**	**45**	**15**	**88**	**99**	**100**	**100**
16	50	5000	55	15	93	100	100	100
17	60	6000	70	15	96	100	100	100
18	80	8000	90	15	99	100	100	100
19	100	10000	150	15	100	100	100	100
20	200	20000	250	15	100	100	100	100
21	300	30000	350	15	100	100	100	100
22	400	40000	450	15	100	100	100	100
23	500	50000	750	15	100	100	100	100
24	1000	100000	1500	15	100	100	100	100
25	2000	200000	2500	15	100	100	100	100
UV Rate constants (m²/J)								
Adenovirus	0.054							
Influenza	0.119							
TB	0.472							
MRSA	0.596							

Note: Boldface indicates normal design range.

the URV scale, which is based on the indicated UV dose. Most UV systems for airstream disinfection will normally fall into the URV 10–15 range. The URV rating system parallels the ASHRAE filter rating system for minimum efficiency reporting value (MERV), and when the recommended MERV filter is coupled with the indicated URV system the removal rates of airborne microbes will be approximately equal across the entire array of microbes. UV disinfection rates for some example microbes in air are shown but do not account for the filter—if the filter were accounted for the removal rates would be higher. Filters rated below MERV 6 are not recommended for use in protecting UV lamps, while MERV 15 filters represent the maximum rating of a filter that would normally be coupled with a UV system. Exceptions to this rule may exist when designing ultraclean operating rooms for hospitals, where HEPA filters might be required.

Once the UV dose is established or selected, predictions can then be made on the UV disinfection rates for any given microbe for which the UV susceptibility, or UV rate constant, is known. The disinfection rates for the UV system can then be combined with the removal rates for the filter to determine the overall removal rates for any given microbe.

Filtration and UVGI are mutually complementary technologies because filtration removes most of the microbes that tend to be resistant to UV, and UV destroys most of the smaller microbes that may penetrate filters. Table 19.2 shows the disinfection rates for all the Class 1 and Class 2 airborne nosocomial pathogens from Table 4.1 for which the UV rate constant is known or could be predicted from the genome. UV rate constants are for air as data are available, otherwise they are for surfaces or water and are averaged where multiple values are available. See Kowalski (2009) for UV rate constant source data.

Some microbes in Table 19.2 for which the UV rate constant is unknown have a predicted rate constant shown based on genomic modeling (Kowalski 2009; Kowalski, Bahnfleth, and Hernandez 2009; Kowalski 2011). Accuracy of these predicted values is expected to be within ±10%. VZV, for which the rate constant is unknown and the genome is also unknown, has the value for varicella virus substituted. For *Clostridium difficile* spores the k value for *Clostridum perfringens* spores is used as a surrogate.

UVGI in-duct air disinfection systems usually consist of lamps, lamp fixtures, or modular lamp arrays located inside ducts or air-handling units (AHUs). UV lamps are often located on the upstream side of the cooling coils and downstream from the mixing box and filters. Assuming the UV lamp performs per design specifications and is properly maintained, the main parameters that impact UV disinfection rates are the relative humidity (RH), air temperature, and air velocity. Increased RH tends to decrease decay rates of bacteria under ultraviolet exposure (Peccia et al. 2001). Air temperature has a negligible impact on microbial susceptibility to UVGI, but low or high air temperature combined with air velocity can impact the UV output of lamps if design operating specifications are exceeded. If a UV lamp

TABLE 19.2

UV Disinfection Rates of Airborne Nosocomial Pathogens

Microbe	Airborne Class	UV (k m²/J)	UV D90 (J/m²)	Disinfection Rate (%)					
				URV 10	URV 11	URV 12	URV 13	URV 14	URV 15
Acinetobacter	2	0.16	14	70	86	94	98	100	100
Adenovirus	2	0.054	43	33	49	61	74	85	91
Aspergillus spores	1	0.00894	258	6	11	14	20	27	33
Blastomyces dermatitidis spores	2	0.01645	140	12	19	25	34	44	52
*Bordetella pertussis**	1	0.0364	63	24	37	47	60	72	81
Clostridium difficile spores	1	0.0385	60	25	38	49	62	74	82
Clostridium perfringens spores	2	0.0385	60	25	38	49	62	74	82
Coronavirus (SARS)	1	0.377	6	94	99	100	100	100	100
Corynebacterium diphtheriae	2	0.0701	33	41	58	71	83	91	96
Coxsackievirus	2	0.111	21	57	75	86	94	98	99
Cryptococcus neoformans spores	2	0.0167	138	12	19	25	34	44	53
Enterobacter cloacae	2	0.03598	64	24	36	47	59	72	80
*Enterococcus**	2	0.0822	28	46	64	76	87	94	98
Fusarium spores	2	0.00855	269	6	10	14	19	26	32
Haemophilus influenzae	2	0.11845	19	59	77	87	95	98	100
*Haemophilus parainfluenzae**	2	0.03	77	20	31	41	53	65	74
Influenza A virus	1	0.119	19	59	77	88	95	98	100
Klebsiella pneumoniae	2	0.04435	52	28	43	54	67	79	86
Legionella pneumophila	1	0.2024	11	78	92	97	99	100	100
Measles virus	1	0.1051	22	55	73	84	93	97	99
Mucor spores	2	0.01012	228	7	12	16	22	30	37
Mumps virus*	1	0.0766	30	44	62	74	85	93	97
Mycobacterium avium	2	0.04387	52	28	42	54	67	78	86
Mycobacterium tuberculosis	1	0.4721	5	97	100	100	100	100	100
Mycoplasma pneumoniae	2	0.2791	8	88	97	99	100	100	100
Neisseria meningitidiś	2	0.1057	22	55	73	84	93	98	99
Nocardia asteroides	2	0.0822	28	46	64	76	87	94	98
Norwalk virus*	2	0.0116	198	8	13	18	25	33	41

Continued

TABLE 19.2 (*Continued*)

UV Disinfection Rates of Airborne Nosocomial Pathogens

Microbe	Airborne Class	UV (k m²/J)	UV D90 (J/m²)	Disinfection Rate (%)					
				URV 10	URV 11	URV 12	URV 13	URV 14	URV 15
Parainfluenza virus*	2	0.1086	21	56	74	85	93	98	99
Parvovirus B19	2	0.092	25	50	68	80	90	96	98
Penicillium spores	2	0.00307	750	2	4	5	7	10	13
Proteus mirabilis	2	0.289	8	89	97	99	100	100	100
Pseudomonas aeruginosa	1	0.5721	4	99	100	100	100	100	100
Reovirus	2	0.01459	158	10	17	23	31	40	48
RSV*	1	0.0917	25	50	68	80	90	96	98
Rhinovirus*	2	0.0142	162	10	16	22	30	39	47
Rhizopus spores	2	0.00861	267	6	10	14	19	26	32
Rotavirus	2	0.02342	98	16	25	34	44	56	65
Rubella virus*	1	0.0037	622	3	5	6	9	12	15
Serratia marcescens	2	0.221	10	81	94	98	100	100	100
Staphylococcus aureus	1	0.5957	4	99	100	100	100	100	100
Staphylococcus epidermis	2	0.09703	24	52	70	82	91	97	99
Streptococcus pneumoniae	1	0.00492	468	4	6	8	12	16	20
Streptococcus pyogenes	1	0.8113	3	100	100	100	100	100	100
VZV (varicella surrogate k)	1	0.1305	18	62	80	90	96	99	100

* UV rate constant is a predicted value based on the complete genome.

is operated at an air velocity above design conditions the UV output will be reduced because of the cooling effect of the air on the lamp.

Air Handling Units

In-duct UV systems serve the purpose of disinfecting an airstream in a building or in the zone that it serves. They generally consist of UV lamp fixtures and ballasts, and rely on existing filters in the air handling unit (AHU) to maintain lamp cleanliness. UV lamp fixtures can be placed almost anywhere in the ductwork or in the AHU. In some UV systems the lamp fixtures and ballasts are installed internally while in others the lamp ballasts are installed external to the ductwork, which will result in lower pressure drops. The pressure losses associated with UV lamps in the airstream are generally minor, especially if the air velocity is within the normal design limits of 2–3 m/s (400–600 fpm), but some internal lamp fixtures may reduce airflow or increase energy demand.

Retrofitting UV lamps in an existing ventilation system may require the upgrading of existing filters. At least a MERV 6–8 filter or better is required for lamp cleanliness, but a MERV 10 filter or higher is recommended to maximize air cleaning. The ideal air velocity is 2–3 mps (400–600 fpm) to ensure UV system effectiveness, and the same is generally true of any filters. Modular UV systems are available for installation in ductwork and some include filters and booster fans.

The performance of any installed UV system should be established by analysis or testing. Complete modular systems will generally have performance specifications that are specified in advance. In any event, the actual disinfection performance will depend on local operating conditions, especially airflow and building volume. Any installed UV system must be built to current electrical standards and this usually requires that wiring be shielded from UV exposure or be made of UV-resistant materials. Any existing ventilation system components should also be protected from UV degradation and their materials verified as UV-resistant. No stray UV rays should exit the duct—this is easy to verify by turning out the lights and looking for the telltale blue glow.

Recirculation Units

UV recirculation units consist of UV lamps and fixtures in a housing containing a fan and a filter to keep the lamps free of dust. Room recirculation units can be used to augment in-duct systems or where in-duct installation is not feasible. The filter should be at least a MERV 6–8 filter, and units with a MERV 10 or higher will have improved air cleaning performance. Some UV recirculation units use HEPA filters, but HEPA filters can have high energy demands without any added benefits. The airflow in recirculation units is often in the range of 1.4–14 m³/min (50–500 cfm) and such units are suitable for small rooms. Many recirculation units are portable and can be positioned on floors or tables. Some units are available for hanging on walls or from ceilings in TB or isolation rooms.

Recirculating air systems will deliver multiple UV doses to airborne microorganisms and this chronic dosing will enhance effectiveness against UV-resistant spores. Because of their compact size, the internal volume of recirculation units often does not allow for extended exposure times, and most manufacturers attempt to make up for this by increasing the total UV wattage. This approach may not meet the proposed requirements of a minimum 0.25-second exposure time (IUVA 2005). Without this minimum residence time, the actual performance may be less than predicted because the exposure time is so short that some UV-dosed microbes remain in the shoulder region of their decay curves.

Recirculation units can be compared in terms of the clean air delivery rate (CADR). For in-duct systems the airflow is preset, and if the ventilation system is designed per ASHRAE guidelines, in-duct systems will deliver an appropriate amount of air to each occupied building zone. For unitary systems, the performance will vary with the zone or room volume in which the unit is placed. Methodologies and test methods have been proposed for measuring the single-pass efficiencies and effectiveness of room air cleaners, and unitary air cleaners should be comparable in terms of performance parameters (Foarde et al. 1999a, 1999b; Hanley, Smith, and Ensor 1995; Janney et al. 2000). Any unit added to a hospital room should be individually evaluated for the room size to ensure it will be effective, and this may require consideration of local air currents.

Upper-Room Systems

Upper-room systems consist of single or multiple UV lamp fixtures hung from ceilings or attached to the walls. They have proven to be effective at controlling various types of infections and reducing airborne concentrations of microbes. They are most often used in TB wards and clinics. A complete upper-room system may include the use of UV-absorbing paints on the walls or ceiling to minimize reflected UV exposure to room occupants. The principle of upper-room system design is to provide maximum irradiance in the upper portion of the room while minimizing UV levels below the UV units to less than ACGIH threshold limit values (TLVs). Upper-room systems can provide a cost-effective alternative to forced air systems where budgets are critical or for buildings that rely on natural ventilation (WHO 1999). Upper-room system effectiveness is largely dependent on the local conditions.

Upper-room systems, also called upper-air systems, create a germicidal zone of UV rays that are confined to the upper portion of a room, known as the stratum or UV zone (see Figure 19.1). Air that passes through this field is disinfected and remixes with lower room air. Any exposed surfaces in the room will also be disinfected. UV exposure levels in the lower room are maintained below the ACGIH 8-hour exposure limit, which is 30 J/m^2 for broadband UV (200–320 nm) or 60 J/m^2 for narrow band UVC (254 nm). Upper-room systems operate continuously in occupied areas and, when properly designed and installed, are safe for patients and HCWs. They can be cost-effective for many types of facilities, including hospital waiting rooms and TB shelters. One of the stated advantages of upper-room systems is that they intercept microbes inside the room where microbes may be generated

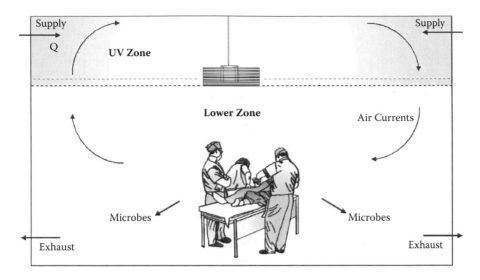

FIGURE 19.1
Air will circulate into and out of the upper-room UV zone while contaminants are released continuously from occupants.

by occupants, thereby controlling infection at the source (First et al. 1999). Continuous upper-room UV disinfection is considered to be the most practical method in resource-limited countries where mechanical systems are often lacking and in cold climates where energy losses from high air turnover rates (high outside air) are prohibitive (WHO 1999).

Upper-room systems have been in use for many decades, and data that have accrued from installations in prior tests in hospitals and other indoor environments have shown them to be effective when properly designed (Dumyahn and First 1999). Upper-room systems are effective against a wide array of airborne viruses and bacteria, including TB, measles, chickenpox, mumps, varicella, and cold viruses. Safety issues have been largely addressed through good design practices, and the history of upper-room systems indicate very few cases of accidental eye and skin burns.

Upper-room UVGI fixtures are available as rectangular wall-mounted units and circular ceiling-hung units. A circular fixture in the center of a room will provide all-around exposure in the upper room. Multiple rectangular fixtures can be placed along the walls, usually in more than one location. Upper-room fixtures may contain one or more UV lamps, and some rectangular fixtures contain grilles and reflectors to direct exiting UV rays in parallel beams. Typical lamp input wattages range from about 18–36 W, with UV lamp output wattages of about 5–10 W. Fixtures should be carefully located so as to provide optimum coverage of the stratum without causing any hazards in the lower room. The cutoff height of the UV zone may be 7–8 feet or higher, depending on available space, well above head height for most people. Some reflected UV from the walls and ceiling may still enter the lower areas of the room, but lower-room levels should be below the ACGIH TLV for 8 hours' exposure.

Koller (1939) proposed the criteria of using one 30-W lamp for every 200 ft², for an 8 foot ceiling, and this has been widely adopted in other guidelines (First et al. 1999; Macher 1992; Riley, Knight, and Middlebrook 1976; CIE 2003; ASHRAE 2007). These criteria are also appropriate for hospital environments.

In terms of overall UV dose, upper-room systems provide a similar dose range to in-duct systems. The effectiveness of an upper-room system is dependent on the air exchange rate between the upper- and lower-room volumes (First et al. 2007). The greatest reduction in infectious microorganisms may occur when the highest rate of air exchange occurs between the upper and lower room areas (CIE 2003).

Riley, Knight, and Middlebrook (1976) demonstrated the disappearance of *Mycobacterium bovis* (BCG) from a room with an upper-air UVGI system in terms of the equivalent air changes per hour (EAC), defined as the slope of the plot of the natural logarithm of the airborne BCG count versus time in hours. First et al. (1999) define the EAC more generally as the number of air exchanges in a well-mixed room that would be required to reduce the number of viable airborne bacteria to the same degree as UV alone, and present the following equation:

$$EAC = -\ln\left(\frac{N_s}{N_0}\right) = kIt = kD \tag{19.3}$$

where
   N_s = microbial population at time t (based on airborne counts)
   N_0 = microbial population at time $t - 0$
   k = microbe rate constant, m²/J
   I = irradiance, W/m²
   t = time, seconds
   D = UV exposure dose, J/m²

Equation (19.3) can be used to compute the EAC for any given nosocomial microorganism for which UV susceptibility data exist. For example, if *Serratia marcescens* (UV rate constant $k = 0.221$ m²/J) were exposed to a UV dose of 54.3 J/m², the EAC would be $(0.221) \times (54.3) = 12$ air changes per hour (ACH). Studies on reduction rates of *Mycobacterium tuberculosis* have shown that equivalent air exchange rates of 10–25 ACH can be achieved (CDC 2005). The dependence of EAC on microbial species, and the lack of data for some microbes, is not a serious drawback provided some representative microbe is used as an indicator microbe.

It is not uncommon for hospitals and other health care facilities to have supply air enter near the ceiling and exit near the floor, and in such cases the entire airflow (Q) will pass through the UV zone. Biocontaminants can come from room occupants, including the patient, as well as from other sources like the floor and equipment, and a downward flow can facilitate air cleaning.

Good air mixing is necessary for upper-room systems to be effective (WHO 1999). When rooms lack adequate air movement, the addition of a mixing fan can be an effective solution (CIE 2003). Relative humidity can also impact system performance because of bacterial response to RH and should be kept at or less than 75%, or typical indoor conditions, but an RH exceeding 80% may reduce system effectiveness against bacteria (Rudnick and First 2007).

UV-absorbing paints containing titanium oxide can be used on ceilings and walls to minimize reflected UV. Usually, painting with flat white or dark paints is sufficient to reduce reflected UV to about 4–5%. If UV levels exceed NIOSH safe limits, all highly UV-reflecting surfaces should be removed or altered.

In-duct air disinfection systems can be modeled with a single-pass efficiency and a clean air delivery rate. For upper-room systems the overall room effectiveness can be defined by the following relation (Beggs and Sleigh 2002):

$$E = 1 - \left(\frac{C_{on}}{C_{off}} \right) \tag{19.4}$$

where

E = room effectiveness, fraction or %
C_{on} = airborne concentration of microbes with system On, cfu/m^3
C_{off} = airborne concentration with system Off, cfu/m^3

The ability of the system to reduce airborne counts of bacteria or fungi is a prime indicator of system performance. In one test of an upper-room system conducted by Xu et al. (2003), an 87 m^3 room was equipped with four 36-W units mounted in the corners and one 72-W unit mounted in the center. The disinfection rates of several microbes in the air were used to determine the effective clean air delivery rate (CADR), which proved to be about 1392 m^3/hr (819 cfm). It was found in a second study that operating the upper-room system in combination with other air cleaners resulted in an additive effect (Kujundzic et al. 2006).

The best test of system performance is the ability to reduce infection rates. Escombe et al. (2009) exposed guinea pigs to exhaust air from a TB ward and 35% of the controls developed tuberculosis, but only 9.5% developed infections when upper-room systems were used, making for a net incidence decrease of 74%. Table 19.3 summarizes airborne infection control results from hospital field trials.

The primary safety hazard of upper-room systems is that occupants may be exposed to direct or reflected UV rays for prolonged periods of time. Measurements should be made to ensure that hazardous UV levels do not exist. Levels of UV irradiance in the lower room should be below those that produce the ACGIH TLV dose after 8 hours of exposure (0.002 W/m^2 for UVC

TABLE 19.3

Result of Field Trials of Upper-Room Systems in Health Care

Location	Infection	Infection Cases (%) Before	Infection Cases (%) UVGI	Decrease (%) Net	Decrease (%) %	Reference
The Cradle, Evanston	Respiratory infection	14.5	4.6	9.9	68	Sauer, Minsk, and Rosentern 1942
St. Luke's Hospital, NY	Respiratory infection	—	—	—	33	Higgons and Hyde 1947
Home for Hebrew Infants, NY	Varicella epidemic	97	0	97	100	Wells 1955
Livermore, CA, Veteran's Hospital	Influenza	19	2	17	89	McLean 1961
Boston Homeless Shelter	Tuberculosis	—	—	—	78	Nardell 1988
North Central Bronx Hospital	TB conversions in staff	2.5	1.0	2.0	60	EPRI 1997
Average Net Decrease	**71%**					

and 0.001 W/m^2 for broadband UV). The ACGIH TLVs for UV depends on the type of UV lamp used in the fixtures because narrow-band UVC lamps (low-pressure mercury lamps) emit most of their UV at 254 nm, while broadband UV lamps (medium-pressure mercury lamps) emit a broader spectrum of UV between 200 and 400 nm. Because the erythemal response curve varies with wavelength, the limits are defined in terms of the UV spectra produced, with the ACGIH TLV being 30 J/m^2 for broadband UV and 60 J/m^2 for narrow band UVC (254 nm). Scientists at the Tuberculosis Ultraviolet Shelter Study (TUSS) are now using an irradiance limit of 0.004 W/m^2 at eye level as their design criteria, which takes into account the body's designed defenses: shading of upper lids and brows, and other factors that limit actual UV exposure (CIE 2003).

Barrier Systems

UV barrier systems are mounted on the sides and/or the overhead portion of a doorway (i.e., each side and overhead) and are intended to irradiate the air that passes between rooms. The lamp fixtures themselves are often of the upper-room type with louvers to constrain the UV rays to minimize the irradiance in the rooms. Though not in common use, UV barrier systems do provide options for the health care industry and can be used to separate patients. UV barrier systems have been used in hospitals to disinfect air transferring from one patient area to another (Buttolph and Haynes 1950; Koller 1965). Irradiance levels within the barrier or curtain must be sufficient to achieve a high disinfection rate in the brief period it takes for air to pass through. Experimental installations have demonstrated that barrier systems can greatly reduce airborne concentrations of bacteria and can reduce rates of respiratory illness (Koller 1965; Wheeler et al. 1945). Wells (1938) found that a UV barrier system was effective in preventing the spread of chickenpox in an isolation ward. Robertson et al. (1939) demonstrated that UV barriers could reduce airborne bacteria by 95%. DelMundo and McKahn (1941) placed UV barrier systems across the cubicles of individual patients and reduced cross-infections from 12.5% to 2.7%. Sommer and Stokes (1942) found UV barrier systems effective at reducing the airborne bacterial concentrations in a hospital ward and that they had some measurable impact on cross-infections. Sauer, Minsk, and Rosenstern (1942) showed that a UV barrier system effectively controlled cross-infections in a nursery. In a number of tests in a children's hospital, Robertson et al. (1943) showed that UV curtains could reduce cross-infections in half or better in most cases.

Because irradiance levels in barrier systems are often high, there is a manifest hazard to personnel who tarry, and personnel may be required to wear protective clothing, skin creams, and eye protection.

UV Air Disinfection Field Trials

Published studies on the ability of UV in-duct air disinfection to reduce microbial contamination in health care facilities have demonstrated major reductions in indoor airborne concentrations of pathogens and allergens. It is often the case in epidemiological studies that natural disease incidence is so low (i.e., about 1–5% on any given day) that achieving statistical significance requires either a large population or a trial period lasting several years. A study by Dionne (1993) on the incidence of respiratory infections at a daycare center showed a negligible reduction in illness but did demonstrate a reduction in airborne concentrations of microbes. Robertson et al. (1943) showed that respiratory infections could be reduced in a children's hospital. Allegra et al. (1997) used a unitary UV system without a filter to reduce airborne concentrations of bacteria in a room and obtained a 93–99% reduction of airborne counts within minutes.

Evidence that UVGI is effective at reducing disease incidence comes mostly from upper-room UV systems. Limited epidemiological data are available on UV air disinfection systems, but what there is confirms that reductions in disease transmission are possible. In 1937 the first application of UVGI to a school ventilation system significantly reduced the incidence of measles, and subsequent applications enjoyed similar success (Wells 1955).

Rosenstern (1948) showed that UV in air conditioning systems could reduce cross-infections by 71% or more in a nursery, although the system simultaneously employed UV barrier systems. Riley and O'Grady (1961) demonstrated the complete elimination of TB bacilli from hospital ward exhaust air using in-duct UV air disinfection. Schneider et al. (1969) applied UV both in ducts and in corridors and effectively controlled pathogens in an isolation unit. A study on mold-sensitized children afflicted with asthma who lived in residential homes showed that UV air disinfection was effective in decreasing respiratory symptoms (Bernstein et al. 2006). For additional studies on UV reduction of respiratory disease incidence see Kowalski (2009).

Combination of UV and Filtration

Filtration and UVGI are mutually complementary technologies because filtration removes most of the pathogens that tend to be resistant to UVGI, and filters remove most of the large bacteria and spores that may be resistant to UV. When the appropriate amount of UV is combined with the right level of filtration, almost any level of air disinfection may be achieved across the entire array of airborne nosocomial pathogens. When a filter is placed in series with a UV system, the removal rates are combined additively based on the penetration of the filter (assumed to come first). That is, the microbes that penetrate the filter will be exposed to a UV dose that will operate on the surviving (penetrating) population.

In mathematical form, we can write the combined total survival for a population passing through a filter and then a UV system as follows:

$$Cr = 1 - (1 - Fr)(Dr) \tag{19.5}$$

where
   Cr = combined removal rate, fractional
   Fr = filtration rate, fractional
   Dr = disinfection rate, fractional

The process is simply repeated if there are three or more filters or UV components in series. This might be the case if there is a prefilter before the primary filter. Equation (19.5) is used to compute the removal rates of all pathogens from Table 19.3 when they are subject to matched MERV and URV ratings. That is, each MERV-rated filter (e.g., MERV 10) is matched with an URV-rated UV system (e.g., URV 10) and the total removal rates are shown in Table 19.4. The filtration rates are taken from Table 8.2 in Chapter 8. Only microbes with a known or estimated UV rate constant are addressed in Table 19.4.

TABLE 19.4

Combined UV and Filter Removal Rates of Airborne Nosocomial Pathogens

MERV/URV Microbe	Disinfection Rate (%)					
	10/10	11/11	12/12	13/13	14/14	15/15
Penicillium spores	99.7	99.7	99.9	99.9	99.9	99.99
Fusarium spores	99.5	99.4	99.9	99.8	99.7	99.97
Rhizopus spores	99.5	99.4	99.9	99.8	99.7	99.97
Mucor spores	99.4	99.3	99.9	99.8	99.7	99.96
Aspergillus spores	99.2	99.1	99.9	99.8	99.7	99.97
Blastomyces dermatitidis spores	99.1	98.9	99.8	99.7	99.6	99.9
Cryptococcus neoformans spores	99.0	98.8	99.8	99.7	99.6	99.9
Rubella virus*	97.8	96.7	96.4	96.7	97.9	97.6
Clostridium perfringens spores	97.8	97.6	99.7	99.4	99.3	99.9
Streptococcus pneumoniae	97.3	96.2	96.2	97.3	99.1	99.6
Norwalk virus*	94.6	92.6	93.5	95.8	98.5	98.6
Rhinovirus*	93.9	92.0	93.5	96.4	98.8	99.1
*Haemophilus parainfluenzae**	92.9	92.6	96.6	99.0	99.3	99.9
Clostridium difficile spores	92.7	92.8	97.4	99.2	99.3	99.9
Reovirus	91.4	87.3	85.8	86.5	90.6	89.7
Enterobacter cloacae	89.3	88.4	93.2	98.0	99.3	99.9
Rotavirus	86.7	80.8	79.0	80.7	87.3	86.5
Mycobacterium avium	84.2	81.9	87.2	95.0	98.9	99.9
*Enterococcus**	79.2	79.4	88.9	97.0	99.0	99.9
Klebsiella pneumoniae	78.5	72.4	74.0	82.7	94.6	97.9
*Bordetella pertussis**	78.4	68.7	63.9	63.2	71.7	74.33
Norcardia asteroides	74.1	72.5	81.7	93.5	98.7	99.9
Adenovirus	72.2	62.2	60.7	65.9	78.8	79.1
Parvovirus B19	70.2	67.1	77.2	89.7	97.3	98.3
Corynebacterium diphtheriae	69.4	63.0	67.2	80.1	94.5	98.1
Staphylococcus epidermis	65.2	61.7	70.6	86.2	97.3	99.4
Coxsackievirus	63.9	60.4	71.2	85.9	96.1	97.2
Acinetobacter	63.6	66.6	81.2	94.6	98.8	99.9
*Neisseria meningitidis**	60.9	56.6	65.1	81.9	95.9	98.9
Mumps virus*	60.6	47.3	43.4	47.0	61.3	64.4
RSV*	54.9	41.4	38.2	43.4	59.7	64.3
Measles virus	50.9	37.6	35.7	42.6	59.8	63.7
Parainfluenza virus*	49.5	36.1	34.2	41.2	59.0	64.2
Influenza A virus	49.2	37.7	39.3	50.3	69.2	71.4
Haemophilus influenzae	47.0	34.8	34.5	44.0	64.8	72.7
VZV (varicella surrogate k)	43.6	31.0	30.8	39.8	58.6	63.4
Serratia marcescens	37.1	37.1	50.1	71.6	91.8	96.8

Continued

TABLE 19.4 (*Continued*)

Combined UV and Filter Removal Rates of Airborne Nosocomial Pathogens

MERV/URV Microbe	Disinfection Rate (%)					
	10/10	11/11	12/12	13/13	14/14	15/15
Legionella pneumophila	35.6	32.5	42.6	62.6	85.9	92.9
Streptococcus pyogenes	34.1	46.9	65.6	86.0	97.5	99.5
Staphylococcus aureus	33.5	45.5	64.0	84.9	97.2	99.4
Proteus mirabilis	26.0	27.3	39.3	60.0	84.1	91.6
Mycobacterium tuberculosis	24.7	33.4	49.4	71.8	92.0	96.9
Mycoplasma pneumoniae	20.7	16.8	23.5	37.3	58.1	63.3
Coronavirus (SARS)	17.8	18.6	28.5	44.6	65.6	68.5
Pseudomonas aeruginosa	17.6	25.3	39.0	60.0	84.1	91.6

* UV rate constant is a predicted value based on the pathogen genome.

References

Allegra, L., Blasi, F., Tarsia, P., Arosio, C., Fagetti, L., and Gazzano, M. (1997). A novel device for the prevention of airborne infections. *J Clinical Microb* 35(7), 1918–1919.

ASHRAE (1999). Chapter 7: Health care facilities; in *ASHRAE Handbook of Applications*, ASHRAE, ed., American Society of Heating, Refrigerating and Air Conditioning Engineers, Atlanta, GA, 7.1–7.13.

———. (2007). *Systems and Equipment: Chapter 16: Ultraviolet Lamp Systems*. American Society of Heating, Refrigerating, and Air-Conditioning Engineers, Atlanta, GA.

Beggs, C. B., and Sleigh, P. A. (2002). A quantitative method for evaluating the germicidal effects of upper room UV lights. *J Aerosol Sci* 33, 1681–1699.

Bernstein, J., Bobbitt, R., Levin, L., Floyd, R., Crandall, M., Shalwitz, R., Seth, A., and Glazman, M. (2006). Health effects of ultraviolet irradiation in asthmatic children's homes. *J Asthma* 43, 255–262.

Buttolph, L. J., and Haynes, H. (1950). Ultraviolet Air Sanitation. *LD-11*, General Electric, Cleveland, OH.

CDC (2005). *Guidelines for Preventing the Transmission of* Mycobacterium tuberculosis *in Health-Care Facilities*. Centers for Disease Control, Atlanta, GA.

CIE (2003). Ultraviolet Air Disinfection. *CIE 155:2003*, International Commission on Illumination, Vienna, Austria.

DelMundo, F., and McKhann, C. F. (1941). Effect of ultra-violet irradiation of air on incidence of infections in an infant's hospital. *Am J Dis Child* 61, 213–225.

Dionne, J.-C. (1993). Assessment of an ultraviolet air sterilizer on the incidence of childhood respiratory tract infections and day care centre indoor air quality. *Indoor Environ* 2, 307–311.

Dumyahn, T., and First, M. (1999). Characterization of ultraviolet upper room air disinfection devices. *Am Ind Hyg Assoc J* 60(2), 219–227.

EPRI (1997). UVGI for TB Infection Control in a Hospital. *TA-107885*, Electric Power Research Institute, Palo Alto, CA.

Escombe, A., Moore, D., Gilman, R., Navincopa, M., Ticona, E., Mitchell, B., Noakes, C., Martinez, C., Sheen, P., Ramirez, R., Quino, W., Gonzalez, A., Friedland, J., and Evans, C. (2009). Upper-room ultraviolet light and negative air ionization to prevent tuberculosis transmission. *PLoS Med* 6(3), 312–322.

First, M., Rudnick, S. N., Banahan, K. F., Vincent, R. L., and Brickner, P. W. (2007). Fundamental factors affecting upper-room ultraviolet germicidal irradiation— part I. Experimental. *J Occup Environ Hyg* 4(5), 321–331.

First, M. W., Nardell, E. A., Chaisson, W., and Riley, R. (1999). Guidelines for the application of upper-room ultraviolet germicidal irradiation for preventing transmission of airborne contagion—Part II: Design and operational guidance. *ASHRAE J* 105, 869–876.

Foarde, K. K., Hanley, J. T., Ensor, D. S., and Roessler, P. (1999a). Development of a method for measuring single-pass bioaerosol removal efficiencies of a room air cleaner. *Aerosol Sci & Technol* 30, 223–234.

Foarde, K. K., Myers, E. A., Hanley, J. T., Ensor, D. S., and Roessler, P. F. (1999b). Methodology to perform clean air delivery rate type determinations with microbiological aerosols. *Aerosol Sci & Technol* 30, 235–245.

Hanley, J. T., Smith, D. D., and Ensor, D. S. (1995). A fractional aerosol filtration efficiency test method for ventilation air cleaners. *ASHRAE Transactions* 101(1), 97.

Higgons, R. A., and Hyde, G. M. (1947). Effect of ultra-violet air sterilization upon incidence of respiratory infections in a children's institution. *New York State J Med* 47(7):15–27.

IUVA (2005). General Guideline for UVGI Air and Surface Disinfection Systems. *IUVA-G01A-2005*, International Ultraviolet Association Ayr, Ontario, Canada.

Janney, C., Janus, M., Saubier, L. F., and Widder, J. (2000). Test Report: System Effectiveness Test of Home/Commercial Portable Room Air Cleaners. *Contract N. SPO900-94-D-0002, Task No. 491*, US Army Soldier, Biological Chemical Command.

Koller, L. R. (1939). Bactericidal effects of ultraviolet radiation produced by low pressure mercury vapor lamps. *J Appl Phys* 10, 624.

_____. (1965). *Ultraviolet Radiation*. John Wiley & Sons, New York.

Kowalski, W. J. (2009). *Ultraviolet Germicidal Irradiation Handbook: UVGI for Air and Surface Disinfection*. Springer, New York.

_____. (2011). Ultraviolet genomic modeling: Current research and applications. *IOA/IUVA World Congress, May 25, 2011*, Paris, France.

Kowalski, W., Bahnfleth, W., and Hernandez, M. (2009). A genomic model for predicting the ultraviolet susceptibility of viruses. *IUVA News* 11(2), 15–28.

Kowalski, W. J., Bahnfleth, W. P., Witham, D. L., Severin, B. F., and Whittam, T. S. (2000). Mathematical modeling of UVGI for air disinfection. *Quantitative Microbiology* 2(3), 249–270.

Kujundzic, E., Matalkah, F., Howard, C. J., Hernandez, M., and Miller, S. L. (2006). UV air cleaners and upper-room air ultraviolet germicidal irradiation for controlling airborne bacteria and fungal spores. *J Occup Environ Hyg* 3, 536–546.

Macher, J. M., Alevantis, L. E., Chang, Y. L., and Liu, K. S. (1992). Effect of ultraviolet germicidal lamps on airborne microorganisms in an outpatient waiting room. *Appl Occup Environ Hyg* 7(8), 505–513.

McLean, R. (1961). The effect of ultraviolet radiation upon the transmission of epidemic influenza in long-term hospital patients. *Am Rev Resp Dis* 83, 36–38.

Memarzadeh, F., Olmsted, R. N., and Bartley, J. M. (2010). Applications of ultraviolet germicidal irradiation disinfection in health care facilities: Effective adjunct, but not stand-alone technology. *Am J Inf Contr* 38(5)(Suppl 1), S13–S24.

Nardell, E. A. (1988). Chapter 12: Ultraviolet air disinfection to control tuberculosis; in *Architectural Design and Indoor Microbial Pollution*, R. B. Kundsin, ed., Oxford University Press, New York, 296–308.

Peccia, J., Werth, H. M., Miller, S., and Hernandez, M. (2001). Effects of relative humidity on the ultraviolet induced inactivation of airborne bacteria. *Aerosol Sci & Technol* 35, 728–740.

Riley, R. L., Knight, M., and Middlebrook, G. (1976). Ultraviolet susceptibility of BCG and virulent tubercle bacilli. *Am Rev Resp Dis* 113, 413–418.

Riley, R. L., and O'Grady, F. (1961). *Airborne Infection*. The Macmillan Company, New York.

Robertson, E. C., Doyle, M. E., Tisdall, F. F., Koller, L. R., and Ward, F. S. (1939). Air contamination and air sterilization. *Am J Dis Child* 58, 1023–1037.

———. (1943). Use of ultra-violet radiation in reduction of respiratory cross-infections in a children's hospital. *JAMA* 121, 908–914.

Rosenstern, I. (1948). Control of air-borne infections in a nursery for young infants. *Am J Dis Child* 75, 193–202.

Rudnick, S. N., and First, M. W. (2007). Fundamental factors affecting upper-room ultraviolet germicidal irradiation—Part II. Predicting effectiveness. *J Occup Environ Hyg* 4(5), 352–362.

Sauer, L. W., Minsk, L. D., and Rosenstern, I. (1942). Control of cross infections of respiratory tract in nursery for young infants. *JAMA* 118, 1271–1274.

Schneider, M., Schwartenberg, L., Amiel, J. L., Cattan, A., Schlumberger, J. R., Hayat, M., deVassal, F., Jasmin, C. L., Rosenfeld, C. L., and Mathe, G. (1969). Pathogen-free isolation unit—Three years' experience. *Brit Med J* 29 March, 836–839.

Sommer, H. E., and Stokes, J. (1942). Studies on air-borne infection in a hospital ward. *J Pediat* 21, 569–576.

Wells, W. F. (1938). Air-borne infections. *Mod Hosp* 51, 66–69.

———. (1955). *Airborne Contagion*. Annals of the National Academy of Sciences, New York Academy of Sciences, New York.

Wheeler, S. M., Ingraham, H. S., Hollaender, A., Lill, N. D., Gershon-Cohen, J., and Brown, E. W. (1945). Ultra-violet light control of airborne infections in a naval training center. *Am J Pub Health* 35, 457–468.

WHO (1999). Guidelines for the Prevention of Tuberculosis in Health Care Facilities in Resource Limited Settings. *WHO/CDS/TB/99.269*, World Health Organization, Geneva.

Xu, P., Peccia, J., Fabian, P., Martyny, J. W., Fennelly, K. P., Hernandez, M., and Miller, S. L. (2003). Efficacy of ultraviolet germicidal irradiation of upper-room air in inactivating airborne bacterial spores and mycobacteria in full-scale studies. *Atmos Environ* 37, 405–419.

20

UVGI Surface Disinfection

Introduction

Ultraviolet germicidal irradiation (UVGI) systems have various applications in the health care industry for the disinfection of equipment and surfaces, including medical equipment disinfection, whole room decontamination, floor disinfection, cooling coil disinfection, and surgical site disinfection. The disinfection of surfaces is perhaps the simplest and most predictable application of ultraviolet irradiation technology. UV is highly effective at achieving sterilization of most types of surfaces and at controlling microbial growth. The sterilization of equipment in the medical industry was one of the first applications of UV technology. These applications are reviewed along with the fundamentals of UV surface disinfection.

UVGI Surface Disinfection Modeling

Modeling UV surface disinfection requires the computation of the surface irradiance at some distance from the UV lamps. The UV dose (or fluence) of microbial populations exposed to UV irradiation is a function of the irradiance multiplied by the exposure time, as follows:

$$D = E_t \cdot I_R \qquad (20.1)$$

where
D = UV exposure dose (fluence), J/m²
E_t = exposure time, sec
I_R = irradiance, W/m²

For surface disinfection, the parameter I_R is the fluence rate on a flat surface. When the UV dose results in a 90% disinfection rate (10% survival), it is known as a D_{90}. The D_{90} value is commonly used as a benchmark for system performance and can be used to assess the survival of individual pathogenic species. Also in common use is the D_{99}, or the dose that results in 99% inactivation. The single-stage decay equation for microbes exposed to UV irradiation is

$$S = e^{-kD}$$

(20.2)

where
  S = survival, fractional
  k = UV rate constant, m^2/J

Any microbial population exposed to UV will decay exponentially over time, but if the exposure time is extended, as is often the case in surface disinfection applications, the dose-response curve will often have two stages. The second stage is sometimes referred to as a tail. The exposed microbial population behaves like two separate populations—one that decays rapidly and another that resists exposure and decays more slowly. Defining the resistant fraction as "f" and the fast decay fraction as the complement, $(1 - f)$ separates the two populations mathematically. We define the first stage (fast decay) rate constant as k_1 and the second stage (slow decay) rate constant as k_2. Note that $k_1 > k_2$. The survival of the two populations is the sum of each decay rate as shown in Equation (20.3) and Figure 20.1 (Kowalski et al. 2000).

$$S = (1 - f)e^{-k_1 D} + fe^{-k_2 D}$$

(20.3)

where
  f = UV resistant fraction (slow decay)
  k_1 = first stage rate constant, m^2/J
  k_2 = second stage rate constant, m^2/J

Most published data on UV inactivation of microbes is limited to the first stage. Single-stage rate constants are accurate predictors of disinfection provided the dose limits of the original studies are not exceeded. Attempting to predict inactivation rates beyond experimental limits amounts to extrapolation. Such extrapolation may be invalid if the population displays a second stage after prolonged periods of exposure. Table 20.1 summarizes the first- and second-stage rate constants and resistant fractions for a number of airborne nosocomial pathogens. Media refers to the UV test media—W for water irradiation and S for surface irradiation. It can be observed that the second-stage rate constant is approximately an order of magnitude more resistant than the first stage, and this might be useful as an approximation for microbes for which no second-stage data exist. It should also be noted that the resistant population is on the order of 1% of the total population, meaning that disinfection rates higher than 99% may be subject to second-

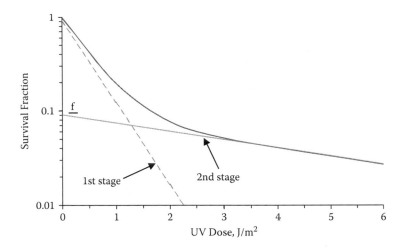

FIGURE 20.1
Graphic illustration of a two-stage decay curve as the summation of the first and second stages, proportioned by the resistant fraction f.

stage decay. This is important in surface disinfection because extended UV exposure will often result in up to six logs of inactivation, and accurate prediction of decay may depend on accounting for the second stage.

In addition, microbial decay often exhibits shoulder effects, in which the initial response to UV exposure hesitates before descending into first-stage exponential decay. In general, this effect is minor or insignificant at the high UV doses produced in most UV surface disinfection applications. In fact, at the prolonged UV exposure times typical of surface disinfection systems (i.e., 30–60 minutes minimum) the shoulder effects would not be noticeable and could be disregarded.

Table 20.1 is arranged in order of the most resistant to the least resistant microbes. It is interesting to note that when the second stage is considered, viruses seem to be the most resistant to UV, instead of fungal spores, although this is not a sufficiently large enough data set to be sure. Figure 20.2 illustrates the two-stage decay rates under continuous UV exposure for the ten microbes listed in Table 20.1, and the numbers correspond to those in Table 20.1.

Equipment Disinfection

Equipment disinfection systems employing UV are available for disinfecting surgical and medical equipment and materials. Medical equipment disinfection units are in common use in medical and dental offices, and they can

TABLE 20.1

Two-Stage Inactivation Rate Constants

No.	Microorganism	Type	Media	k original (µW/cm²)	k₁ (µW/cm²)	Two-Stage Curve				Reference
						(1 – f) Susc.	k₂ (µW/cm²)	f Resist.		
1	Adenovirus Type 2	Virus	W	0.00470	0.00778	0.99986	0.00500	0.00014		Rainbow and Mak 1973
2	Reovirus	Virus	W	0.00853	0.014	0.92	0.00430	0.08		McClain and Spendlove 1966
3	Coxsackievirus A-9	Virus	W	0.015900	0.01600	0.9807	0.01250	0.0193		Hill et al. 1970
4	Fusarium oxysporum	Fungi	W	0.0142	0.0155	0.999	0.00370	0.001		Asthana and Tuveson 1992
5	Penicillium italicum	Fungi	W	0.0114	0.017	0.996	0.00500	0.004		Asthana and Tuveson 1992
6	Rhizopus nigricans spores	Fungi	S	0.0133	0.0285	0.92	0.00203	0.08		Kowalski 2001
7	Mycobacterium tuberculosis	Bacteria	W	0.031	0.04	0.997	0.01150	0.003		Boshoff et al. 2003
8	Streptococcus pyogenes	Bacteria	S	0.061600	0.28700	0.8516	0.01670	0.1484		Lidwell and Lowbury 1950
9	Staphylococcus aureus	Bacteria	W	0.08531	0.15	0.9998	0.00700	0.0002		Chang et al. 1985
10	Legionella pneumophila	Bacteria	S	0.44613	0.45	0.999	0.14000	0.001		Knudson 1985

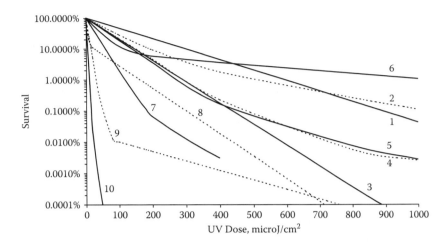

FIGURE 20.2
Two-stage decay of microbes under continuous UV exposure. Numbers correspond to microbes in Table 20.2.

sterilize every type of equipment from reusable needles to scalpels (Philips and Novak 1955). Such systems often consist of an enclosure or cabinet lined with reflective aluminum in which UV lamp fixtures are installed. Many systems mount lamp fixtures on multiple sides or rely on internal reflective surfaces to obtain complete exposure of items.

Biological safety cabinets are commonly used in hospitals and laboratories to disinfect medical instruments, surgical tools, vials, containers, and other items. Items to be disinfected are placed inside these cabinets and exposed to UV for designated periods of time, typically about 30–60 minutes. UV lamps are typically located inside the cabinets over the work surfaces. Some cabinets have interlocking switches to deactivate UV lamps when the cabinet is opened. Another safety option is the use of fluorescent labels inside the cabinet that will glow if the UV lamp is on. Levels of UV irradiance in these systems can vary widely, but a typical cabinet might produce between 0.25 and 10 W/m².

Lightweight battery-operated or rechargeable handheld UV disinfection units (hand wands or portable handheld UV units) are available for surface disinfection applications. These units emit UV light that is focused downward, usually with parabolic reflectors. Handheld units can be used in health care settings for room surface disinfection, kitchens, toilets, mattresses, clothes, and other items. Surface irradiation times are typically on the order of 10–20 seconds and such units usually include timers or other controls. These units must be used with extreme caution and users should be trained in the hazards and wear eye protection and protective clothing. Users must also ensure that for short-duration exposures the UV beam is directed away from the user and not toward any reflective UV surfaces.

Because the UV reflectivity of local room surfaces is often unknown and reflected UV rays are not visible, users must take special precautions. Users of such devices should be familiar with the ACGIH threshold limit values (TLVs) before attempting to disinfect surfaces with these devices, and should be aware that invisible reflected UV rays can cause severe eye damage and erythema (CIE 2003; ACGIH 2004; NIOSH 1972).

UV only disinfects surfaces and has no penetrating power, and its effectiveness may be limited when microbial contamination exists within dust, dirt, grease, or in shadowed crevices (Burgener 2004). As a result, UV is not recommended as the sole means of disinfection but only when used in conjunction with cleaning by chemical means (NSF 2004; NIH 1995). A study by Birch (2000) in which ultraviolet disinfection cabinets were compared against traditional means of disinfecting stored instruments found that, although microbial contamination levels were greatly reduced by UV cabinets, there was typically some residual level of contamination that resisted sterilization. UV cabinets can be used to achieve high levels of disinfection provided care is taken to clean instruments of dirt and grease prior to irradiation. Instruments can be rotated during the disinfection process to help ensure that there are no areas left unexposed. UV irradiation of polymerase chain reaction (PCR) supplies and equipment has been used to decontaminate DNA using irradiation times of up to 8 hours. Cone and Fairfax (2009) present a protocol for verifying the disinfection of DNA in PCR products. In a study by Fitzwater (1961) of a prototype tray for irradiating surgical instruments during operations, some 86% of inoculated plates placed in the UV tray grew no colonies, while 100% of the plates placed openly in the room during operations grew colonies. UV levels outside the tray were within safe limits.

Whole Room Disinfection

Entire rooms can be disinfected with continuous exposure from naked UV lamps. Permanently installed UV room disinfection systems generally consist of UV lamp fixtures mounted on ceilings or walls. These installations require periodic cleaning and personnel protection or shutoff controls during room occupation. Portable UVGI systems are also commonly available that can be moved into place temporarily to decontaminate room surfaces or equipment. Whole room or area disinfection systems are of two types, portable and permanent. Portable UV systems can be used to disinfect rooms that have become contaminated with pathogens or body fluids. Figure 20.3 shows two examples of such UV systems.

Permanently installed area disinfection systems can be used to eliminate surface contamination in open areas, either for remediation or for prevention of potential hazards, but they are generally only used in unoccupied

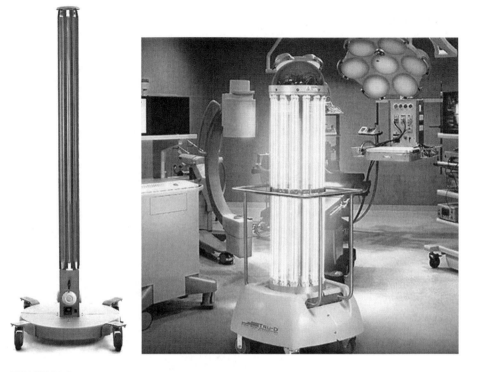

FIGURE 20.3
Portable UV area disinfection units for decontaminating entire hospital rooms. (Left photo of Surface Sanitizer provided courtesy of UVDI, Valencia, CA. Right photo of Tru-D disinfection unit provided courtesy of Lumalier, Memphis, TN.)

areas. They provide options for hospitals and laboratories where contamination potential exists or where there is a need to keep all surfaces sterilized. Permanent UV units are sometimes used in such applications to continually control mold growth, especially where water damage has occurred.

After-hours UVGI systems are permanent fixtures used in open areas to disinfect surfaces like walls and floors during unoccupied periods. They consist of UV lamp fixtures coupled with a control unit and can be engaged by timers to operate overnight, and they may include automatic controls to disengage when doors are opened or movement is detected (i.e., radar detectors). The controls can also be set to operate continuously and to turn off when anyone enters the area.

Rooms may remain contaminated after patients are removed. Patients placed in rooms previously occupied by other patients with MRSA, VRE, or *Clostridium difficile* are at significant risk for acquiring these microbes from contaminated room surfaces (Huang, Datta, and Platt 2006; Drees et al. 2008; Shaughnessey et al. 2008). Rutala, Gergen, and Weber (2010) demonstrated a room decontamination system using a model Tru-D ultraviolet disinfection unit (see Figure 20.3, right) to disinfect surfaces of MRSA, VRE, *Acinetobacter*, and *Clostridium difficile* spores. Inoculations of *Clostridium* spores on various surfaces were reduced by 99.8% within 50 minutes of UV exposure, while the three vegetative bacteria were reduced by 99.9% in approximately 15 minutes. Both directly exposed surfaces and surfaces exposed only to reflected radiation were disinfected. Another hospital study on the Tru-D device found that it reduced surface levels of MRSA and VRE by 93% and reduced *Clostridium difficile* spores by 80% (Nerandzic et al 2010).

A study by Anderson et al. (2006) compared UV area disinfection with chemical cleaning in a hospital. Four ceiling-mounted units and nine wall-mounted units were placed in three areas: a patient room, a bathroom, and an anteroom. Microbial counts on UV-exposed surfaces decreased by 93% after about 40 minutes. Cleaning and disinfection with chloramines were used in combination, and the results are shown in Figure 20.4. Significant levels of disinfection were achieved with UV alone, but the combination of cleaning, UV, and chloramines produced the best results.

Schneider et al. (1969) describes a pathogen-free isolation unit in a hospital in which the corridors surrounding the unit were irradiated for twelve hours a day by five UV lamps. This approach, in combination with rigorous disinfection protocols, succeeded in reducing infections among chemotherapy patients.

The UV irradiance produced by area disinfection systems are typically far above levels that can be safely tolerated by occupants even for short periods of time. Attention must be paid to their safe use, and procedural controls are necessary to ensure that no hazards are created for HCWs. Because levels below NIOSH/ACGIH TLVs are still capable of disinfection, it is possible for

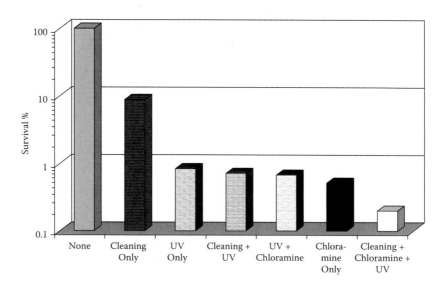

FIGURE 20.4
Comparison of UV and chemical disinfection methods. Based on data from Anderson et al. (2006).

systems that operate at low levels of UV irradiance to have room occupancy of 1–8 hours. This is true of upper-room systems that produce UV levels of up to 0.002 W/m² in the lower room as a by-product of air disinfection.

Vehicles can also be disinfected with UV area disinfection units. Ambulances are subject to various hazards that may include spills of contaminated fluids and soiling with pathogens. Ambulances are often scrubbed down with detergents and disinfectants by procedure between pickups. Interior UV disinfection systems are available that can be turned on during unoccupied periods to disinfect all exposed surfaces of the vehicle.

Floor Disinfection

The floor is a primary source of contaminants because dust, airborne fungal spores, and bacteria tend to gravitate to the floor, and are easily stirred up by activity (Lidwell and Lowbury 1950). Airborne concentrations of pathogens are typically higher near the floor. Lower-room UVGI systems create a UV field in the lower 1–2 feet of floor space in the same way that upper-room systems do for ceilings, except that the intent is to disinfect the floor surface. Fomites on floors can be resuspended in the air and irradiating the floor to destroy settled microbes should reduce airborne contamination levels, although this approach has never been studied in detail. The fixtures used for lower-room systems are virtually identical to wall-mounted upper-room systems. Exposure hazards are minimal provided personnel use appropriate legwear. Floor and lower wall surfaces must be nonreflective for UV so that stray UV rays do not reflect upward. Lower-room UV systems could be used in operating rooms, hospital hallways, and in general wards where airborne spores and bacteria may tend to settle. Lower-room UV systems can be designed to radiate directly on the floor surfaces. UV exposure above the specified height (0.5 m) is subject to the same TLVs as upper-room systems. No exposure hazards will be created for HCWs when levels of irradiance at the floor are below the ACGIH TLV, or 0.001 W/m² (for broadband UV). Lower-room UV systems may be located in doorways and entryways in hospitals where traffic brings in environmental spores and bacteria. Any such installations, however, should be coupled with motion detectors and warning signs to prevent hazards to visitors.

Willmon, Hollaender, and Langmuir (1948) installed lower-room UV systems in conjunction with upper-room systems in a naval barracks. They found that suppressing dust with oil coatings alone could reduce the infection rate, indicating a contribution to respiratory infections from floor dust, but that the highest reduction in infection rates occurred when lower-room UV was used. Both Wheeler et al. (1945) and Miller et al. (1948) used lower-room UV systems at a naval training center to reduce respiratory infections, but they employed upper-room systems simultaneously and so the effectiveness of the lower-room systems could not be isolated.

Lower-room UV systems also have potential applications in operating rooms because most SSI bacteria will settle toward the floor and may be stirred up again by activity. Such systems will keep the floor and lower air (below about 18 inches) virtually sterile and turn the most contaminated portion of the room, the floor, into the cleanest area. UV levels above 18 inches would be below ACGIH/NIOSH 8-hour limits, and upper body coverings and eyewear would not necessarily be required. The same type of system might be useful in controlling environmental spores that are subject to airborne transport in hospital environments, especially *Clostridum difficile* spores and *Aspergillus* spores. These spores will alternately settle to floors and be re-aerosolized throughout the course of a day, and spread far and wide due to the fact that spores survive well indoors. Irradiating the floor at a low level (i.e., below NIOSH TLVs) continuously for 24 hours a day and seven days a week will eradicate the hardiest spores and should reduce overall airborne transport. Such units could also be placed in patients' rooms where patients infected with *C. difficile* are disseminating spores. Figure 20.5 shows the results of analysis of a 12-foot square room with low-wattage UV

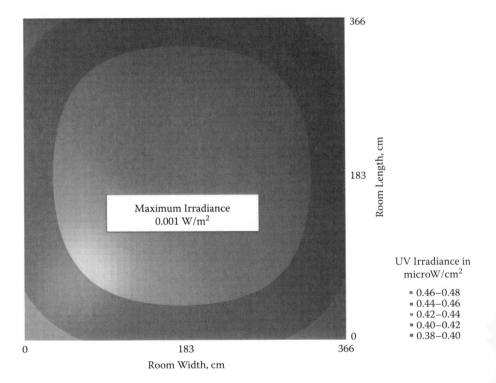

FIGURE 20.5
Irradiance contours on a 12-foot square floor with UV lamps located at floor level in the corners.

lamps located on the floor in all four corners. The contours show that the central area of the room has a maximum irradiance below the NIOSH limit of 0.001 W/m² and would present no hazards to personnel, but it would continuously disinfect the floor of spores and bacteria.

Overhead Surgical Site UV Systems

Overhead surgical site UV systems use ultraviolet lights suspended over a patient to directly disinfect the surgical site during procedures. Although they have had a proven record of success since the 1930s, they have not been widely adopted due mainly to concerns about UV hazards and proper operating parameters. Successful long-term demonstration studies with statistically significant results have generated renewed interest (Ritter, Olberding, and Malinzak 2007).

Overhead UV systems consist of UV lamp fixtures either hanging over the operating table or installed in the ceiling in recessed troffers like normal lighting fixtures, as shown in Figure 20.6, where the UV lamp fixtures are located in the ceiling directly above the operating table (and above the hanging lighting fixture). Overhead surgical site systems require operating room personnel to wear protective eyewear, clothing, and skin creams. Variable controls and ON/OFF switches are provided such that the operating room personnel may adjust the UV irradiance before, during, and after procedures.

FIGURE 20.6
Overhead surgical site UV system with UV lamps in the four recessed troffers above the normal lighting fixture in the foreground. (Image provided courtesy of Dr. Merrill Ritter of the St. Francis Hospital, Mooresville, IN.)

Overhead UVGI systems have been used to successfully reduce airborne bacterial concentrations and surgical site infections in some operating rooms since at least 1936 (Hart and Sanger 1939; Kraissl, Cimiotti, and Meleney 1940; Overholt and Betts 1940; DelMundo and McKhann 1941; Brown et al. 1996; Goldner et al. 1980). One study showed that UVGI reduced airborne microbial concentrations to below 10 cfu/m^3 in the operating room (Berg, Bergman, and Hoborn 1991; Berg-Perier, Cederblad, and Persson 1992). Moggio et al. (1979) demonstrated a 49% decrease in airborne bacteria with an overhead UV system. Lowell and Kundsin (1980) report on an overhead UV system that produced a 99–100% decrease in aerosolized *E. coli* and resulted in a 54% decrease in airborne bacteria during procedures. Duke University has successfully used overhead UVGI systems since 1940 to keep levels of orthopedic infections low (Goldner and Allen 1973). Wright and Burke (1969) reported a reduction from 5.3% to 0.7% in postoperative sepsis following craniotomy, and a reduction from 4.1% to 0.3% for laminectomy after a 36-month field trial at the Massachusetts General Hospital. Table 20.2 summarizes the various field trials that have been performed in operating rooms, including all

TABLE 20.2

Results of Overhead UV Field Trials in Operating Rooms

Location	Infection/ Operation	Infection Cases (%)		Decrease (%)		Reference
		Before	After	Net	%	
Duke University Hospital	SSI	5	1	4	80	Kraissl, Cimiotti, and Meleney 1940
NE Deaconess Hospital	SSI	13.8	2.7	11.1	80	Overholt and Betts 1940
Infant and Children's Hospital, Boston	SSI	12.5	2.7	9.8	78	Del Mundo and McKhann 1941
Montreal Neurological Institute	SSI	1.1	0.36	0.7	67	Woodhall, Neill, and Dratz 1949
MA General Hospital	Craniotomies	5.3	0.70	4.6	87	Wright and Burke 1969
MA General Hospital	Laminectomies	4.1	0.30	3.8	93	Wright and Burke 1969
Duke University Hospital	Hip arthroplasty infection	5	0.5	5	90	Lowell and Kundsin 1980
Brigham Hospitals	Hip and knee	3.5	0.89	3	75	Young 1991
Watson Clinic, FL	Mediastinitis	1.4	0.23	1.2	84	Brown et al. 1996
St. Francis Hospital	SSI	1.77	0.57	1.2	68	Ritter, Olberding, and Malinzak 2007
Average Reduction	**80%**					

those equipped with overhead UV systems, and these show a net average reduction of approximately 80%.

Overhead UV systems typically produce a UV irradiance of about 0.25–0.3 W/m^2 (25–30 $\mu W/cm^2$) at operating table height. They require UV-proof clothing, gloves, goggles or visors, and protective skin creams (Young 1991). The overhead UV system implemented by Ritter, Olberding, and Malinzak (2007) achieved significant reductions in surgical site infections over a ten-year period as compared with a similar period before. The overhead UV system consisted of eight UV lamp fixtures in the ceiling that produced an irradiance of 25 $\mu W/cm^2$ at the operating table. The Ritter system reduced the surgical site infection rate from 1.77% to 0.5%.

The level of irradiance produced by overhead UV systems is capable of inactivating all major surgical site infection (SSI) bacteria. Table 20.3 summarizes the inactivation rates predicted for this system at table height for all major SSI bacteria and fungi (see Chapter 12) for which UV surface rate constants are known (see Kowalski 2009 for UV rate constant source data). Disinfection levels above 99% (survival below 1%) are shown in bold because results beyond these points are dubious without considering the second stage of decay. The duration of exposure, being the duration of the operation, may typically be 1–2 hours or less depending on the procedure. The UV rate constant for *Clostridium difficile* spores is based on the surrogate *Clostridium perfringens*, which will have a similar decay rate.

Figure 20.7 illustrates the inactivation rates of SSI microbes produced by 25 $\mu W/cm^2$ based on Table 20.3. It is clear that most bacteria, and particularly the most hazardous bacteria (*Streptococcus* and *Staphylococcus*) are reduced to below 1% within the first few minutes. Most surgical procedures take about 1–2 hours, whereas a six log reduction is achieved for all but spores within about 30 minutes. In an actual OR, because levels would rarely exceed a few thousand cfu/m^3, this level of inactivation should provide operating conditions that approach sterility. If the airstream passing over the surgical site is sterile and not carrying bacteria from the surgeon, then sterility might be maintained indefinitely. Figure 20.8 illustrates the irradiance contour produced on an operating table by an overhead UV lamp. Although the peak irradiance reached 300 $\mu W/cm^2$, the average irradiance is approximately 0.25 W/m^2 across the table.

The Ritter, Olberding, Malinzak (2007) study did not report airborne concentrations of bacteria in the ORs, but it is likely that this system significantly reduced airborne levels of bacteria even with personnel present. It is probable that this UV system could inhibit both airborne transport and survival of bacteria on other OR surfaces, including on personnel and on floors.

UV irradiance levels need not be as high as 0.25 W/m^2 to adequately disinfect the bacteria in Table 20.3 (excepting *Aspergillus* spores, which should not appear in the operating room if the ventilation system and filters are working properly). In fact, levels below NIOSH TLVs can have considerable impact on the SSI bacteria. Table 20.4 shows the survival rates of SSI microbes at

TABLE 20.3

Surface Disinfection of SSI Pathogens at 0.25 W/m² Constant Exposure

SSI Pathogen	AS	SE	Eb	CD	Klebsiella	GAS	MRSA	VRE	PA	Serratia	Proteus
Airborne Class	1	2	2	1	2	1	1	2	1	2	2
UV k (m²/J)	0.00401	0.01433	0.03598	0.0385	0.04435	0.06161	0.07132	0.0822	0.12802	0.1769	0.18288
Time (min)											
0	100	100	100	100	100	100	100	100	100	100	100
0.5	97	90	76	75	72	63	59	54	38	27	25
1	94	81	58	56	51	40	34	29	15	7	6
2	89	65	34	32	26	16	12	8	2	5.E-01	4.E-01
3	83	52	20	18	14	6	4	2	3.E-01	3.E-02	3.E-02
4	79	42	12	10	7	2	1	1	5.E-02	2.E-03	2.E-03
5	74	34	7	6	4	1	5.E-01	2.E-01	7.E-03	2.E-04	1.E-04
6	70	28	4	3	2	4.E-01	2.E-01	6.E-02	1.E-03	1.E-05	7.E-06
7	66	22	2	2	1	2.E-01	6.E-02	2.E-02	1.E-04	9.E-07	5.E-07
8	62	18	1	1	5.E-01	6.E-02	2.E-02	5.E-03	2.E-05	6.E-08	3.E-08
10	55	12	5.E-01	3.E-01	1.E-01	1.E-02	2.E-03	4.E-04	5.E-07	3.E-10	1.E-10
30	16	2.E-01	9.E-06	3.E-06	2.E-07	9.E-11	1.E-12	9.E-15	1.E-23	3.E-33	2.E-34
60	3	3.E-04	9.E-13	9.E-14	5.E-16	8.E-23	1.E-26	7.E-31	9.E-49	7.E-68	3.E-70
120.0	7.E-02	6.E-10	7.E-27	8.E-29	2.E-33	7.E-47	2.E-54	6.E-63	8.E-99	5.E-137	1.E-141

Note: As = *Aspergillus* spores; SE = *S. epidermis*; Eb = *Enterobacter*; PA = *P. aeruginosa*; CD = *C. difficile*.

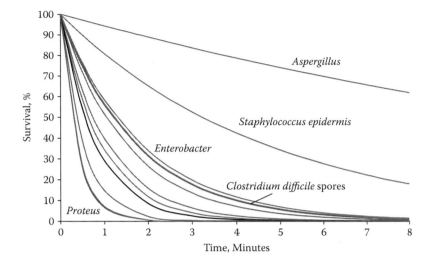

FIGURE 20.7
Survival of airborne SSI microbes on surfaces (per Table 20.3). Dark line represents *C. difficile*.

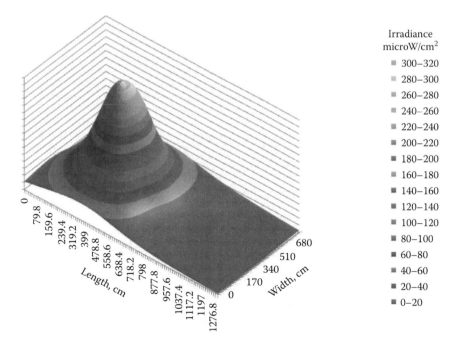

FIGURE 20.8
Irradiance at operating table height from an overhead UV system producing average irradiance of 0.25 μW/cm².

TABLE 20.4

Surface Disinfection Rates of SSI Bacteria at NIOSH 2-Hour Exposure Limit

SSI Bacteria	S. epidermis	Enterobacter	Klebsiella	GAS	MRSA	VRE	Pseudomonas	Serratia	Proteus
Airborne Class	2	2	2	1	1	2	1	2	2
UV k (m²/J)	0.01433	0.03598	0.04435	0.06161	0.07132	0.0822	0.12802	0.1769	0.18288
Time (min)									
0	100	100	100	100	100	100	100	100	100
1	99	98	98	97	97	96	94	92	91
2	99	96	96	94	93	92	88	84	83
4	97	93	92	88	87	85	77	70	69
8	94	87	84	78	75	72	60	49	48
10	93	84	80	74	70	66	53	41	40
15	90	76	72	63	59	54	38	27	26
30	81	58	52	40	34	29	15	7	7
45	73	45	37	25	20	16	6	2	2
60	65	34	27	16	12	9	2	1	**4.E-01**
120	42	12	7	3	1	1	**5.E-02**	**3.E-03**	**2.E-03**
240	18	1	**5.E-01**	**6.E-02**	**2.E-02**	**5.E-03**	**2.E-05**	**7.E-08**	**3.E-08**
360	8	**2.E-01**	**4.E-02**	**2.E-03**	**3.E-04**	**4.E-05**	**1.E-08**	**2.E-12**	**6.E-13**
480	3	**2.E-02**	**2.E-03**	**4.E-05**	**4.E-06**	**3.E-07**	**5.E-12**	**4.E-17**	**1.E-17**

0.0083 W/m² (the NIOSH two-hour TLV for UVC). Disinfection levels above 99% (survival below 1%) are shown in gray because results beyond these points are dubious, as in Table 20.3. Note that even at levels of UV irradiance safe for human occupation significant reductions are possible for most SSI bacteria. If a particular microbe like MRSA were being targeted in the operating room, it seems entirely possible that all exposed surfaces could be kept highly disinfected during any procedure, and that the air might simultaneously be kept disinfected.

Overhead UV systems can have two operating modes—Constant UV Mode and Decontamination Mode. The constant mode operates continuously during surgery but can be under internal control by the surgeon or other personnel. Decontamination Mode would serve as an area disinfection system and would operate when the OR is unoccupied to disinfect room surfaces. It would be engaged after the OR was cleaned or scrubbed with disinfectants to achieve the highest possible levels of disinfection. Decontamination Mode can be controlled by a timer such that it operates for about 30–60 minutes but would disengage automatically whenever anyone entered the room. Control can be linked to a door switch or a motion detector to shut off automatically when any HCWs enter.

Cooling Coil Disinfection

The disinfection of cooling coils with UV is an approach that provides energy savings as well as improving air quality in hospitals by disinfecting the coils of microbial contamination (Keikavousi 2004). Cooling coil disinfection systems perform surface disinfection, but because they are installed in air-handling units, they simultaneously perform some air disinfection. UV cooling coil disinfection systems have enjoyed widespread success because of favorable economics and energy savings (Scheir and Fencl 1996). Cooling coil disinfection systems have been demonstrated to save energy with short payback periods, typically less than two years (Shaughnessy, Levetin, and Rogers 1999; Levetin et al. 2001). Because biofilms on cooling coils reduce heat transfer coefficient and increase the pressure loss on the air side, keeping these surfaces clean maintains peak design performance.

Cooling coil disinfection systems consist of UV lamp fixtures located around cooling coils. They operate continuously and disinfect the surface of the cooling coil fins, which are prone to accumulate dust and environmental microbes, including fungal spores, from the outdoor air or the return air. The continuous UV exposure is sufficient to sterilize the cooling coil surfaces of even the hardiest fungal and bacterial spores. These systems also disinfect any exposed areas inside the air-handling units like drain pans and air filters. The energy savings alone can justify installation of such systems but the removal of biocontamination from the coils, and preventing them from entering the hospital environment, provides an added benefit to health care facilities. Because air handling units with cooling coils usually have filters installed upstream of the coils, they may not require the addition of filters to protect the lamps. The existing filters should be rated at least MERV 6–8.

The microorganisms that accumulate on cooling coils are primarily spores, including fungal spores and bacterial spores. Common environmental fungal spores like *Aspergillus*, *Penicillium*, and *Cladosporium*, and bacterial spores like *Clostridium*, may be found on contaminated coils. For example, *Cladosporium* spores have a UV rate constant of 0.00384 m^2/J (Luckiesh 1946). Under constant exposure at 1 W/m^2 (100 $\mu W/cm^2$), the survival rate will be about one in a million (0.0001%) after one hour, per the classic single-stage exponential decay equation. If mathematical sterilization is assumed to be six logs of reduction it can be seen that even at 0.10 W/m^2 (10 $\mu W/cm^2$) the surface will be sterilized within about 10 hours. Even if the population decay involves two stages, sterilization will eventually be achieved. Coils may be irradiated on one or both sides. If only one side of the coil is exposed, the biofilm on the coil surface is destroyed by UV exposure over time and the UV light is able to reflect its way through the coil more efficiently, increasing irradiance levels on the opposite side. The higher the irradiance on the upstream side, the more UV will penetrate to the opposite face. In general, a mean irradiance of approximately 1–10 W/m^2 (100–1000 $\mu W/cm^2$) on the upstream side should provide eventual sterilization on both sides of the cooling coil. The minimum acceptable irradiance can be on the order of 0.5–1 W/m^2 (50–100 $\mu W/cm^2$).

Radiometer measurements can be used to confirm irradiance levels. A better indicator of the effectiveness of UV is cooling system performance, because the elimination of surface contamination should restore heat exchange efficiency and airflow to original design operating values. Parameters that could be measured before and after a UV installation to verify system performance include pressure drop across the coil, coil total airflow, coil entering wet bulb temperature, coil leaving dry bulb temperature, entering and exiting chilled water temperature, chilled water flow rate, and fan motor amperage.

References

ACGIH (2004). Threshold Limit Values and Biological Exposure Indices. American Conference of Governmental Industrial Hygienists, Cincinnati, OH.

Anderson, B. M., Banrud, H., Boe, E., Bjordal, O., and Drangsholt, F. (2006). Comparison of UV C light and chemicals for disinfection of surfaces in hospital isolation units. *Infect Contr Hosp Epidemiol* 27(7), 729–734.

Asthana, A., and Tuveson, R. W. (1992). Effects of UV and phototoxins on selected fungal pathogens of citrus. *Int J Plant Sci* 153(3), 442–452.

Berg, M., Bergman, B. R., and Hoborn, J. (1991). Ultraviolet radiation compared to an ultra-clean air enclosure. Comparison of air bacteria counts in operating rooms. *JBJS* 73(5), 811–815.

Berg-Perier, M., Cederblad, A., and Persson, U. (1992). Ultraviolet radiation and ultra-clean air enclosures in operating rooms. *J Arthroplasty* 7(4), 457–463.

Birch, R. (2000). A study into the efficacy of ultraviolet disinfection cabinets for storage of autoclaved podiatric instruments prior to use, in comparison with current practices. University College, Northampton, England.

Boshoff, H. I. M., Reed, M. B., Barry, C. E., and Mizrahi, V. (2003). DnaE2 polymerase contributes to in vivo survival and the emergence of drug resistance in *Mycobacterium tuberculosis*. *Cell* 113, 183–193.

Brown, I. W., Moor, G. F., Hummel, B. W., Marshall, W. G., and Collins, J. P. (1996). Toward further reducing wound infections in cardiac operations. *Ann Thorac Surg* 62(6), 1783–1789.

Burgener, J. (2004). Position paper on the use of ultraviolet lights in biological safety cabinets. *Applied Biosafety* 11(4), 228–230.

Chang, J. C. H., Ossoff, S. F., Lobe, D. C., Dorfman, M. H., Dumais, C. M., Qualls, R. G., and Johnson, J. D. (1985). UV inactivation of pathogenic and indicator microorganisms. *Appl & Environ Microbiol* 49(6), 1361–1365.

CIE (2003). Ultraviolet Air Disinfection. *CIE 155:2003*, International Commission on Illumination, Vienna, Austria.

Cone, R., and Fairfax, M. (2003). Protocol for ultraviolet irradiation of surfaces to reduce PCR contamination. *PCR Methods and Appl* 3, S15–S17.

DelMundo, F., and McKhann, C. F. (1941). Effect of ultra-violet irradiation of air on incidence of infections in an infant's hospital. *Am J Dis Child* 61, 213–225.

Drees, M., Snydman, D. R., Schmid, C. H., Barefoot, L., Hansjosten, K., Vue, P. M., Cronin, M., Nasraway, S. A., and Golan, Y. (2008). Prior environmental contamination increases the risk of acquisition of vancomycin-resistant enterococci. *Clin Infect Dis* 46, 678–685.

Fitzwater, J. (1961). Bacteriological effect of ultraviolet light on a surgical instrument table. *Pub Health Rep* 76(2), 97 103.

Goldner, J. L., and Allen, B. L. (1973). Ultraviolet light in orthopedic operating rooms at Duke University. *Clin Ortho* 96, 195–205.

Goldner, J. L., Moggio, M., Beissinger, S. F., and McCollum, D. E. (1980). Ultraviolet light for the control of airborne bacteria in the operating room; in *Airborne Contagion*, Annals of the New York Academy of Sciences, R. B. Kundsin, ed., NYAS, New York, 271–284.

Hart, D., and Sanger, P. W. (1939). Effect on wound healing of bactericidal ultraviolet radiation from a special unit: Experimental study. *Arch Surg* 38(5), 797–815.

Hill, W. F., Hamblet, F. E., Benton, W. H., and Akin, E. W. (1970). Ultraviolet devitalization of eight selected enteric viruses in estuarine water. *Appl Microb* 19(5), 805–812.

Huang, S. S., Datta, R., and Platt, R. (2006). Risk of acquiring antibiotic-resistant bacteria from prior room occupants. *Arch Int Med* 166, 1945–1951.

Keikavousi, F. (2004). UVC: Florida hospital puts HVAC maintenance under a new light. *Engin Sys* March, 60–66.

Knudson, G. B. (1985). Photoreactivation of UV-irradiated *Legionella pneumophila* and other *Legionella* species. *Appl & Environ Microbiol* 49(4), 975–980.

Kowalski, W. J. (2001). Design and Optimization of UVGI Air Disinfection Systems (PhD thesis). The Pennsylvania State University, State College.

_____. (2009). *Ultraviolet Germicidal Irradiation Handbook: UVGI for Air and Surface Disinfection.* Springer, New York.

Kowalski, W. J., Bahnfleth, W. P., Witham, D. L., Severin, B. F., and Whittam, T. S. (2000). Mathematical modeling of UVGI for air disinfection. *Quantitative Microbiology* 2(3), 249–270.

Kraissl, C. J., Cimiotti, J. G., and Meleney, F. L. (1940). Considerations in the use of ultra-violet radiation in operating rooms. *Ann Surg* 111, 161–185.

Levetin, E., Shaughnessy, R., Rogers, C. A., and Scheir, R. (2001). Effectiveness of germicidal UV radiation for reducing fungal contamination within air-handling units. *Applied & Environ Microbiol* 67(8), 3712–3715.

Lidwell, O. M., and Lowbury, E. J. (1950). The survival of bacteria in dust. *Annual Review of Microbiology* 14, 38–43.

Lowell, J., and Kundsin, R. (1980). Ultraviolet Radiation: Its Beneficial Effect on the Operating Room Environment and the Incidence of Deep Wound Infection Following Total Hip and Total Knee Arthroplasty. *A-810*, American Ultraviolet Company, Murray Hill, NJ.

Luckiesh, M. (1946). *Applications of Germicidal, Erythemal and Infrared Energy.* Van Nostrand, New York.

McClain, M. E., and Spendlove, R. S. (1966). Multiplicity reactivation of reovirus particles after exposure to ultraviolet light. *J Bact* 92(5), 1422–1429.

Miller, W. R., Jarrett, E. T., Willmon, T. L., Hollaender, A., Brown, E. W., Lewandowski, T., and Stone, R. S. (1948). Evaluation of ultra-violet radiation and dust control measures in control of respiratory disease at a naval training center. *J Infect Dis* 82, 86–100.

Moggio, M., Goldner, J. L., McCollum, D. E., and Beissinger, S. F. (1979). Wound infections in patients undergoing total hip arthroplasty. Ultraviolet light for the control of airborne bacteria. *Arch Surg* 114(7), 815–823.

Nerandzic, M., Cadnum, J., Pulz, M., and Donskey, C. (2010). Evaluation of an automated ultraviolet radiation device for decontamination of *Clostridium difficile* and other healthcare-associated pathogens in hospital rooms. *BMC Infect Dis* 10, 197.

NIH (1995). *Primary Containment for Biohazards: Selection, Installation and Use of Biological Safety Cabinets.* National Institute of Health, US Department of Health and Human Services, US Government Printing Office, Washington, DC.

NIOSH (1972). Occupational Exposure to Ultraviolet Radiation. *HSM 73-110009*, National Institute for Occupational Safety and Health, Cincinnati, OH.

NSF (2004). International Standard, American National Standard: Class II (Laminar Flow) Biohazard Cabinetry. *NSF/ANSI 49-2004A*, National Sanitation Foundation, Ann Arbor, MI.

Overholt, R. H., and Betts, R. H. (1940). A comparative report on infection of thoracoplasty wounds. *J Thoracic Surg* 9, 520–529.

Phillips, G. B., and Novak, F. E. (1955). Applications of germicidal ultraviolet in infectious disease laboratories. *Appl Microb* 4, 95–96.

Rainbow, A. J., and Mak, S. (1973). DNA damage and biological function of human adenovirus after UV irradiation. *Int J Radiat Biol* 24(1), 59–72.

Ritter, M., Olberding, E., and Malinzak, R. (2007). Ultraviolet lighting during orthopaedic surgery and the rate of infection. *J Bone Joint Surg* 89, 1935–1940.

Rutala, W. A., Gergen, M. F., and Weber, D. J. (2010). Room decontamination with UV radiation. *Inf Contr Hosp Epidemiol* 31(10), 1025–1029.

Scheir, R., and Fencl, F. B. (1996). Using UVC technology to enhance IAQ. *HPAC* February.

Schneider, M., Schwartenberg, L., Amiel, J. L., Cattan, A., Schlumberger, J. R., Hayat, M., deVassal, F., Jasmin, C. L., Rosenfeld, C. L., and Mathe, G. (1969). Pathogen-free isolation unit—Three years' experience. *Brit Med J* 29 March, 836–839.

Shaughnessey, M., Micielli, R., Depestel, D., Daryl, D., Arndt, J., Strachan, C. L., Welch, K. B., and Chenoweth, C. E. (2008). Evaluation of hospital room assignment and acquisition of *Clostridium difficile* associated diarrhea (CDAD). *Annual Interscience Conference on Antimicrobial Agents and Chemotherapy/Infections Disease Society of America 46th Annual Meeting*, Washington, DC.

Shaughnessy, R., Levetin, E., and Rogers, C. (1999). The effects of UV-C on biological contamination of AHUs in a commercial office building: Preliminary results. *Indoor Environment '99*, 195–202.

Wheeler, S. M., Ingraham, H. S., Hollaender, A., Lill, N. D., Gershon-Cohen, J., and Brown, E. W. (1945). Ultra-violet light control of airborne infections in a naval training center. *Am J Pub Health* 35, 457–468.

Willmon, T. L., Hollaender, A., and Langmuir, A. D. (1948). Studies of the control of acute respiratory diseases among naval recruits. *Am J Hyg* 48, 227–232.

Woodhall, B., Neill, R., and Dratz, H. (1949). Ultraviolet radiation as an adjunct in the control of post-operative neurosurgical infection. Clinical experience 1938–1948. *Ann Surg* 129, 820–825.

Wright, R., and Burke, J. (1969). Effect of ultraviolet radiation on post-operative neurosurgical sepsis. *J Neurosurg* 31, 533–537.

Young, D. P. (1991). Ultraviolet Lights for Surgery Suites. St. Francis Hospital, Mooresville, IN.

21

Alternative Air Cleaning Technologies

Introduction

In addition to ventilation dilution, filtration, and UVGI, there are several alternative technologies that can reduce the levels of airborne pathogens in hospital environments, or that can be used to disinfect hospital surfaces. The most promising of these technologies are addressed here, including ozonation, photocatalytic oxidation, plasma or corona technologies, pulsed light, passive solar exposure, ionization, vegetation air cleaning, and antimicrobial coatings. Microbiological disinfection data or epidemiological data demonstrating infection reduction in hospitals is quite limited for these developmental technologies. Some of these technologies may be considered Green Building Technologies, in which active, passive, or natural technologies are applied, or sustainable materials and renewable energy resources are incorporated into hospital design. These include passive solar exposure, vegetation air cleaning, and material selectivity. Of course, any technology that reduces energy consumption while promoting improved air quality may be considered green. In most cases it is easy to demonstrate via engineering that energy consumption is reduced, but the proof that airborne biocontamination or infection rates are reduced is often more difficult. Gas phase filtration technologies such as carbon absorption are not addressed here because gases and odors are not infectious hazards, even if they come from microbiological sources. See Kowalski (2006) for specific details on gas phase filtration technologies.

PCO Systems

Photocatalytic oxidation (PCO) systems employ UV light to activate surfaces coated with titanium dioxide. Titanium dioxide (TiO_2), under exposure to light, reacts with various organic compounds, destroying them and inactivating microorganisms. The light source is typically a UV lamp, and the

FIGURE 21.1
Recirculating unit employing PCO technology. (Image courtesy of Zander Scientific, Inc., Vero Beach, FL.)

disinfection is produced by a combination of the PCO and the UV. The substrate on which the TiO_2 is located is often a screen or other material that acts like a filter to bring the contaminants in close contact with the activated TiO_2. In addition to destroying microbes, PCO systems are often used to remove volatile organic compounds and odors from the air. PCO systems create reactive radicals, such as hydroxyl radicals, and decompose organic molecules to form carbon dioxide, water, and mineral acids as final products (Goswami 2003; Zhao and Yang 2003). There are many good references in the literature that readers may consult for additional and more detailed technical information (Goswami, Trivedi, and Block 1997; Jacoby, Maness, and Wolfrum 1998; Hodgson et al. 2007).

PCO technology can also be applied to surfaces exposed to visible light. Although the photocatalytic effect is not as pronounced under visible light as it is under UV exposure, extended exposure to visible light could still provide high levels of disinfection. Any surface subject to biocontamination or to which fomites may adhere, such as doorknobs, handrails, light switches, etc., could be coated with TiO_2 and the surface would continuously be disinfected under light from the sun or from light fixtures. Figure 21.1 shows an example of a unitary PCO system.

Ozone

Ozone has been used for disinfecting water for decades, but the use of ozone for air disinfection has not yet been fully developed due to safety

TABLE 21.1

Ozone Disinfection of Airborne Nosocomial Pathogens

Test Microbe	Ozone (ppm)	RH (%)	Time (min)	Survival (%)	Reference
Fusarium oxysporum	0.1	35–75	240	2	Hibben and Stotzky (1969)
Rhizopus stolonifer	0.1	35–77	240	43	Hibben and Stotzky (1969)
Aspergillus niger	0.1	35–76	240	84	Hibben and Stotzky (1969)
Staphylococcus aureus	0.3–0.9	—	240	0.5	Dyas, Boughton, and Das (1983)
Proteus	0.3–0.9	—	240	0.9	Dyas, Boughton, and Das (1983)
Serratia	0.3–0.9	—	240	3.2	Dyas, Boughton, and Das (1983)
Aspergillus fumigatus	0.3–0.9	—	240	8	Dyas Boughton, and Das (1983)
Pseudomonas aeruginosa	0.3–0.9	—	240	31	Dyas Boughton, and Das (1983)
Staphylococcus epidermis	0.47	60–75	60	1	Heindel, Streib, and Botzenhart (1993)
Streptococcus salivarius	0.6	60–75	10	2	Elford and van den Eude (1942)
Staphylococcus aureus	300	18–21	1.5	0.001	Kowalski, Bahnfleth, and Whittam (1998)
Penicillium chrysogenum	3–9	90	1380	0.1	Foarde, Van Osdell, and Steiber (1997)

issues. Ozone systems that produce low levels of ozone in indoor air for air quality control are currently being marketed, but the effectiveness of such approaches have not been proven to outweigh the potential health hazards, and the use of ozone generators in indoor environments is not recommended (Steiber 1995). Ozone has promise as a means of disinfecting surfaces and is useful for decontamination of unoccupied rooms and buildings (Kowalski 2003; Masaoka et al. 1982). Ozone is very reactive and can produce a variety of by-products in humid air, including various hydroxyl radicals. These products are unstable and ultimately break down into water and oxygen. Bacterial cell constituents are typically high-molecular-weight compounds with a variety of low-energy bonds that are dissociated under reaction with ozone (Langlais, Reckhow, and Brink 1991).

Various researchers have studied the biocidal properties of airborne ozone for the disinfection of air or surfaces and a summary of these results is shown in Table 21.1 for airborne nosocomial pathogens.

The ozone dose is measured as the concentration (ppm) multiplied by the time of exposure. High RH, about 90–95%, tends to maximize the disinfectant rate of ozone. Other studies report that RH below about 50% has negligible or limited effects (Elford and van den Eude 1942). A high-speed ozone sterilizing device has been developed that produces a high concentration of ozone of 20,000–30,000 ppm (Masuda et al. 1990). For reference bacteria, this ozonizer produces the standard sterilizing effect of killing one million cells in 3–5 minutes. A single run, including processing, lasts 12–14 minutes as compared with approximately 30 minutes for a conventional autoclave.

Ozone has been evaluated as a means of disinfecting hatcheries, and levels as high as 9091 ppm produced bacterial reductions of 4–7 logs (Whistler and Sheldon 1989). It was found that ozone was almost as effective as formaldehyde in reducing levels of *Pseudomonas, Proteus, Aspergillus, Staphylococcus,* and *Streptococcus.*

Khurana (2003) investigated ozone as a method of controlling microbial growth in air conditioning systems. The findings indicated that frequent treatment with ozone at 9 ppm was sufficient to prevent microbial growth. Levels as high as 45 ppm used for 15 minutes a day were able to inhibit microbial growth, and ozone doses of 600–2400 ppm-min, when level was 11 ppm or higher, were effective at preventing microbial growth.

Ionization

Ionization occurs when electrons are stripped from or added to atoms, leaving a temporary charge imbalance. Charged atoms tend to agglomerate or gather together in clumps. Clumping of dust particles and airborne microbes can cause them to precipitate out of the air and settle on interior surfaces including floors, walls, and air filters. By removing airborne particles, dust, and bacteria, ionization of indoor air can reduce the airborne concentration of bioaerosols as well as dust particles (Makela et al. 1979; Phillips, Harris, and Jones 1964). Some reports indicate that ions can also kill bacteria directly or inhibit microbial growth (Krueger 1985; Krueger and Reed 1976; Krueger, Smith, and Go 1957; Phillips, Harris, and Jones 1963). Negative air ionization has been shown to reduce the incidence of respiratory infection transmission, but the effect is somewhat species dependent and can be impacted by relative humidity (Estola, Makela, and Hovi 1979; Happ, Harstad, and Buchanan 1966). Ion generators are commercially available and energy-efficient units can produce controlled outputs of specific ions on demand, while minimizing the formation of undesirable by-products such as ozone (Daniels 2000).

Some success in reducing infections has been reported in burn wards and dental offices (Gabbay 1990). Figure 21.2 summarizes results from Makela et al. (1979), who found that bacterial aerosols in patient rooms of a burn and plastic surgery unit could be reduced with negative air ionization. Variations in the bacterial levels were associated with activities in the room and the relative humidity in the rooms was low, which may have enhanced the effect. The average for two days of monitoring indicated a significant reduction in airborne levels of *Staphylococcus.*

Ozone concentrations of 0.03 ppm combined with negative ionization was effective in killing 98% of *S. aureus* on plates after 72 hours of exposure (Li et al. 1989). Arnold and Mitchell (2002) report on a laboratory test in which negative air ionization effectively decreased the survival levels of bacteria

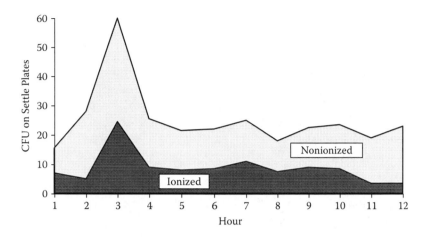

FIGURE 21.2
Reduction of airborne *Staphylococcus aureus* in a patient waiting room with negative ionization.
Based on data from Makela et al. (1979).

on stainless steel, with a reduction efficiency of 99.8%. An epidemiological study is currently being conducted at the University of Leeds on the use of ionization in hospital wards (Chard 2005; McDowell 2003). Initial reports indicated that after the first year the nosocomial agent *Acinetobacter* was completely eliminated from the ward air and new infections were reduced to zero.

Plasma and Corona Systems

Plasma is an intermediate state of matter between gas and solid, consisting of ions, electrons, and neutral particles. Plasma systems can be used to disinfect or sterilize air, water, and surfaces. Plasma generators create reactive oxygen species including atomic oxygen, oxygen free radicals, and hydroxyl radicals in humid airstreams. Plasma has been used for the disinfection of water (Manolache et al. 2001). Experiments have demonstrated the ability of plasma systems to destroy microorganisms (Kelley-Wintenberg et al. 1999). Montie, Kelly-Wintenberg, and Roth (2000) report test results using a plasma source called the One Atmosphere Uniform Glow Discharge Plasma (OAUGDP), which operates at atmospheric pressure in air and produces antimicrobial active species. OAUGDP exposures have been effective at reducing microbial populations of Gram-negative bacteria, Gram-positive bacteria, bacterial endospores, yeast, and bacteriophages on a variety of surfaces. The nature of the surface impacts the inactivation rate, with microorganisms on poly propylene being most vulnerable, followed by glass, and cells embedded in agar. Results showed at least a 5 $\log_{10}$ cfu reduction in bacteria after exposure for 50–90 seconds. Oxygen radicals are believed to be responsible for some of the lethal cell damage. In a test on medical equipment disinfection, plasma was used for the inactivation of microbiological cultures (Karelin et al. 2001). Park et al. (2003) used microwave-induced argon plasma for the sterilization of microorganisms. Experimental results for six species of bacteria and fungi indicated that all six species were fully sterilized within 20 seconds. Bergeron et al. (2007) report on the application of a nonthermal plasma system in a pediatric unit in which levels of opportunistic fungi were reduced by 75%.

Corona discharge systems are similar to plasma systems, and they involve the generation of a localized corona in or around electrical conductors (Vincent 1995). In corona discharge systems a high-voltage alternating current is maintained across wires or other media that generate electrons and sparking does not occur if the potential is not too high (Nunez et al. 1993). These free electrons induce chemical reactions and ionization in the passing airstream, which may ionize and dissociate molecules, and generate free radicals. These products, in turn, induce the breaking of chemical bonds in organic molecules.

Pulsed Light

Pulsed light, also called pulsed white light (PWL) involves the pulsing of a high-power xenon lamp for about 0.1–3 milliseconds for some types of systems (Johnson 1982), or about 100 microseconds to 10 milliseconds for other types (Wekhof 2000). The spectrum of light produced resembles the spectrum of sunlight but is briefly about 20,000 times as intense (Bushnell et al. 1997). The spectrum of PWL includes a large component of ultraviolet light, which is responsible for most of the biocidal effects. The ability of pulsed light to penetrate materials more deeply than UVGI allows its use in various applications where material translucence would limit the effectiveness of UVGI, such as sterilizing medicine bottles. PWL is currently being used for medical equipment sterilization and in the pharmaceutical packaging industry where translucent aseptically manufactured bottles and containers are sterilized in a once-through light treatment chamber (Wallen et al. 2001; Bushnell et al. 1998). In food industry applications, Hillegas and Demirci (2003) have used pulsed light to inactivate *Clostridium* in clover honey. Jun, Irudayaraj, and Geisner (2004) have demonstrated the inactivation of *Aspergillus* spores after treatment of corn meal.

In some cases, only two or three pulses are sufficient to completely eradicate bacteria and fungal spores. Two pulses at 0.75 J/cm² each were sufficient to reduce plate cfu cultures of *Staphylococcus aureus* by more than 7 logs

(Dunn et al. 1997). Spores of *Aspergillus niger* were completely inactivated from an initial 6–8 logs of cfu with 1–3 pulses (Bushnell et al. 1998). Table 21.2 shows the results of a number of studies on pulsed light disinfection. The Dose represents the total PWL dose while the UV Dose is based on the estimated percentage of light that is UV.

PWL inactivates pathogens more rapidly than UVGI, but in terms of energy consumption it is less efficient (Kowalski 2009). The effect of PWL is primarily due to UV, but there is a secondary effect that occurs at high power levels that involves flash heating and cell rupture. In higher power PWL systems, spores experience cavitation due to sudden expansion and evaporation of internal cell water and other constituents. This effect can occur when power levels are increased far above those normally used. This effect operates even if the UVC component is filtered out. Pulsed light with the UV removed produces pulsed light that is not necessarily hazardous to humans but is still biocidal for bacteria and spores (Wekhof, Trompeter, and Franken 2001). The overheating results primarily from the UVA content and under high doses the disintegration effect is dominant.

UV-filtered pulsed light systems may be able to destroy bacterial cells on skin surfaces without necessarily harming skin cells due to the fact that skin cells are packed in a matrix, giving them protection against sudden overpressure. Even if skin cells were damaged, they may recover from natural regeneration processes. This unique ability has promising potential applications in the medical industry where it could be used in operating rooms to control surgical site infections. PWL hand disinfection is also feasible, and one such hand disinfection unit is currently available in Europe.

Passive Solar Exposure

Passive exposure to solar irradiation as a means of destroying airborne pathogens is based on the fact that sunlight contains some ultraviolet radiation and is lethal to microorganisms (Beebe 1959; Fernandez and Pizarro 1996). Sunlight contains some trace levels of UVA and UVB (Webb 1991). Sunlight is one of the primary reasons that most human pathogenic microorganisms die off rapidly in the outdoor air (Harper 1961; Mitscherlich and Marth 1984). Unfiltered sunlight can destroy about 99% of *Staphylococcus aureus* cells within 70 minutes (El-Adhami, Daly, and Stewart 1994). Based on the empirical data for filtered light in Figure 21.3, it is estimated that *S. aureus* would be sterilized by Perspex filtered sunlight in about 240 minutes, while it would take about 625 minutes to sterilize with sunlight in which all the UVB was removed. Sunlight in which all UVB and UVA was removed showed no significant disinfection within the first 120 minutes.

TABLE 21.2

Pulsed Light Disinfection of Airborne Nosocomial Pathogens

Microbe	Medium	Dose (J/m²)	Est. % UV	UV Dose (J/m²)	Survival frac	Disinfection (%)	Reference
Staphylococcus aureus	Plates	15000	25	3750	1.0E-07	100.0000	Dunn 1996
	Solution	3500	25	875	1.0E-08	100.0000	Dunn 2000
	Plates	20000	100	8000	1.6E-07	100.0000	Wekhof 2000
	Plates	168000	100	168000	1.3E-05	99.9987	Krishnamurthy, Demirci, and Irudayaraj 2003
Staphylococcus epidermis	Air	1558	100	1558	4.8E-04	99.9516	UVDI 2002
Aspergillus niger spores	Solution	53000	25	13250	1.0E-06	99.9999	Dunn 2000
	Plastic	45000	56	25200	1.6E-01	83.6346	Bushnell et al. 1998
	Plastic	45000	65	29250	1.0E-01	89.9741	Bushnell et al. 1998
	Plastic	45000	79	35550	2.6E-03	99.7368	Bushnell et al. 1998
	Glass	11000	30	3300	3.2E-06	99.9997	Wekhof, Trompeter, and Franken 2001
Aspergillus versicolor spores	Air	1781	100	1781	2.2E-03	99.7811	UVDI 2002

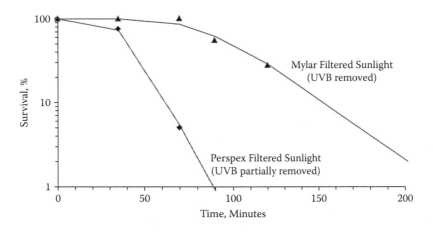

FIGURE 21.3
The effect of filtered sunlight on *Staphylococcus aureus*. Based on data from El-Adhami, Daly, and Stewart (1994).

Open air and sunlight are one way in which ancient Greek physicians dealt with cases of consumption and other diseases. Generous amounts of sunlight will tend to destroy spores over time and can be effective in destroying mold and mold spores in rugs, blankets, furniture, etc., after only a few days of exposure. The common practice of exposing mattresses, blankets, and rugs to open air in the spring could be applied by maximizing solar exposure indoors. Although indoor solar exposure is limited by transmittance through glass and total window area, given sufficient time almost any material can be disinfected with sunlight. The use of atriums and extensive fenestration in hospitals may be of some benefit in controlling environmental microbes that find their way into hospitals and contaminate floors.

Vegetation Air Cleaning

Large amounts of living vegetation can act as a natural biofilter, removing or reducing levels of airborne microbes (Darlington, Dixon, and Pilger 1998; Rautiala et al. 1999). The surface area of large amounts of vegetation may absorb or adsorb microbes or dust. The oxygen generation of the plants may have an oxidative effect on microbes. The presence of symbiotic microbes in soil such as *Streptomyces* may have some disinfection effects on the air. Natural plant defenses against bacteria may also operate against mammalian pathogens. Finally, gardens and vegetation may have an effect on the psyche of occupants that may stimulate a sense of well-being. This effect did not go unnoticed in hospitals through the ages and it is not uncommon for them to include "healing gardens" as part of their architectural landscape (Gerlach-Spriggs, Kaufman, and Warner 1998). A number of specific house plants have been identified that may contribute to passive cleaning of indoor air pollutants, but their effects on airborne microbes are unknown (Wolverton 1996). Winter gardens may act as buffer zones in moderating the indoor climate and creating healthy indoor environments (Watson and Buchanan 1993).

Although house plants are often considered a source of potential fungal spores, a study by Rautiala et al. (1999) indicated that no significant increase in airborne concentrations of fungi or surface samples occurred when they were added to indoor environments. In a vegetation air cleaning system air flows through areas filled with vegetation or through entire greenhouses before entering the ventilation system. Such vegetation areas may include water or waterfalls, which can have an effect on local ionization levels. One downside to growing large amounts of vegetation indoors is that the potting soil may include potentially allergenic fungi.

Material Selectivity

Selecting appropriate materials for use in healthy indoor environments can preempt health hazards during the lifetime of a building. The removal of gypsum that has become contaminated with mold provides one example of an expensive and unhealthy problem that could be avoided by careful material selection prior to construction. The selection of appropriate materials for healthy building construction must, of course, be done during the design phase. The drive to produce cheaper buildings faster has resulted in the adoption of a variety of construction materials and practices that have contributed to unhealthy buildings. Among these developments is the excessive use of cheap wood and cellulose-based products, synthetic materials with hazardous potential, and fragile materials subject to damage and moisture accumulation, and that provide substrates for microbial growth. A return to the use of natural materials like stone, glass, lime or mud plasters, adobe or rammed earth, bricks, tiles, and other natural materials can improve both the aesthetics and healthiness of a building (Pearson 1998).

Carpets and rugs provide a substrate that may both collect spores and grow mold and mildew under moist conditions. Because mold spores tend to settle over time, they inevitably collect in carpets and are further ground in by foot traffic, needing only wet spills to foster germination. Because such floor coverings may do more harm than good in a hospital environment, the use of carpets and rugs in such facilities is questionable. If floor coverings serve the purpose of protecting feet from cold floors then one obvious solution is to use radiant floor heating systems. Another alternative is to use

TABLE 21.3

Healthy Replacements for Common Building Materials

Commonly Used Materials	Suggested Replacement Materials
Wooden walls, structural wood products	Concrete, stone, brick, steel, glass
Wood lumber	Plastic lumber, steel beams
Nonsustainable wood products	Certified sustainable wood products (i.e., Forest Stewardship Council [FSC] certified wood product), recycled wood products
Arsenic-treated wood	Any safely treated wood product or substitute
PVC plastic, vinyl	Any PVC-free materials
Carpeting, rugs	Radiant floor heating, natural linoleum, rubber and synthetics, cork or bamboo flooring, wood flooring
Alkyd oil-based paint	Water-based paint
Solvent-based adhesives	Non-VOC or low-VOC producing adhesives
Paints and sealants	Surfaces requiring no sealants or painting: natural stone, ceramic tile, stucco, raw plaster, waxed concrete, glass
Concrete formwork, precast concrete	High fly-ash (i.e., 15%) concrete
Fabrics	Natural fibers, low-VOC materials, low-dust
Furnishings/furniture	Low dust-producing synthetics or leather

healthy materials like natural linoleum, natural cork and rubber, or soft synthetic materials that do not absorb spores.

Table 21.3 provides a select summary of green building materials that have been proposed or used as replacements for more hazardous materials (Spiegel and Meadows 1999). Although some of the material replacements represent choices based on volatile organic compounds (VOCs) rather than microbiological concerns, they all contribute to healthier buildings.

Various other sources of information on practical and healthy building materials and good construction practices are available, and the reader should consult these documents for more detailed information (Berman 2001; Watson and Buchanan 1993). In particular, the "Green Healthcare Design Guidance Statement" from ASHE (2002) may prove useful for hospital facility design.

References

Arnold, J. W., and Mitchell, B. W. (2002). Use of negative air ionization for reducing microbial contamination on stainless steel surfaces. *J Appl Poultry Res* 11, 179–186.

ASHE (2002). Green healthcare design guidance statement. Green Building Committee of American Society for Healthcare Engineering, Chicago, IL.

Beebe, J. M. (1959). Stability of disseminated aerosols of *Pastuerella tularensis* subjected to simulated solar radiations at various humidities. *J Bacteriol* 78, 18–24.

Bergeron, V., Reboux, G., Poirot, J. L., and Laudinet, N. (2007). Decreasing airborne contamination levels in high-risk hospital areas using a novel mobile air-treatment unit. *Inf Contr Hosp Epidemiol* 28(10), 1181–1186.

Berman, A. (2001). *Your Naturally Healthy Home: Stylish, Safe, Simple.* St. Martin's Press, New York.

Bushnell, A., Clark, W., Dunn, J., and Salisbury, K. (1997). Pulsed light sterilization of products packaged by blow-fill-seal techniques. *Pharm Eng* 17(5), 74–84.

Bushnell, A., Cooper, J. R., Dunn, J., Leo, F., and May, R. (1998). Pulsed light sterilization tunnels and sterile-pass-throughs. *Pharm Eng* March/April, 48–58.

Chard, A. (2005). *New Weapon to Fight Hospital Infections.* University of Leeds, Leeds, UK.

Daniels, S. L. (2000). Applications of air ionization for control of VOCs and PNx. *Air & Waste Management Association 94th Annual Conference & Exhibition*, Orlando, Florida, http://www.precisionair.com/news/iaq.pdf

Darlington, A., Dixon, M. A., and Pilger, C. (1998). The use of biofilters to improve indoor air quality: The removal of toluene, TCE, and formaldehyde. *Life Support Biosph Sci* 5(1), 63–69.

Dunn, J. (1996). Pulsed light and pulsed electric field for food and eggs. *Poult Sci* 75(9), 1133–1136.

_____. (2000). *Pulsed Light Disinfection of Water and Sterilization of Blow/Fill/Seal Manufactured Aseptic Pharmaceutical Products.* Automatic Liquid Packaging, Woodstock, IL.

Dunn, J., Bushnell, A., Ott, T., and Clark, W. (1997). Pulsed white light food processing. *Cereal Foods World* 42(7), 510–515.

Dyas, A., Boughton, B. J., and Das, B. C. (1983). Ozone killing action against bacterial and fungal species; microbiological testing of a domestic ozone generator. *J Clin Path* 36, 1102–1104.

El-Adhami, W., Daly, S., and Stewart, P. R. (1994). Biochemical studies on the lethal effects of solar and artificial ultraviolet radiation on *Staphylococcus aureus. Arch Microbiol* 161, 82–87.

Elford, W. J., and van den Eude, J. (1942). An investigation of the merits of ozone as an aerial disinfectant. *J Hygiene* 42, 240–265.

Estola, T., Makela, P., and Hovi, T. (1979). The effect of air ionization on the air-borne transmission of experimental Newcastle disease virus infections in chickens. *J Hyg* 83, 59–67.

Fernandez, R. O., and Pizarro, R. A. (1996). Lethal effect induced in *Pseudomonas aeruginosa* exposed to ultraviolet-A radiation. *Photochem & Photobiol* 64(2), 334–339.

Foarde, K. K., Van Osdell, D. W., and Steiber, R. S. (1997). Investigation of gas-phase ozone as a potential biocide. *Appl Occup Envrion Hyg* 12(8), 535–541.

Gabbay, J. (1990). Effect of ionization on microbial air pollution in the dental clinic. *Environ Res* 52(1), 99.

Gerlach-Spriggs, N., Kaufman, R. E., and Warner, S. B. J. (1998). *Restorative Gardens: The Healing Landscape.* Yale University Press, New Haven, CT.

Goswami, D. Y. (2003). Decontamination of ventilation systems using photo catalytic air cleaning technology. *J Solar Energy Engrg* 125(3), 359–365.

Goswami, D. Y., Trivedi, D. M., and Block, S. S. (1997). Photocatalytic disinfection of indoor air; in *Transactions of the ASME—Solar Engineering*, ASME, ed., 92–96.

Happ, J. W., Harstad, J. B., and Buchanan, L. M. (1966). Effect of air ions on submicron T1 bacteriophage aerosols. *Appl Microb* 14, 888–891.

Harper, G. J. (1961). Airborne micro-organisms: Survival tests with four viruses. *Journal of Hygiene* 59, 479–486.

Heindel, T. H., Streib, R., and Botzenhart, K. (1993). Effect of ozone on airborne micro-organisms. *Zbl Hygiene* 194, 464–480.

Hibben, C. R., and Stotzky, G. (1969). Effects of ozone on the germination of fungus spores. *Canadian J Microbiol* 15, 1187–1196.

Hillegas, S. L., and Demirci, A. (2003). Inactivation of *Clostridium sporogenes* in clover honey by pulsed UV-light treatment. *ASAE 2003 International Meeting*, Las Vegas, NV.

Hodgson, A. T., Destaillats, H., Sullivan, D. P., and Fisk, W. J. (2007). Performance of ultraviolet photo catalytic oxidation for indoor air cleaning applications. *Indoor Air* 17, 305–316.

Jacoby, W. A., Maness, P. C., and Wolfrum, E. J. (1998). Mineralization of bacterial cell mass on a photocatalytic surface in air. *Environ Sci & Technol* 32(17), 2650–2653.

Johnson, T. (1982). Flashblast: The light that cleans. *Popular Science* July, 82–84.

Jun, S., Irudayaraj, J., and Geisner, D. (2004). Pulsed UV-light treatment of corn meal for inactivation of *Aspergillus niger*. *Int J Food Science & Technol*, 21:883–888.

Karelin, V. I., Buranov, S. N., Voevodin, S. V., Voevodina, I. A., Matvey, T. N., Repin, P. B., and Selemir, V. D. (2001). Study of high-voltage diffused gas discharge effect on microbiological cultures. *IEEE International Conference on Plasma Science*, Las Vegas, NV.

Kelly-Wintenberg, K., Montie, T., Hodge, A., Gaskins, J., Roth, J. R., and Chen, Z. (1999). Mechanism of killing microorganisms by a one atmosphere uniform glow discharge plasma. *IEEE International Conference on Plasma Science*, Monterey, CA.

Khurana, A. (2003). *Ozone treatment for prevention of microbial growth in air conditioning systems*, PhD thesis, University of Florida.

Kowalski, W. J. (2003). *Immune Building Systems Technology*. McGraw-Hill, New York.

———. (2006). *Aerobiological Engineering Handbook: A Guide to Airborne Disease Control Technologies*. McGraw-Hill, New York.

———. (2009). *Ultraviolet Germicidal Irradiation Handbook: UVGI for Air and Surface Disinfection*. Springer, New York.

Kowalski, W. J., Bahnfleth, W. P., and Whittam, T. S. (1998). Bactericidal effects of high airborne ozone concentrations on *Escherichia coli* and *Staphylococcus aureus*. *Ozone Science & Engineering* 20(3), 205–221.

Krishnamurthy, K., Demirci, A., and Irudayaraj, J. (2003). Paper #03-037: Inactivation of *Staphylococcus aureus* using pulsed UV treatment. *NABEC 2003 Northeast Agricultural Biological Engineering Conference*, Storrs, CT.

Krueger, A. P. (1985). The biological effect of air ions. *Int J Biometeorol* 29, 205–206.

Krueger, A. P., and Reed, E. J. (1976). Biological impact of small air ions. *Science* 193(Sep), 1209–1213.

Krueger, A. P., Smith, R. F., and Go, I. G. (1957). The action of air ions on bacteria. *J Gen Physiol* 41, 359–381.

Langlais, B., Reckhow, D. A., and Brink, D. R. (1991). *Ozone in Water Treatment: Application and Engineering*. Lewis Publishers, Chelsea, MI.

Li, J., Wang, X., Yao, H., Yao, Z., Wang, J., and Luo, Y. (1989). Influence of discharge products on post-harvest physiology of fruit. *International Symposium on High Voltage Engineering* 6, 1–4.

Makela, P., Ojajarvi, J., Graeffe, G., and Lehtimaki, M. (1979). Studies on the effects of ionization on bacterial aerosols in a burns and plastic surgery unit. *J Hyg* 83, 199–206.

Manolache, S., Somers, E. B., Wong, A. C. L., Shamamian, V., and Denes, F. (2001). Dense medium plasma environments: A new approach for the disinfection of water. *Environ Sci Technol* 18, 3780–3785.

Masaoka, T., Kubota, Y., Namiuchi, S., Takubo, T., Ueda, T., Shibata, H., Nakamura, H., Yoshitake, J., Yamayoshi, T., Doi, H., and Kamiki, T. (1982). Ozone decontamination of bioclean rooms. *Appl & Environ Microb* 43(3), 509–513.

Masuda, S., Kiss, E., Ishida, K., and Asai, H. (1990). Ceramic-based ozonizer for high speed sterilization. *IEEE T Ind Appl* 26(1), 36–41.

McDowell, N. (2003). Air ionizers wipe out hospital infections. NewScientist.com news service. www.newscientist.com/article.ns?id=dn3228

Mitscherlich, E., and Marth, E. H. (1984). *Microbial Survival in the Environment.* Springer-Verlag, Berlin.

Montie, T. C., Kelly-Wintenberg, K., and Roth, J. R. (2000). Overview of research using the one atmosphere uniform glow discharge plasma (OAUGDP) for sterilization of surfaces and materials. *IEEE Trans on Plasma Sci* 28(1), 41–50.

Nunez, C. M., Ramsey, G. H., Ponder, W. H., Abbott, J. H., Hamel, L. E., and Kariher, P. H. (1993). Corona destruction: An innovative control technology for VOCs and air toxics. *J Air & Waste Mgt* 43, 242–247.

Park, B., Lee, D. H., Park, J.-C., Lee, I.-S., Lee, K.-Y., Hyun, S. O., Chun, M. S., and Chung, K. H. (2003). Sterilization using a microwave-induced argon plasma system at atmospheric pressure. *Phys Plasmas* 10(11), 4539–4544.

Pearson, D. (1998). *The New Natural House Book: Creating a Healthy, Harmonious and Ecologically Sound Home.* Fireside, New York.

Phillips, G., Harris, G. J., and Jones, M. W. (1963). The effect of ions on microorganisms. *Int J Biometerol* 8, 27–37.

Phillips, G., Harris, G. J., and Jones, M. V. (1964). Effect of air ions on bacterial aerosols. *Intl J of Biometeorol* 8, 27–37.

Rautiala, S., Haatainen, S., Kallunki, H., Kujanpaa, L., Laitinen, S., Miihkinen, A., Reiman, M., and Seuri, M. (1999). Do plants in office have any effect on indoor air microorganisms? *Indoor Air 99: Proceedings of the 8th International Conference on Indoor Air Quality and Climate*, Edinburgh, Scotland, 704–709.

Spiegel, R., and Meadows, D. (1999). *Green Building Materials: A Guide to Product Selection and Specification.* John Wiley & Sons, New York.

Steiber, R. S. (1995). Ozone Generators in Indoor Air Systems. *EPA-600/R-95-154*, Environmental Protection Agency, Washington, DC.

UVDI (2002). Report on pulsed light disinfection of microorganisms prepared by K. Foarde and Research Triangle Institute. Ultraviolet Devices Incorporated, Valencia, CA.

Vincent, J. H. (1995). *Aerosol Science for Industrial Hygienists.* Pergamon, New York.

Wallen, R. D., May, R., Reiger, K., Holoway, J. M., and Cover, W. H. (2001). Sterilization of a new medical device using broad-spectrum pulsed light. *Biomed Instr & Tech* 35(5), 323–330.

Watson, D., and Buchanan, G. (1993). *Designing Healthy Buildings.* American Institute of Architects, Washington, DC.

Webb, A. R. (1991). Solar ultraviolet radiation in southeast England: The case for spectral measurements. *Photochem Photobiol* 54(5), 789–794.

Wekhof, A. (2000). Disinfection with flashlamps. *PDA J of Pharmaceutical Science and Technology* 54(3), 264–267.

Wekhof, A., Trompeter, I.-J., and Franken, O. (2001). Pulsed UV-disintegration, a new sterilization mechanism for broad packaging and medical-hospital applications. *Proceedings of the First International Congress on UV-Technologies*, Washington, DC.

Whistler, P. E., and Sheldon, B. W. (1989). Biocidal activity of ozone versus formaldehyde against poultry pathogens inoculated in a prototype setter. *Poultry Sci* 68, 1068–1073.

Wolverton, B. C. (1996). *How to Grow Fresh Air: 50 Plants That Purify Your Home or Office*. Penguin Press, Baltimore, MD.

Zhao, J., and Yang, X. (2003). Photo catalytic oxidation for indoor air purification: A literature review. *Building Environ* 38(5), 645–654.

Appendix: Database of Airborne Nosocomial Pathogens

Pathogen	Group	Mean Dia. μm	Type	BSL	ID50	LD50	IR Days	Incub Days	Peak Infection	Annual Cases	Duration Days	Source	Toxins	Untreated Fatality Rate
Absidia	FS	3.536	NC	2	NA	NA	none	–	NA	rare	–	E	N	–
Acinetobacter	B	1.225	E	2	Unk	NA	–	–	NA	147	–	E G1 W	N	–
Acremonium	FS	2.449	NC	1-2	NA	–	none	–	NA	–	–	E I	Y	–
Actinomyces israelii	B	0.901	E	2	Unk	–	–	–	NA	rare	–	H A	N	–
Adenovirus	V	0.079	C	2	32-150	none	0.51-0.75	1-10	3-4 days	common	8-19	H G	N	NA
Aeromonas	B	2.098	NC	2	Unk	–	none	Unk	NA	–	–	E W S	Y	–
Alcaligenes	B	0.775	E	2	Unk	–	–	–	NA	rare	–	H S W	N	–
Alternaria alternata	FS	11.225	NC	1	NA	–	none	–	NA	–	–	E I	Y	–
Arthrinium phaeospermum	FS	5.000	NC	2	NA	NA	none	–	NA	–	–	E	N	NA
Aspergillus	FS	3.354	NC	2	9643-58154	–	none	3-30 days	NA	rare	–	E I	Y	NA
Aureobasidium pullulans	FS	4.899	NC	1	NA	NA	none	–	NA	–	–	E I	N	NA
Bacillus anthracis	BS	1.118	NC	2	1300-10000	28000	none	2-3 days	NA	rare	7-21	A S	N	5-20%
Bacillus cereus	BS	1.118	NC											
Bacteroides fragilis	B	3.162	E	2	Unk	–	none	Unk	–	rare	–	H	N	–
Blastomyces dermatitidis	FS	12.649	NC	2	11000	Unk	none	weeks	NA	rare	–	E	N	–
Bordetella pertussis	B	0.245	C	2	(4)	(1314)	high	7-10 days	7-14 days	6,564	28-42	H	N	–
Botrytis cinerea	FS	6.708	NC	1	NA	–	none	–	NA	rare	–	E S	N	–
Brucella	B	0.566	NC	3	1300	–	–	5-60 days	–	98	–	A S	N	<2%
Burkholderia cenocepacia	B	0.707	NC	1	Unk	Unk	–	Unk	–	–	–	E	N	–
Burkholderia mallei	B	0.674	NC	3	3200	–	none	1-14 days	NA	–	14-21	E A	N	–

Organism														
Burkholderia pseudomallei	B	0.494	NC	3	Unk	–	none	2 days min.	NA	rare	–	EASW	N	–
Candida albicans	FS	4.899	E	1	Unk	NA	–	variable	–	rare	–	HWS	N	NA
Cardiobacterium	B	0.612	E	2	Unk	–	–	–	NA	rare	–	H	N	–
Chaetomium globosum	FS	5.455	NC	1	Unk	–	none	–	NA	–	–	ESI	Y	NA
Chlamydia pneumoniae	B	0.548	C	2	Unk	–	0.5	7 days	–	rare	28	H	N	–
Chlamydophila psittaci	B	0.283	NC	2	Unk	–	none	5–15 days	NA	33	7–28	A	N	<6%
Cladosporium	FS	8.062	NC	2	Unk	NA	none	–	NA	–	–	E	Y	–
Clostridium botulinum	B	1.975	NC	2	Unk	–	–	12–36 hours	–	–	–	E	Y	–
Clostridium difficile	B	2.000	NC											
Clostridium perfringens	B	5.000	NC	2	10 per g	–	none	6–24 hrs.	NA	10,000	1	EHAS	Y	–
Coccidioides immitis	FS	3.464	NC	3	100–1350	–	none	1–4 weeks	13 days	rare	28	ES	N	90%
Coronavirus	V	0.110	C	2	Unk	–	0.34–0.5	2–5 days	3–4 days	1,700,000	8–18	H	N	none
Corynebacterium diphtheriae	B	0.698	C	2	Unk	–	varies	2–5 days	–	10	10	H	Y	5–10%
Coxiella burnetii	B/R	0.283	NC	3	10	–	none	9–18 days	NA	rare	–	A	N	<1%
Coxsackievirus	V	0.027	C	2	18–67	–	0.53–0.64	1–4 days	3–5 days	common	2–10	HCG	N	none
Crimean-Congo														
Cryptococcus neoformans	FY	4.899	NC	2	1000	NA	none	Unk	NA	high	–	EI	N	–
Cryptosporidium parvum														
Cryptostroma corticale	FS	3.742	NC	2	NA	–	–	–	–	rare	–	E	N	–
Curvularia lunata	FS	11.619	NC	1	Unk	NA	–	–	–	rare	–	E	N	–
Drechslera	FS	69.282	NC	2	NA	NA	NA	NA	NA	rare	–	E	N	NA
Ebola	V		C											

Continued

Appendix: Database of Airborne Nosocomial Pathogens

Pathogen	Group	Mean Dia. μm	Type	BSL	ID50	LD50	IR Days	Incub Days	Peak Infection	Annual Cases	Duration Days	Source	Toxins	Untreated Fatality Rate
E. coli O157:H7	B													
Echovirus	V	0.024	C	2	Unk	–	0.43–0.80	2–14 days	3–4 days	common	8–18	H	N	–
Emericella nidulans	FS	3.240	NC	1	NA	–	none	–	NA	–	–	E	Y	–
Enterobacter cloacae	B	1.414	E	1	Unk	–	–	Unk	–	rare	–	H E S W	Y	–
Enterococcus	B	1.414	E	1–2	Unk	–	–	–	–	rare	–	H	N	–
Enterococcus faecalis	B	0.707	E	1	Unk	–	–	–	–	–	–	C	N	–
Epicoccum purpurascens	FS	17.321	NC	1	NA	NA	none	NA	NA	–	–	E I	Y	–
Eurotium	FS	5.612	NC	1	NA	–	none	–	NA	–	–	E I	Y	–
Exophiala	FS	2.121	NC	2	NA	NA	none	–	NA	–	–	E S W G	N	–
Francisella tularensis	B	0.200	NC	3	10–100	–	none	1–14 days	NA	rare	21–42	A W	N	5–15%
Fugomyces cyanescens														
Fusarium	FS	11.225	NC	1	NA	–	none	–	NA	–	–	E I	Y	–
Haemophilus influenzae	B	0.285	C	2	Unk	–	0.2–0.5	2–4 days	3–4 days	1,162	10–14	H N	N	–
Haemophilus parainfluenzae	B	1.732	E	2	Unk	–	–	14–30 days	NA	common	–	H	N	–
Hantaan virus	V	0.096	NC	3	Unk	–	none	14–30 days	NA	44	30–90	A	N	5–15%
Helicobacter pylori	B													
Helminthosporium	FS	11.577	NC	1	NA	NA	none	–	NA	–	–	E I	N	NA
Histoplasma capsulatum	FS	2.236	NC	3	10	40000	none	4–22 days	NA	common	84	E	N	–
HPV	V													
Influenza A virus	V	0.098	C	2	20–790	–	0.2–0.83	2–3 days	3–4 days	2,000,000	10–21	H A	N	low
Junin virus	V	0.122	NC	4	Unk	10–100000	low	2–14 days	7 days	2,000	7–21	A	N	10–50%

												E S H I W	Y	
Klebsiella pneumoniae	B	0.671	E	2	Unk	-	-	-	NA	1,488	-	A	Y	-
Lassa virus	V	0.122	C	4	15	2-200000	high	7-14 days	-	-	14-21	E W	N	10-50%
Legionella pneumophila	B	0.520	NC	2	<129	140000	<0.01	2-10 days	NA	1,163	7-21	E H	N	39-50%
Listeria monocytogenes	B	0.707	NC	2	Unk	<1000	-	-	NA	363	21	E H	N	-
Lymphocytic choriomeningitis	V	0.087	NC	2-3	Unk	<1000	na	8-13 days	3-7 days	rare	-	A	N	<1%
Machupo	V	0.120	NC	4	Unk	<1000	-	7-16 days	-	-	28	A	N	5-30%
Marburg virus	V	0.039	C	4	Unk	-	-	7 days	-	rare	-	H A	N	25%
Measles virus	V	0.158	C	2	0.2 units	-	0.85	7-18 days	9-11 days	500,000	10-14	H	N	25%
Micromonospora faeni	BS	0.866	NC	1	Unk	Unk	none	-	NA	rare	-	F I	N	0-20%
Moraxella	B	1.225	E	2	Unk	-	-	-	NA	rare	-	H	N	-
Mucor	FS	7.071	NC	1	Unk	-	none	-	NA	rare	-	E G	N	-
Mumps virus	V	0.164	C	2	Unk	-	0.6-0.85	14-28 days	3-10 days	10,000	39	H	N	-
Mycobacterium abcessus	B													
Mycobacterium avium	B	1.118	NC	2	Unk	-	none	-	NA	rare	-	E W	N	none
Mycobacterium chelonae	B													
Mycobacterium fortuitum	B													
Mycobacterium kansasii	B	0.637	NC	2	Unk	-	none	-	NA	rare	-	W A	N	-
Mycobacterium marinum	B													
Mycobacterium tuberculosis	B	0.637	C	3	1-10	-	0.33	4-12 weeks	varies	20,000	-	H	N	-
Mycobacterium ulcerans	B													

Continued

Pathogen	Group	Mean Dia. μm	Type	BSL	ID50	LD50	IR Days	Incub Days	Peak Infection	Annual Cases	Duration Days	Source	Toxins	Untreated Fatality Rate
Mycoplasma pneumoniae	B	0.177	E	2	100	-	-	6–23 days	NA	rare	/months	H	N	-
Neisseria meningitidis	B	0.775	E	2	1Unk10	-	varies	2–10 days	2–4 days	3,308	7–21	H	N	50%
Nocardia asteroides	BS	1.118	NC	2	Unk	-	none	-	NA	rare	-	E S G	N	10%
Nocardia brasiliensis	BS	1.414	NC	2	Unk	-	none	-	NA	rare	-	E S G	N	10%
Norwalk virus	V	0.029	C	2	Unk	NA	high	10–60 hours	1–2 days	181,000	4–Mar	E W	N	0%
Paecilomyces variotii	FS	2.828	NC	1	Unk	NA	none	-	NA	-	-	E1W	Y	-
Paracoccidioides brasiliensis	FS	4.472	NC	2	8000000	Unk	none	-	NA	rare	>6 months	E S	N	-
Parainfluenza virus	V	0.194	C	2	Unk	>1.5	0.2–0.75	1–3 days	3–4 days	common	7–21	H	N	low
Parvovirus B19	V	0.022	C	2	0.5 ml of ser	-	0.3–0.8	4–20 days	-	rare	28	H	N	-
Penicillium	FS	3.262	NC	2	Unk	NA	none	-	NA	rare	-	E I	Y	NA
Phialophora	FS	1.470	NC	2	Unk	-	none	-	NA	-	-	E S I W	N	-
Phoma	FS	3.162	NC	1	NA	NA	none	-	NA	-	-	E S I	Y	NA
Pneumocystis carinii	FS	2.000	C	1	Unk	-	none	4–8 weeks	NA	rare	-	E H	N	-
Proteus mirabilis	B	0.494	E	2	Unk	-	-	-	-	-	-	H	Y	-
Pseudallescheria boydii														
Pseudomonas aeruginosa	B	0.494	NC	1	Unk	-	none	2–3 days	2–4 days	2,626	7–21	E G 1 W	Y	-
Reovirus	V	0.075	C	2	Unk	-	-	3–4 days	-	-	7–10	H	N	-
Respiratory Syncytial Virus	V	0.190	C	2	160–640	-	0.5–0.9	4–5 days	1–3 days	common	10–14	H	N	-
Rhinovirus	V	0.023	C	2	1–5	-	0.38–0.89	2–4 days	3–4 days	common	3–7	H	N	none

Pathogen														
Rhizomucor pusillus	FS	4.183	NC	1	Unk	NA	none	–	NA	rare	–	ES	N	NA
Rhizopus	FS	6.928	NC	2	Unk	NA	none	–	NA	rare	–	E	Y	NA
Rhodoturula	FS	13.856	NC	1	NA	NA	none	–	NA	–	–	ESWC	N	NA
Rickettsia prowazeki	B	0.600	VB	3	10	–	–	1–2 weeks	–	–	2 weeks	HA	N	10–40%
Rubella virus	V	0.061	C	2	10–60	–	0.3–0.8	12–23 days	10–14 days	3,000	28–31	H	N	–
Saccharopolyspora rectivirgula	BS	1.342	NC	2	Unk	–	none	–	NA	–	–	FI	N	0–20%
Salmonella typhi	B	0.806	FB	2	100,000	–	–	5–21 days	–	2,000,000	27–Nov	H	Y	10–40%
SARS virus	V	0.110	C	4	low	–	high	2–7 days	3–4 days	8,000	8–18	HA	N	2–15%
Scedosporium spp.	FS													NA
Scopulariopsis	FS	5.916	NC	2	Unk	NA	none	–	NA	–	–	ES	N	NA
Serratia marcescens	B	0.632	E	1	Unk	–	–	–	NA	479	–	EIW	Y	uncommon
Sporothrix schenckii	FS	6.325	NC	2	Unk	NA	none	1–12 weeks	NA	rare	–	ES	N	–
Stachybotrys chartarum	FS	5.623	NC	1–2	Unk	–	none	–	NA	–	–	EI	Y	–
Staphylococcus aureus	B	0.866	E	2	varies	–	–	4–10 days	NA	2,750	–	HG	N	–
Staphylococcus epidermis	B	0.866	E	1	Unk	–	–	–	–	common	–	HG	N	–
Streptococcus pneumoniae	B	0.707	C	2	Unk	–	0.1–0.3	1–5 days	2–10 days	500,000	7–28	H	N	5–40%
Streptococcus pyogenes	B	0.894	C	2	Unk	–	–	1–5 days	2–10 days	213,962	7–28	H	N	3%
Thermoactinomyces sacchari	BS	0.855	NC	2	Unk	Unk	none	–	NA	–	–	F	N	0–20%
Thermoactinomyces vulgaris	BS	0.866	NC	1	Unk	Unk	none	–	NA	rare	–	FI	N	0–20%
Thermomonospora viridis	BS	0.520	NC	1	Unk	Unk	none	–	NA	–	–	FI	N	0–20%

Continued

Pathogen	Group	Mean Dia. µm	Type	BSL	ID50	LD50	IR Days	Incub Days	Peak Infection	Annual Cases	Duration Days	Source	Toxins	Untreated Fatality Rate
Trichoderma	FS	4.025	NC	1	NA	NA	none	–	NA	–	–	E S	Y	NA
Trichophyton	FS	4.899	NC	2	Unk	–	none	–	NA	–	–	E	N	none
Trichosporon	FY	8.775	NC	3	Unk	NA	none	–	NA	–	–	E S W	N	none
Ulocladium	FS	14.142	NC	1	NA	NA	none	–	NA	–	–	E S	N	NA
Ustilago	FS	5.916	NC	1	Unk	NA	none	Unk	NA	–	–	E	N	NA
Vaccinia virus	V	0.224	NC	2	Unk	–	none	–	NA	rare	–	F	N	none
Varicella–zoster virus	V	0.173	C	2	Unk	–	0.75–0.96	11–21 days	2–4 days	common	25	H	N	rare
Variola (smallpox)	V	0.173	C	4	10–100 (?)	–	0.3–0.9	12–14 days	14–17 days	0	22–24	H	N	10–40%
Verticillium	FS	4.796	NC	1	NA	NA	–	–	–	–	–	E S	N	NA
Wallemia sebi	FS	2.958	NC	1	NA	NA	none	–	NA	–	–	E I	N	NA
Yersinia pestis	B	0.707	C	3	3000	–	varies	2–6 days	–	4	6	A	Y	50%

Note: E = Environment, I = Indoor, H = Humans, S = Soil, A = Animals, F = Farms, W = Water, G = Sewage, C = Feces, F = Fungi, FY = Fungal yeast, FS = Fungal spore, B = Bacteria, BS = Bacterial spore, and R = Rickettsia.

Index